PROGRESS IN
Nucleic Acid Research and Molecular Biology
Volume 36

PROGRESS IN
Nucleic Acid Research and Molecular Biology

edited by

WALDO E. COHN
Biology Division
Oak Ridge National Laboratory
Oak Ridge, Tennessee

KIVIE MOLDAVE
University of California
Santa Cruz, California

Volume 36
Transposable Elements in Mutagenesis
and Regulation of Gene Expression

ACADEMIC PRESS, INC.
Harcourt Brace Jovanovich, Publishers

San Diego   New York   Berkeley   Boston
London   Sydney   Tokyo   Toronto

COPYRIGHT © 1989 BY ACADEMIC PRESS, INC.
All Rights Reserved.
No part of this publication may be reproduced or transmitted in any form or by any means, electronic or mechanical, including photocopy, recording, or any information storage and retrieval system, without permission in writing from the publisher.

ACADEMIC PRESS, INC.
San Diego, California 92101

United Kingdom Edition published by
ACADEMIC PRESS LIMITED
24-28 Oval Road, London NW1 7DX

LIBRARY OF CONGRESS CATALOG CARD NUMBER: 63-15847

ISBN 0-12-540036-5 (alk. paper)

PRINTED IN THE UNITED STATES OF AMERICA
89 90 91 92   9 8 7 6 5 4 3 2 1

Contents

ABBREVIATIONS AND SYMBOLS xi
ARTICLES TO APPEAR IN VOLUME 37 xiii
INTRODUCTION xv

I. Transposable Elements in *Drosophila*

Drosophila Foldback Elements, Primate L1 Elements, and Transgenic Mice   3
S. Potter, B. Heineke, S. Kaur, G. Jones, J. Lloyd, J. McNeish, M. Mucenski, W. Scott, N. Smyth-Templeton, J. Stock, and K. Sulik

Transposable Elements in Natural Populations of *Drosophila*   25
Brian Charlesworth

The *hobo* Element of *Drosophila melanogaster*   37
William M. Gelbart and Ronald K. Blackman

Molecular Biology of *Drosophila* P-Element Transposition   47
Rhonda F. Doll, Paul D. Kaufman, Sima Misra, and Donald C. Rio

The Use of Molecularly Tagged P Elements
to Monitor Spontaneous and Induced
Frequencies of Transposon Excision
and Transposition   59
   Christopher Osgood and Sonya Seward

II. Transposable Elements in *Drosophila* (Continued)

Asymmetrical Exchanges and Chromosomal
Rearrangements in *Drosophila* (Abstract
only, see p. 325)
   B. H. Judd, E. A. Montgomery, S.-M. Huang,
   and C. H. Langley

Spread of P Transposable Elements
in Inbred Lines of *Drosophila melanogaster*   71
   Christine R. Preston and William R. Engels

Suppressible Insertion-Induced Mutations
in *Drosophila*   87
   Zuzana Zachar and Paul M. Bingham

Identifying and Cloning *Drosophila* Genes
by Single P Element Insertional Mutagenesis   99
   Lynn Cooley, Celeste Berg, Richard Kelley,
   Dennis McKearin, and Allan Spradling

Molecular Lesions Induced by I-R Hybrid
Dysgenesis in *Drosophila melanogaster*   111
   Isabelle Busseau, Alain Pelisson, Michèle Crozatier,
   Chantal Vaury, and Alain Bucheton

III. Regulation of Gene Expression

X-Chromosome Inactivation as a System of Gene Dosage Compensation to Regulate Gene Expression   119
    Mary F. Lyon

The Developmental Regulation of Albumin and α-Fetoprotein Gene Expression   131
    Sally A. Camper, Roseline Godbout, and Shirley M. Tilghman

A Methylation Mosaic Model for Mammalian Genome Imprinting   145
    Carmen Sapienza, Thu-Hang Tran, Jean Paquette, Ross McGowan, and Alan Peterson

Insertional Mutations in Transgenic Mice   159
    Frank Costantini, Glenn Radice, James J. Lee, Kiran K. Chada, William Perry, and Hyeung Jin Son

IV. Structure and Function of Repetitive and Unusual Sequences

The L1 Family of Repetitive Sequences in Mammals (Abstract only, see p. 327)
    M. H. Edgell, D. D. Loeb, R. Shehee, M. B. Comer, N. C. Casavant, and C. A. Hutchinson, III

Repetitive Sequences in the Human Genome (Abstract only, see p. 329)
    R. K. Moyzis

Transposition of Intracisternal
A-Particle Genes   173
  Kira Lueders and Edward Kuff

Use of Variable Number of Tandem Repeat
(VNTR) Sequences for Monitoring
Chromosomal Instability   187
  Paul M. Kraemer, Robert L. Ratliff,
  Marty F. Bartholdi, Nancy C. Brown,
  and Jonathan L. Longmire

V. Retroviruses

A Retroviral Insertion in the Dilute (*d*) Locus
Provides Molecular Access to This Region
of Mouse Chromosome 9   207
  Nancy A. Jenkins, Marjorie C. Strobel,
  Peter K. Seperack, David M. Kingsley,
  Karen J. Moore, John A. Mercer,
  Liane B. Russell, and Neal G. Copeland

Spontaneous Germ-Line Ecotropic Murine
Leukemia Virus Infection: Implications
for Retroviral Insertional Mutagenesis
and Germ-Line Gene Transfer   221
  Neal G. Copeland, Leslie F. Lock, Sally E. Spence,
  Karen J. Moore, Deborah A. Swing, Debra J. Gilbert,
  and Nancy A. Jenkins

The Specific Consequences of c-*fos*
Expression in Transgenic Mice   235
  Ulrich Rüther and Erwin F. Wagner

Mouse Endogenous Retroviral Long-Terminal-Repeat (LTR) Elements and Environmental Carcinogenesis 247

Wen K. Yang, L. -Y. Ch'ang, C. K. Koh, F. E. Myer, and M. D. Yang

VI. Molecular Analysis of Chromosomal Translocation and Gene Insertion

Molecular Genetics of Lymphoid Tumorigenesis 269

F. G. Haluska, Y. Tsujimoto, G. Russo, M. Isobe, and C. M. Croce

Molecular Analysis of Chromosome Breakpoints 281

John Groffen, André Hermans, Gerard Grosveld, and Nora Heisterkamp

Homologous Recombination in Mammalian Somatic Cells 301

Raju S. Kucherlapati

Gene Transfer into Primates and Prospects for Gene Therapy in Humans 311

Kenneth Cornetta, Robert Wieder, and W. French Anderson

Addendum

Abstracts 325

INDEX.. 331

Abbreviations and Symbols

All contributors to this Series are asked to use the terminology (abbreviations and symbols) recommended by the IUPAC-IUB Commission on Biochemical Nomenclature (CBN) and approved by IUPAC and IUB, and the Editors endeavor to assure conformity. These Recommendations have been published in many journals (*1, 2*) and compendia (*3*) and are available in reprint form from the Office of Biochemical Nomenclature (OBN); they are therefore considered to be generally known. Those used in nucleic acid work, originally set out in section 5 of the first Recommendations (*1*) and subsequently revised and expanded (*2, 3*), are given in condensed form in the frontmatter of Volumes 9–33 of this series. A recent expansion of the one-letter system (*5*) follows.

SINGLE-LETTER CODE RECOMMENDATIONS[a] (*5*)

Symbol	Meaning	Origin of symbol
G	G	Guanosine
A	A	Adenosine
T(U)	T(U)	(ribo)Thymidine (Uridine)
C	C	Cytidine
R	G or A	puRine
Y	T(U) or C	pYrimidine
M	A or C	aMino
K	G or T(U)	Keto
S	G or C	Strong interaction (3 H-bonds)
W[b]	A or T(U)	Weak interaction (2 H-bonds)
H	A or C or T(U)	not G; H follows G in the alphabet
B	G or T(U) or C	not A; B follows A
V	G or C or A	not T (not U); V follows U
D[c]	G or A or T(U)	not C; D follows C
N	G or A or T(U) or C	aNy nucleoside (i.e., unspecified)
Q	Q	Queuosine (nucleoside of queuine)

[a] Modified from *Proc. Natl. Acad. Sci. U.S.A.* **83**, 4 (1986).
[b] W has been used for wyosine, the nucleoside of "base Y" (wye).
[c] D has been used for dihydrouridine (hU or H_2 Urd).

Enzymes

In naming enzymes, the 1984 recommendations of the IUB Commission on Biochemical Nomenclature (*4*) are followed as far as possible. At first mention, each enzyme is described *either* by its systematic name *or* by the equation for the reaction catalyzed *or* by the recommended trivial name, followed by its EC number in parentheses. Thereafter, a trivial name may be used. Enzyme names are not to be abbreviated except when the substrate has an approved abbreviation (e.g., ATPase, but not LDH, is acceptable).

REFERENCES

1. *JBC* **241**, 527 (1966); *Bchem* **5**, 1445 (1966); *BJ* **101**, 1 (1966); *ABB* **115**, 1 (1966), **129**, 1 (1969); and elsewhere.† General.
2. *EJB* **15**, 203 (1970); *JBC* **245**, 5171 (1970); *JMB* **55**, 299 (1971); and elsewhere.†
3. "Handbook of Biochemistry" (G. Fasman, ed.), 3rd ed. Chemical Rubber Co., Cleveland, Ohio, 1970, 1975, Nucleic Acids, Vols. I and II, pp. 3–59. Nucleic acids.
4. "Enzyme Nomenclature" [Recommendations (1984) of the Nomenclature Committee of the IUB]. Academic Press, New York, 1984.
5. *EJB* **150**, 1 (1985). Nucleic Acids (One-letter system).†

Abbreviations of Journal Titles

Journals	*Abbreviations used*
Annu. Rev. Biochem.	ARB
Annu. Rev. Genet.	ARGen
Arch. Biochem. Biophys.	ABB
Biochem. Biophys. Res. Commun.	BBRC
Biochemistry	Bchem
Biochem. J.	BJ
Biochim. Biophys. Acta	BBA
Cold Spring Harbor	CSH
Cold Spring Harbor Lab	CSHLab
Cold Spring Harbor Symp. Quant. Biol.	CSHSQB
Eur. J. Biochem.	EJB
Fed. Proc.	FP
Hoppe-Seyler's Z. Physiol. Chem.	ZpChem
J. Amer. Chem. Soc.	JACS
J. Bacteriol.	J. Bact.
J. Biol. Chem.	JBC
J. Chem. Soc.	JCS
J. Mol. Biol.	JMB
J. Nat. Cancer Inst.	JNCI
Mol. Cell. Biol.	MCBiol
Mol. Cell. Biochem.	MCBchem
Mol. Gen. Genet.	MGG
Nature, New Biology	Nature NB
Nucleic Acid Research	NARes
Proc. Natl. Acad. Sci. U.S.A.	PNAS
Proc. Soc. Exp. Biol. Med.	PSEBM
Progr. Nucl. Acid. Res. Mol. Biol.	This Series

†Reprints available from the Office of Biochemical Nomenclature (W. E. Cohn, Director).

Articles to Appear in Volume 37

Polynucleotide–Protein Cross-Links Induced by Ultraviolet Light and Their Use for Structural Investigation of Nucleoproteins
  EDWARD I. BUDOWSKY AND GULNARA G. ABDURASHIDOVA

Regulation of Collagen Gene Expression
  PAUL BORNSTEIN AND HELENE SAGE

Left-handed Z-DNA and Genetic Recombination
  JOHN A. BLAHO AND ROBERT D. WELLS

Polycistronic Animal Virus mRNAs
  CHARLES E. SAMUEL

Mammalian β-Glucuronidase: Genetics, Molecular Biology, and Cell Biology
  KENNETH PAIGEN

Structure and Function of Signal Recognition Particle RNA
  CHRISTIAN ZWIEB

Eukaryotic DNA Polymerases α and β: Conserved Properties and Interactions, from Yeast to Mammalian Cells
  PETER M. J. BURGERS

Structure and Regulation of the Multigene Family Controlling Maltose Fermentation in Budding Yeast
  MARCO VANONI, PAUL SOLLITTI, MICHAEL GOLDENTHAL, AND JULIUS MARMUR

Introduction

The symposium on Transposable Elements in Mutagenesis and Regulation of Gene Expression marked the fortieth anniversary of a series of symposia begun in the Spring of 1948 by Alexander Hollaender shortly after he became the Director of the Biology Division at the Oak Ridge National Laboratory.* This symposium is dedicated and named in honor of the late Alexander Hollaender. It was held in Gatlinburg, Tennessee, in the setting and spirit of scientific excellence that he established when he originated the Biology Division's annual conferences.

The conference opened with dedications to Hollaender presented eloquently by Virginia White, his long-time Administrative Assistant, and Alvin Weinberg, Director of the Oak Ridge National Laboratory when Hollaender was Director of its Biology Division. Frank Ruddle presented a special evening lecture on The Physical and Functional Organization of Mammalian Genomes prior to a poster session contributed by conference attendees.

This symposium focused on current research to reveal, at the molecular level, how transposable elements cause mutations and regulate gene expression in *Drosophila*, mice, and humans. Integration of transposable elements into the DNA of germinal cells gives rise to a variety of heritable effects, ranging from minor alterations in developmental phenotypes through early lethality. Neoplasia is often associated with the integration of transposable elements into the DNA of somatic cells. Attempts are being made to integrate normal genes into the DNA of somatic cells with defective genes to mitigate inherited disorders. Research in these areas is presented and discussed in this collection of papers. In a few cases, the authors had already published their work elsewhere. For these, the preconference abstracts are included to indicate the actual composition of the entire conference.

The organizing committee, consisting of R. A. Popp, coordinator, J. L. Epler, R. J. Preston, E. M. Rinchik, and W. -K. Yang, thanks the Council for Research Planning in Biological Sciences, the National Institutes of Environmental Health Sciences, Martin Marietta Energy Systems Inc., and the Department of Energy for their generous support. We also thank all of the participants, as well as Norma Cardwell, ORNL Conference Office, Angela Barnes, Conference Secretary, and Waldo Cohn, editor, for their valuable contributions toward the success of the conference.

RAYMOND A. POPP
Biology Division
Oak Ridge National Laboratory
Oak Ridge, Tennessee

*Three earlier "Gatlinburg Symposia" have been published in this Series: Volume 19, mRNA; Volume 26, DNA: Protein Interactions; and Volume 29, Genetic Mechanisms in Carcinogenesis.

I. Transposable Elements in *Drosophila*

Drosophila Foldback Elements, Primate L1 Elements, and Transgenic Mice   3
S. POTTER, B. HEINEKE, S. KAUR, G. JONES, J. LLOYD, J. MCNEISH, M. MUCENSKI, W. SCOTT, N. SMYTH-TEMPLETON, J. STOCK, AND K. SULIK

Transposable Elements in Natural Populations of *Drosophila*   25
BRIAN CHARLESWORTH

The *hobo* Element of *Drosophila melanogaster*   37
WILLIAM M. GELBART AND RONALD K. BLACKMAN

Molecular Biology of *Drosophila* P-Element Transposition   47
RHONDA F. DOLL, PAUL D. KAUFMAN, SIMA MISRA, AND DONALD C. RIO

The Use of Molecularly Tagged P Elements to Monitor Spontaneous and Induced Frequencies of Transposon Excision and Transposition   59
CHRISTOPHER OSGOOD AND SONYA SEWARD

Drosophila Foldback Elements, Primate L1 Elements, and Transgenic Mice

S. Potter[1]
B. Heineke*
S. Kaur
G. Jones
J. Lloyd†
J. McNeish
M. Mucenski
W. Scott
N. Smyth-Templeton
J. Stock and
K. Sulik‡

Children's Hospital Research Foundation, Cincinnati, Ohio 45229
*Molecular Biology and Biochemistry, Wesleyan University, Middletown, Connecticut 06457
†Molecular Genetics, Biochemistry and Microbiology, University of Cincinnati College of Medicine, Cincinnati, Ohio 45267, and
‡Department of Anatomy, University of North Carolina School of Medicine, Chapel Hill, North Carolina 27514

I. Drosophila Foldback Elements

The *Drosophila* genome is rich with a variety of types of transposable elements. Retrovirus-like elements, including members of the *copia* family, were the first eukaryotic transposons to be molecularly cloned (1). *In situ* hybridizations to polytene chromosomes (2) and molecular analysis (3) demonstrated that these were indeed mobile elements. Subsequently, the "foldback" elements, with their long inverted terminal repeats (4), the P elements, with their short

---

[1] Speaker.

inverted terminal repeats (5), and the L1-type elements, with no terminal repeats (6), were cloned and characterized. These elements are not only structurally distinct, but also transpose by different mechanisms and have different impacts upon the genome.

In many ways, the *Drosophila* system is nearly ideal for the study of transposable elements. The short generation time, the small genome size, the presence of polytene chromosomes, the availability of a variety of strains, and the enormous power of the available genetics all contribute greatly toward a better understanding of transposable elements in this organism. Indeed, the transposable elements of *Drosophila* are currently the best understood of all animal transposons, and this is likely to continue. We are beginning to achieve a true molecular understanding of transposition mechanisms for several types of *Drosophila* transposable elements.

The foldback elements of *Drosophila* have a particularly interesting structure and exhibit a wide range of biological activities. These elements were originally "found" by simply hypothesizing that the genomes of eukaryotes would carry transposable elements with inverted repeats, as were known to exist in bacteria. A genomic DNA library screen using "snapback" sequences as a probe revealed the presence of a family of elements with a heterogeneous construction (4). The termini of the elements are conserved, but about 90% carry internal deletions of variable size.

The inverted terminal repeats of the foldback elements display a most unusual sequence organization (Fig. 1) (7). Although the outer termini are conserved, their sequence appears unremarkable except for the presence of scattered multiple copies of a 10-base-pair repeat. At about 300 bp from the termini, this 10-bp sequence grows by the addition of another 10 bp of different sequence, and these (now 20-bp) repeats are only separated by short stretches of variable sequence. At about 500 bp from the ends, this 20-bp repeat grows again by the addition of yet another 11 different bp, generating a 31-bp repeat. Contiguous repeat copies of this sequence then extend toward the center of the element. However, the many copies of this 31-bp sequence are not identical (8). There are, in fact, five distinct variations, which appear in periodic order, resulting in a higher level repeat of about 155 bp. Finally, at the junctions between the central loop sequences and these inverted repeats are distinctly different perfect inverted repeats of 308 bp (9). The function of this rather bizarre and complex sequence organization is uncertain, although testable models have been proposed (8).

The foldback elements appear to maintain a rather high steady-

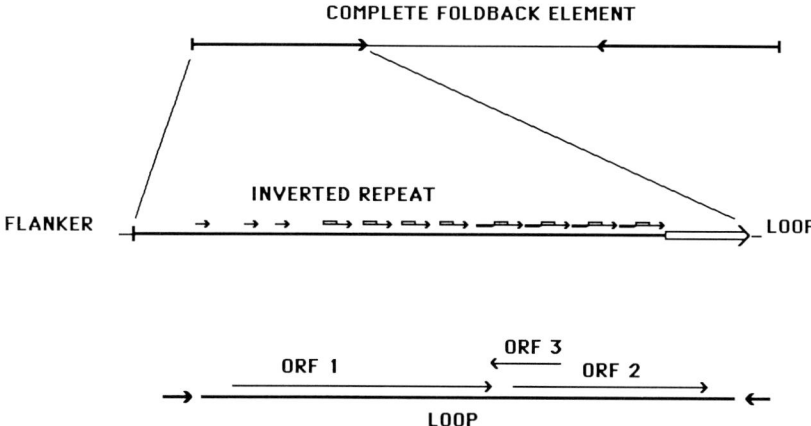

FIG. 1. A complete "foldback" element with a loop flanked by two inverted repeats is shown at the top. The structure of the inverted repeats is shown in the middle. Near the outer edge are sporadic copies of a 10-bp repeat, which then grows to a 20-bp repeat by the addition of 10 different base-pairs. This becomes a 31-bp repeat by the addition of yet another 11 different base-pairs, and at the boundary with the loop is found a distinct 308-bp repeat. Many more copies of the 10-, 20-, and 31-bp repeats are present than are shown. The loop with the three open reading frames (ORFs) is shown at the bottom.

state level of activity, unlike the P elements, which are extremely active, but only in bursts, and unlike the retrovirus-like elements that in many respects are relatively inactive. For example, the foldback elements can precisely excise, restoring wild-type DNA sequences, at a high frequency (10); in certain mutant stocks, these excision rates have remained constant for decades. It has also been observed that DNA rearrangements often occur within the vicinity of a foldback element, some of which result from imprecise excision events (11, 12). Some of these DNA alterations associated with particular elements appear to be similar in structure and to occur in a fixed percentage of the DNA from a variety of *Drosophila* strains (Potter, unpublished observations). Furthermore, it is particularly interesting to note that two foldback elements can sometimes move cooperatively, carrying large intervening blocks of DNA to new genomic positions (13, 14).

Of key interest is the elucidation of the molecular mechanisms by which the foldback elements transpose, excise, and control copy number. By analogy to plant elements, bacterial elements, and *Drosophila* P elements, it seems reasonable to consider that the foldback elements themselves might encode proteins that play important roles in these processes. However, the inverted repeat regions of the elements apparently carry no coding regions, and eight of the first nine elements analyzed apparently lacked central loop (potential coding) regions altogether. The one remaining element, designated FB4, that did carry a definite loop was sequenced and did carry a significantly long open reading frame (ORF). However, the structure of the loop-inverted repeat junctions was ambiguous, suggesting that the loop of FB4 might represent another transposable element inserted into a foldback element (8). Further studies demonstrated that this was the case, as the loop sequences of FB4 were found to lack the foldback inverted repeats at other genomic positions (15). The FB4 loop therefore represents a member of a distinct family of elements (designated HB elements); FB4 is a composite structure consisting of one of these HB elements inserted into a foldback element.

The HB elements are of considerable interest in their own right. First, their structure, with short inverted terminal repeats, places them in the same category as P elements, although there is no significant sequence similarity. This structure suggests that certain genetic crosses might trigger the movement of HB elements, as is the case for the similar P elements and *hobo* elements. This could be useful for the purpose of insertional mutagenesis since the various element families (HB, P, *hobo*) might display somewhat different target specificities. Thus, some genes might be targeted more efficiently by one family. Furthermore, S. Henikoff originally pointed out to us that the HB element ORF could encode a protein with dramatic amino-acid-sequence similarity to that encoded by the Tcl (transposable) element of *C. elegans*.

It is indeed remarkable that two families of transposable elements in two such phylogenetically distant species could be so similar. This strong sequence similarity suggests that the Tcl and HB elements are related and raises two possibilities. First, perhaps the common ancestor of *D. melanogaster* and *C. elegans* carried these elements, which were then passed "vertically," from generation to generation, in highly conserved form. Alternatively, perhaps the elements can spread "horizontally," somehow crossing even wide species barriers. Viral transduction, for example, is one possible mechanism for trans-species infection. Supporting the "vertical" explanation is the observation that the HB elements exhibit extreme DNA-sequence divergence from element to element, suggesting that the family is relatively ancient, so that individual elements in the genome have had ample opportunity to accumulate numerous mutations. Supporting the "horizontal spread" hypothesis, however, are observations suggesting that the P elements also might have been recently introduced into the *D. melanogaster* genome (*16, 17*). This indicates that cross-species transfer may indeed be possible.

There is, in addition, a theoretical basis for thinking that once a new active element enters a population, even if only as a rare isolated event, it might nevertheless spread rapidly throughout the population, despite possible harmful consequences (*18*). That is, the ability of a transposable element to replicate preferentially and to move copies of itself to new genomic locations gives it something of an infectious nature once introduced into a population. In a sexually reproducing population, the chromosomes are constantly mixed at each generation, as sperm meets egg, and chromosomes with transposable elements gain access to chromosomes without them. Replicative transposition can thus "contaminate" new chromosomes, resulting in a rapid spread of the elements, even if individuals carrying the elements are seriously reduced in fitness (*18*). In any case, the HB elements are quite interesting. They are either extremely ancient phylogenetically, or somehow capable of crossing wide species boundaries.

Returning to the foldback elements, it was eventually shown that about 10% of them do carry a well-conserved loop sequence that is unique to foldback elements, and nowhere else in the genome (*15*). These putative "complete" elements might then encode functions necessary for the activities of the foldback family.

Sequence analysis of such a complete foldback element revealed several interesting features (*9*). Three large ORFs encoding possible proteins of 633, 403, and 182 amino acids were found. The first two encode basic proteins that display significant, although short, homologies to proteins that interact with DNA. The first ORF has a 9 out of 13 amino-acid match to *Dpn*II DNA methylase from *Streptococcus pneumoniae;* the second ORF has a 8 out of 9 match with the β' subunit of RNA polymerase III from *Saccharomyces cerevisiae*. The third ORF is unusual in that it is largely embedded in the opposite strand of ORF two, and it shows homology to the E2 protein of Shope papillomavirus (10 of 14) and to the ns72 protein of the Middleburg virus (8 of 12). These sequence similarities are certainly interesting and suggestive of possible viral relationships and enzymatic functions for the encoded proteins. It must be emphasized, however, that these possible relationships remain very speculative at the moment.

In order to better understand the possible coding functions of the "complete" foldback elements, we have initiated efforts to purify the encoded proteins and directly analyze them (*9*). A decapeptide from the first hydrophilic peak of the first ORF was synthesized, coupled to keyhole limpet cyanin, and used to raise antibodies in guinea pigs. The purified antibody was then used to accomplish an immunoaffinity purification of protein from *Drosophila* embryos. This protein migrated as a single band on a sodium dodecyl sulfate/polyacrylamide gel, with an apparent mass of 71,000 Da, which closely matches the predicted value.

The antibodies were also used for immunofluorescent and immunoperoxidase studies designed to identify the location of the protein in adult *Drosophila*. Interestingly, the protein was found only in egg chambers. This suggests a germ-line specificity of expression for the gene, reminiscent of that previously observed for the P elements of *Drosophila* (19), and compatible with the observation that foldback insertions generally revert in germ-line, not somatic, cells (11).

As the foldback element proteins are purified, they will first be tested for possible specific binding to various portions of the foldback elements themselves. Possible functions could be suggested by such specific binding. Furthermore, with purified proteins in hand, it will be possible to test directly for possible enzymatic activities. It should become possible to work out the transposition and copy control mechanisms for these elements using procedures analogous to those that have already been applied successfully to the study of some bacterial transposable elements.

II. Primate L1 Elements

The L1 category of transposable elements is rather unusual. Unlike most other mobile elements, they do not carry either direct or inverted terminal repeats that are a part of the element. Instead, at their 3' ends, they carry dA-rich DNA that is reminiscent of the poly(A) "tail" of mRNA, suggesting a relationship between these elements and processed pseudogenes (20, 21). This notion is further supported by the observation that the elements are often lacking variable portions of the 5' end, as if the reverse transcription that generated them was often incomplete. When it was shown that the mammalian L1 elements carried long ORFs (22–24), this also appeared compatible with the model stating that the elements were actually pseudogenes derived from one or a few functional genes.

A distinguishing feature of the L1 elements, however, is that they encode a protein similar to reverse transcriptase (25). This suggests that they do not play a purely passive role in their own movement, as presumably most pseudogenes do, but rather facilitate the reverse transcription and transposition process themselves. This point, that the L1 elements are indeed transposable elements and not pseudogenes, is made dramatically by the I elements of *Drosophila*. These elements have a fairly standard L1 element construction and encode a protein with significant similarity to that encoded by mammalian L1 elements (26). The I elements of *Drosophila* are remarkable in that they are responsible for the I-R system of hybrid dysgenesis in *Drosophila*. That is, certain genetic crosses induce the movement of these elements at high rates. This ability to transpose rapidly in certain circumstances more closely resembles the properties associated with P elements than pseudogenes, and indicates that the category of L1 elements represents active, autonomous transposons, and not passive pseudogenes.

At least some of the L1 elements of the human genome are apparently still active and not mere relics of past transposition events. Several recent insertions of L1 elements into the human clotting-factor VIII (27) and *myc* (28) genes have now been documented.

Because the L1 elements encode a protein with some homology to reverse transcriptase, it is generally assumed that they transpose by a mechanism employing first transcription and then reverse transcription and chromosomal integration of the L1 element sequences. An inherent difficulty with this general scheme results from the observations suggesting that the L1 elements are transcribed by RNA polymerase II, which recognizes an upstream promoter. New copies of the element, derived from transcripts, would then lack the upstream regulatory sequences and would therefore not be transcribed and would remain inactive.

There are several possible solutions to this problem. First, perhaps there are indeed only a very few active elements with appropriate upstream regulatory regions, and the rest of the family of sequences is transpositionally inert. Second, perhaps the L1 elements are actually transcribed by an RNA polymerase that recognizes an "internal" promoter that could travel with the RNA transcript. This could be RNA polymerase III or even a yet-to-be-characterized, possibly self-encoded, RNA polymerase activity. The available data certainly do not exclude this possibility.

Yet another model has been proposed for a class of L1 elements carrying tandem repeats at the 5' termini (25). In this case, it was suggested that the tandem repeats could serve as "portable" promoters. That is, an upstream copy of this repeat sequence could serve to initiate transcription of downstream sequences, including downstream copies of the tandem repeat. The resulting new copy of the element, following reverse transcription and integration, would still carry some copies of the tandem repeat promoter sequence and would remain transcriptionally active. The tandem repeats could then be replenished in number by unequal crossing-over or by slipped mispairing during DNA replication.

The portable promoter model is attractive in that it resolves the question of how L1 elements transcribed by polymerase II, which lack the elegant structure of the retrovirus genome, could transpose to give new functional elements. Nevertheless, the model fails to explain why L1 elements of primates and *Drosophila* apparently lack the 5' terminal tandem repeats. Of course, it is possible that there are multiple solutions to this problem, and that while the mouse L1 elements use 5' tandem repeats, the L1 elements of other organisms may use other mechanisms. It is also possible that the primate genome, for example, carries a few L1 elements that do indeed carry 5' tandem repeats and that these are, in fact, the only active members of the family that can transpose. It is also possible that the model is simply wrong and that the repeats do not serve as portable promoters of RNA polymerase II.

Another question of key interest concerns the mechanism of concerted evolution of the L1 elements. The various copies of L1 elements of the human genome, for example, are about 90% homologous to each other. The mouse genome carries L1 elements that are also highly homologous to each other. But even though both of these species carry similar L1 elements, suggesting that these L1 sequences were present in a common ancestor, it is nevertheless observed that the mouse elements are much less homologous to the human elements than they are to each other. This argues that the elements were not simply passed down from the common ancestor, accumulating random mutations with time, for this would generate a situation in which the mouse elements would be as similar to human elements as to each other. Instead, it would seem that some mechanism exists that causes the elements of a species to evolve as a single unit, staying more similar to each other than they otherwise would. Although the concerted evolution of L1 elements has been documented and some possible mechanisms (such as gene conversion or recent family amplification) have been described and discussed (29, 30), nevertheless a satisfactory answer to the question of how this concerted evolution is accomplished has not appeared.

In performing a phylogenetic screen of the primate genome, looking for hypervariable repetitive sequences that had changed rapidly or had been recently acquired in evolutionary time, we found that the 5' regions of the L1 elements of humans and the prosimian primate, *Galago garnetti* (the African bushbaby), do not cross-hybridize under normal stringency (*31*). In order to better understand the nature of this 5' variation, a detailed analysis of the *galago* L1 5' region was conducted. Eighteen *galago* L1 elements were partially sequenced, and the 5' portions, presumably important in intiating the transcription and transposition of these elements, were well defined.

Several interesting observations were made (*32*). First, two very distinct *galago* L1 element subfamilies were found. The first 1.2 kb of the 5' portions of these subfamilies differ dramatically, with only limited, scattered sequence homology. The presence of these two subfamilies, with related, yet quite different, 5' sequences, places important constraints on the possible mechanisms of concerted evolution. It is also interesting to note that the scattered blocks of homology observed between these two subfamilies, and between these subfamilies and the L1 elements of the related primate, the Loris (*Nycticebus coucang*), indicate that the 5' regions of L1 elements evolve by wholesale rearrangement as well as by single-base substitutions.

The 5' end point of one of the *galago* L1 subfamilies was precisely defined by the identification of the 5' base beyond which the sequence appeared random, presumably due to the presence of flanking DNA. It is interesting to note that the first 16 bases of the three elements of this subfamily sequenced in this region are identical. The most striking feature of the sequence of this subfamily, however, is the presence of multiple copies of a tandem repeat. The tandem repeats are about 73 bp in length and are present in six to eight copies in the sequenced elements. The individual copies of the repeat are not identical, but are between 67% and 85% similar to a consensus. Twelve bases have been found to be invariant in the analyzed repeats (*32*).

It is also interesting to note that there is a detectable pattern of variability in the repeats. The first repeat of one element most closely resembles the first repeat of another element. Likewise, the second repeats of various elements tend to be more similar to each other than to other copies of the repeat. Indeed, it was possible to identify six different types of the repeat, and in those elements with more than six repeats, the sequence was compatible with the extra repeats resulting from an unequal crossover event.

Furthermore, the *galago* repeats were observed to carry a region with a highly significant similarity to the mouse L1 tandem repeats previously discussed. A 17-base span of DNA from the *galago* repeat matches the mouse repeat at 15 bases. This is particularly meaningful since two very short sequences, one of 73 bases and one of 208 bases, are being compared. A given 10-base sequence, for example, would be found by chance only once in a million-base stretch of random sequence. The occurrence of a 15 of 17 base match in these short tandem repeats therefore strongly suggests a relationship and perhaps a shared function for these elements. Moreover, the most conserved portion of the *galago* repeats (in comparing the various sequenced copies of the *galago* element) shows the homology to the mouse repeats.

Despite the discovery of these tandem repeats in a primate genome, and despite their sequence similarity with the mouse tandem repeats, it nevertheless seems unlikely that they serve the function that has been suggested for the mouse elements. The primate repeats cannot be portable promoters because they are not at the 5' terminus of the L1 elements, but rather are about 700 bases in from the end (32). Any transcript initiating from a repeat would therefore lack the 5' nonrepeat sequence of the L1 element and would be incomplete.

In sumary, the data suggest that these 5'-region tandem repeats represent a conserved structural motif in primates as well as rodents. The sequence similarity between the *galago* and mouse repeats indicates a common function, although the *galago* elements are subterminal and therefore are not likely to be serving as portable promoters. Of course, it remains possible that the tandem repeats of the mouse are multifunctional, sharing one purpose with the *galago* repeats as well as providing portable promoters. Nevertheless, the portable promoter model is clearly not a general solution to the problem of L1 element mobility, since many families of L1 element sequences in primates and *Drosophila* lack these structures. Indeed, the evidence suggests that elements without 5'-terminal tandem repeats can, by some unknown mechanism, generate complete copies of themselves. Perhaps an internal promoter recognized by RNA polymerase III or some other polymerase (self-encoded?) remains to be found. Alternatively, the elements might possess some mechanism for generating complete copies from RNA polymerase II transcripts that originate within the L1 elements, as the retroviruses are capable of doing. Of course, the L1 elements are structurally extremely different from the retroviruses, and any such mechanism would necessarily be quite novel.

III. Insertional Mutagenesis in Transgenic Mice

Transgenic mice can be generated by injecting DNA into the male pronucleus of a fertilized egg (33). In some cases, the foreign DNA will integrate into a chromosomal position, and the resulting mouse that develops after implantation into a surrogate mother will carry the transgene sequences in every cell of its body and transmit these sequences in Mendelian fashion to its offspring (34, 35). Generally, the injected DNA integrates at a single chromosomal position as a "head-to-tail" arrangement of a variable number of copies of the injected sequences.

In some cases (about 5–15% of transgenic mice), the foreign DNA will disrupt an endogenous gene and result in a detectable mutation. Such mutations are particularly useful for many reasons. First, the phenotype of the mutant line of mouse directly reveals the biological effect of altering the gene. That is, one can observe the *in vivo* impact of disrupting the gene and thereby gain important information concerning the significance and possible function of the gene. Second, the availability of a mutant line enables one to continuously generate more mutant mice, and makes it possible to employ the tools of embryology. That is, the mutant phenotype can be analyzed at different developmental times, thus revealing the possible influence of the mutation on the developmental program. The disrupted gene is marked by the presence of the foreign transgene tag. In principle, this should render molecular retrieval of the gene straightforward, and thereby allow one to employ all of the very powerful tools of recombinant DNA technology in the analysis of the gene and its function. In summary, the great power of the "transgene tagging" procedure of insertional mutagenesis derives from its combined approach. The tools of genetics, embryology, and recombinant DNA technology can all be brought to bear on the problem, with each aspect of the analysis complementing the others and leading to a more thorough understanding.

An especially powerful feature of this combined approach is the ability to examine developmental defects in insertional-mutant mice that have been "rescued" with *in vitro* manipulated versions of the wild-type gene. That is, the wild-type version of the disrupted gene can be used after its isolation to rescue otherwise mutant mice to confirm that the cloned gene was responsible for the observed phenotype. Then the wild-type gene can be manipulated *in vitro* in a systematic fashion affecting both regulatory and coding sequences, and then used again to attempt rescue of mutant mice. Embryological analysis of the resulting double transgenics (original disruption plus modified wild-type gene) will allow a functional construct dissection of both regulatory and coding regions of the gene.

The chief disadvantages of this approach are that it is laborious to generate transgenic mice and screen them for mutations, and the procedure is probably fairly random in targeting genes. It may well be true that most of the genes interrupted will be of the "housekeeping" or structural protein variety. Therefore, if one is particularly interested in identifying genes of considerable developmental consequence, genes that directly play a key role in the process of development, a great amount of brute force or random luck will be necessary to succeed.

Because most insertional mutations will be recessive in nature, it is necessary to mate siblings of a single transgenic line (derived from a single transgenic parent) in order to generate homozygotes (with two copies of the interrupted gene) that might display the mutant phenotype. The genotypes of the mice are determined by quantitative Southern blot analysis of DNAs derived from cut tails. By using the transgene sequences as a probe, it is possible to identify those mice with zero, one, or two copies of the transgene concatemer (string of repeats). The data accumulated by us and others suggest that about 5–15% of transgenic lines will carry an insertional mutation. The literature and our own data suggest that most of these insertional mutations result in prenatal lethality, which means that no mice with two copies of the transgene are born. We have now identified two insertions that result in such prenatal lethality, and these are currently under further investigation to better understand the nature of the defects. Initial characterization of one of these suggests a developmental arrest at around day 8 of embryonic development.

We have also been fortunate in the identification of an insertional mutation that appears to have disrupted a gene that plays a very important role in development (36). Homozygotes, with two copies of the disrupted gene, are born with a most unusual phenotype. They appear to lack hind limbs, suggesting the name "legless"; their forelimbs show an interesting pattern of abnormality, and their brains and craniofacial features are malformed. Figure 2 is a photograph comparing mutant and wild-type newborn mice.

The hind limb abnormalities are remarkably uniform. Over 60 homozygous *legless* mice have now been analyzed, and the hind limbs appear severely truncated in all cases. A skeletal analysis, as

FIG. 2. A newborn mouse with the *legless* (*lgl*) newborn mutant phenotype (A) is shown next to a wild-type littermate (B).

illustrated in Fig. 3, shows that the hind limbs terminate abruptly at the end of the femur. That is, the femur is indeed present and normal, but the leg simply stops at the end of the femur, with no distal structures present.

The forelimbs, in contrast, show variable expressivity. The pattern of abnormalities observed, however, is quite interesting. In a few cases, the forelimbs actually appear normal, but more typically, structures at the anterior and distal portions of the limb will be missing. For example, digits are almost always missing, and if only one digit is missing, then it is the most anterior one, the thumb. If two digits are missing, they are the two most anterior, the thumb and forefinger, and so on, as shown in Fig. 4. It is also common for the anterior long bone of the forearm, the radius, to be reduced in size or entirely absent. However, the posterior long bone of the forearm, the ulna, which resides immediately behind the radius, is always present and normal. In the most severe phenotype observed, the ulna is normal, but the entire radius and paw are missing.

Figure 5 is a diagrammatic representation of the patterns of forelimb malformations observed in several of the legless mice. It is

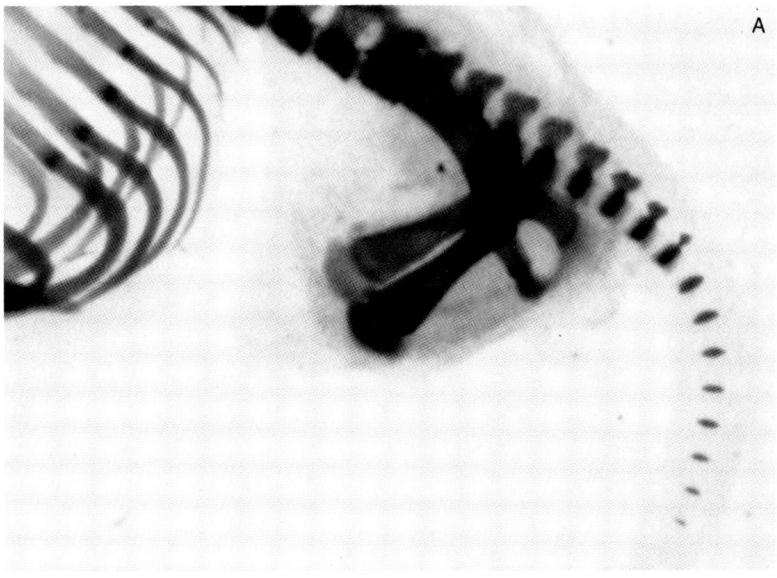

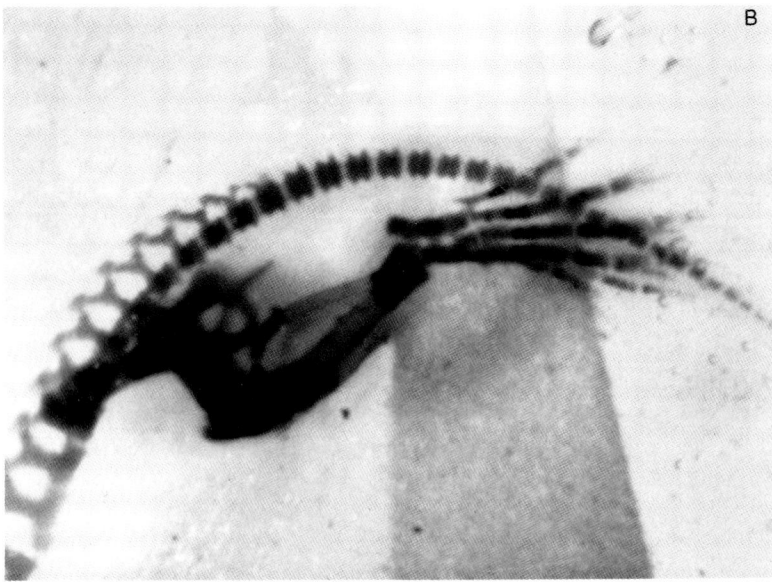

Fig. 3. Skeletal stains of hind limb regions of *lgl* (A) and wild-type (B) mice are shown. Only the femurs are present in mutant hind limbs.

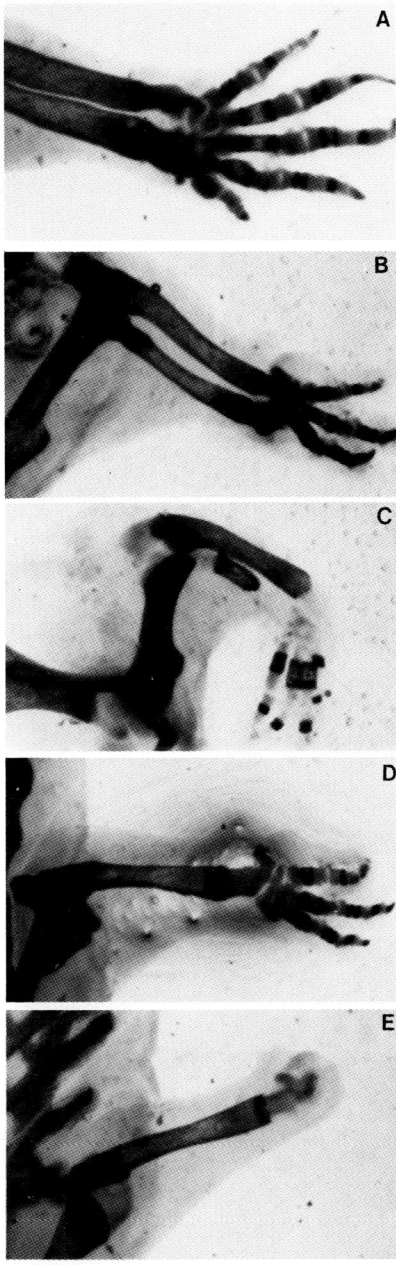

FIG. 4. Forelimb malformations of *lgl* mice: At one end (A) is a wild-type forelimb, and at the other (E) is the most extreme malformation observed, with missing radius and paw, although the ulna is present and normal. Intermediate manifestations (B-D) are also shown.

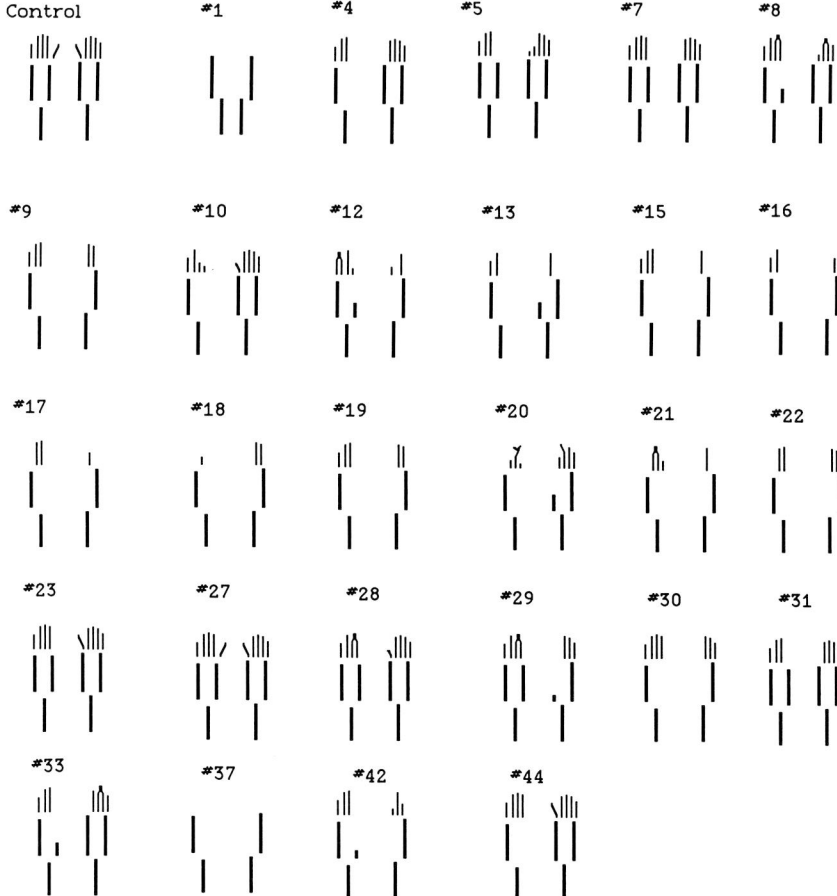

FIG. 5. Shown is the range of forelimb malformations observed in *lgl* mice. The small vertical lines represent digits, with the two lines below representing the ulna (outside) and radius (inside). Control (top left) and 27 mutant phenotypes are shown.

interesting to note that there is a tendency for a single mouse to have forelimbs that are similar in the severity of their malformation. Because the transgenic mice are generated from F_1 hybrid eggs and sperm coming from C57BL/6 × C3H/He hybrid mice, it is clear that the mutants resulting from interbreeding siblings will carry variable genetic backgrounds, and this could cause part of the observed variable expressivity. This is being tested by the generation of a C57BL/6 congenic line carrying the legless (*lgl*) mutation. Nevertheless, even single mice can carry two forelimbs with quite different abnormalities, suggesting that there are other factors at work in addition to genetic background.

The brains of the newborn *lgl* mice are also aberrant. As shown in Fig. 6, the mutant brains almost always lack olfactory lobes, the prominent, most anterior portion of the brain. Furthermore, the next most anterior part of the brain, the cerebrum, is generally malformed and carries darkened necrotic blebs, which in many cases are actually externally visible in the forehead region of the newborn mutants. It is interesting that these defects of the mutant brain involve missing and defective anterior structures, as do the forelimb defects. This provides a common link between these two manifestations of the phenotype, and suggests possible roles for the disrupted gene in the formation of these anterior structures.

The legless mice die within 2 days of birth, probably because of the brain defects, although we are not yet certain of the exact cause of lethality. Although simple measures, such as removing wild-type siblings from the litter, have failed to prolong survival, we have not yet attempted more difficult procedures, such as forced feeding.

We have begun to analyze the embryology of the mutant mice in order to begin to better understand the nature of the defects present.

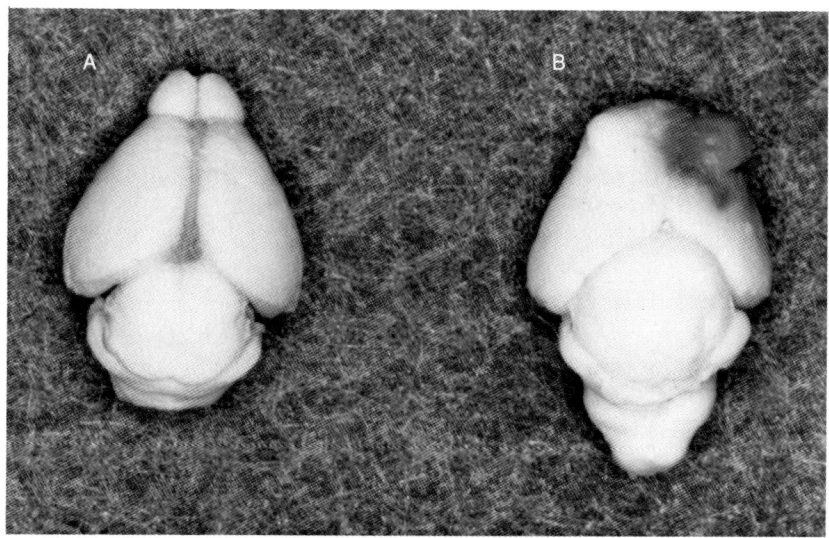

FIG. 6. Shown is a comparison of the brains of wild-type (A) and (B) newborns. The anterior (top) region of the mutant brain is seriously malformed, with missing olfactory lobes and a misshapen cerebrum, which includes necrotic bleb.

In examining the formation of the hind limbs, we have found that at days 11 and 12 of development, when limb buds are just beginning to grow into hind limbs, serious defects are already present (36). At day 10, although the mutant hind limb buds appear normal, there is nevertheless abnormal mesenchymal cell death occurring, which presumably plays an early causal role in the malformation.

Initial observations have also shown a very abnormal pattern of brain development in the mutant mice. At the 20-somite stage of development, the cells of the anterior neuroepithelium of the mutant brain are more loosely aggregated than normal, there is an aberrant plug of apparently deteriorated tissue, and once again there is within the neuroepithelium itself evidence of abnormal cell death. Moreover, the neuropore of the mutant brains is very late to close, sometimes remaining open to the 50-somite stage and beyond.

Several experiments show that this rather remarkable constellation of defects is the result of the disruption of an endogenous gene by the insertion of the transgene, not some other cause. Sixteen matings between heterozygous mice were set up and terminated prior to parturition. Of 118 progeny collected, 30 exhibited the mutant phenotype. That is, the litter sizes were normal, and about 25% of the progeny were mutant, as would be expected for a recessive insertional mutation. DNA was prepared from these mice and carefully quantitated by fluorimetry. Southern-blot analysis, using transgene sequences as a probe, then enabled us to determine which mice carried zero, one, or two copies of the transgene concatemer. The uniform result was that the mice with the mutant phenotype invariably had two copies of the concatemer (and therefore two copies of the disrupted gene), while mice with only one or zero copies were of the wild-type phenotype. These observations are quite consistent with the presence of a recessive insertional mutation.

In attempting to clone the DNA flanking the transgene concatemer, the first step in the isolation and identification of the disrupted gene, we have encountered a rather remarkable and unexpected phenomenon. For reasons that remain uncertain, the transgene sequences of the *lgl* mice are very resistant to cloning. The first efforts to isolate these sequences involved the use of the λ-phage vector EMBL3. About a dozen libraries of DNA from *lgl* mice that had been partially digested with *Sau*3A1, or in a few cases *Mbo*I, and size-selected, were constructed and directly screened without amplification. Packaging extracts from Amersham and Stratagene proved to be the most efficient. λ arms were made by us or purchased from Stratagene. A variety of cells was used for plating the phage, including

VCS 257, Le392, and DH1. In total, over five million phage were screened, using the transgene sequences as a probe, and no true positives were found. This is particularly remarkable, since the transgene concatemer is estimated to be about 200–400 kb in size, which means several hundred positive clones should have been found. It should also be mentioned that dozens of faintly hybridizing clones that fell into overlapping groups upon restriction analysis were found, and we suspect that these may represent endogenous mouse heat-shock genes that show faint homology to the *Drosophila* heat-shock genes in the transgene. [The transgene consists of a *Drosophila* heat-shock gene, a herpes virus thymidine kinase gene, and pBR322 sequences in a construct made by H. Pelham and designated PHT (37)].

We concluded at this point that some unknown feature of the construct was selected against in the process of λ cloning, and we proceeded to attempt cloning in pBR322 in a $recA^-$ derivative of *Escherichia coli* MC1061, named NM554. One million clones, again using *Sau*3A1 partials, were made and screened, this time with one true positive. The positive is interesting because it carried an insert of only about 200 bp (all from within the concatemer) even though the library was made with size-selected DNA of 4–8 kb; two dozen randomly selected clones that were analyzed to test the library all carried inserts in this size range. The data suggest that this small DNA segment, which presumably represented a trace contaminant of our size-selected DNA, was cloned because its small size enabled it to escape some very strong selection against the transgene DNA.

We subsequently made a number of small plasmid libraries using a variety of *E. coli* strains that were believed to be particulary deficient in restriction systems that might be removing the transgene sequences. These libraries were screened by Southern blots against total plasmid DNA preparations; none carried the transgene sequences.

Vigorous attempts at plasmic rescue have also been to no avail. Indeed, the data strongly suggest that the combination of sequences present in this transgene concatemer is somehow modified during passage in the mouse in a manner that causes it to be selectively resistant to molecular cloning by conventional methods. It is clear that this phenomenon is not a general feature of all transgenes, since some have been cloned without remarkable difficulty (38).

We are currently employing a highly modified version of the polymerase chain reaction in order to generate synthetic copies of the genomic DNA. This synthetic DNA is yielding transgene-carrying clones at the expected frequency. With the flanking sequences in hand, it will be possible to begin a molecular analysis of the biological basis of the *lgl* mutant phenotype.

References

1. D. J. Finnegan, G. M. Rubin, M. W. Young, and D. S. Hogness, *CSHSQB* **42**, 1053 (1978).
2. E. Strobel, P. Dunsmuir, and G. M. Rubin, *Cell* **17**, 429 (1979).
3. S. S. Potter, W. J. Brorein, P. Dunsmuir, and G. Rubin, *Cell* **17**, 415 (1979).
4. S. S. Potter, M. Truett, M. Phillips, and A. Maher, *Cell* **20**, 639 (1980).
5. G. M. Rubin, M. G. Kidwell, and P. M. Bingham, *Cell* **29**, 987 (1982).
6. I. B. Dawid, E. O. Long, P. P. DiNocera, and M. L. Pardue, *Cell* **25**, 399 (1981).
7. M. A. Truett, R. S. Jones, and S. S. Potter, *Cell* **24**, 753 (1981).
8. S. S. Potter, *Nature (London)* **297**, 201 (1982).
9. N. Smyth-Templeton and S. S. Potter, (submitted for publication).
10. M. Collins and G. M. Rubin, *Nature (London)* **303**, 259 (1983).
11. R. Levis and G. M. Rubin *Cell* **30**, 543 (1982).
12. S. Potter, *MGG* **188**, 107 (1982).
13. G. Ising and C. Ramel, in "The Genetics and Biology of *Drosophila*" (M. Ashburner and E. Novitski, eds.), p. 947. Academic Press, London, 1976.
14. R. Paro, M. L. Goldberg, and W. J. Gehring, *EMBO J.* **2**, 853 (1983).
15. H. L. Brierly and S. S. Potter, *NARes* **13**, 485 (1985).
16. M. G. Kidwell, *Genet. Res.* **33**, 205 (1979).
17. M. G. Kidwell, in "The Genetics and Biology of *Drosophila*" (M. Ashburner, H. L. Carson, and J. N. Thompson, Jr., eds.), Vol. 3C. Academic Press, London, 1982.
18. D. Hickey, *Genetics* **101**, 519 (1982).
19. F. A. Laski, D. C. Rio, and G. M. Rubin, *Cell* **44**, 7 (1986).
20. M. R. Singer, *Int. Rev. Cytol.* **76**, 67 (1982).
21. J. H. Rogers, *Int. Rev. Cytol.* **93**, 187 (1985).
22. L. Manuelidas, *NARes* **10**, 3211 (1982).
23. S. S. Potter, *PNAS* **81**, 1012 (1984).
24. S. L. Martin, C. F. Voliva, F. H. Burton, M. H. Edgell, and C. A. Hutchinson III, *PNAS* **81**, 2308 (1984).
25. D. D. Loeb, R. W. Padgett, S. C. Hardies, W. R. Shehee, M. B. Comer, M. H. Edgell, and C. A. Hutchinson III, *MCBiol* **6**, 168 (1986).
26. D. H. Fawcett, C. K. Lister, E. Kellett, and D. J. Finnegan, *Cell* **47**, 1007 (1986).
27. H. H. Kazazian, Jr., L. Wong, H. Youssoufian, A. F. Scott, D. G. Phillips, and S. E. Antonarakis, *Nature (London)* **332**, 164 (1988).
28. B. Morse, P. G. Rotherg, V. J. South, J. M. Spandorfer, and S. M. Astrin, *Nature (London)* **335**, 87 (1988).
29. S. M. D. Brown and G. Dover, *JMB* **150**, 441 (1981).
30. S. L. Martin, C. F. Voliva, S. C. Hardies, M. H. Edgell, and C. A. Hutchinson III, *Mol. Biol. Evol.* **2**, 127 (1985).
31. J. A. Lloyd, A. N. Lamb, and S. S. Potter, *Mol. Biol. Evol.* **4**, 85 (1987).
32. J. A. Lloyd and S. S. Potter, *NARes* **16**, 6147 (1988).
33. J. W. Gordon, G. A. Seagnos, D. J. Plotkin, J. A. Barbosa, and F. H. Ruddle, *PNAS* **77**, 7380 (1980).

34. F. Constantini and E. Lang, *Nature (London)* **294**, 92 (1981).
35. T. A. Stewart, E. F. Wagner, and B. Mintz, *Science* **217**, 1046 (1982).
36. J. McNeish, W. Scott, and S. Potter, *Science* **241**, 837 (1988).
37. H. R. B. Pelham, *Cell* **30**, 517 (1982).
38. R. P. Woychik, T. A. Stewart, L. G. Davis, P. D'Eustachio, and P. Leder, *Nature (London)* **318**, 36 (1985).
39. L. J. Mizrokhi, S. G. Georgieva, and Y. V. Ilyin, *Cell* **54**, 685 (1988).

NOTE ADDED IN PROOF. Mizrokhi *et al.* (39) report that RNA polymerase II recognizes an internal promoter in the *Drosophila* L1 element *jockey*, initiating transcription at or near the beginning of the element.

Transposable Elements in Natural Populations of *Drosophila*

BRIAN CHARLESWORTH

Department of Ecology and
Evolution, University of Chicago,
Chicago, Illinois 60637

Models and data relating to the problem of the maintenance of transposable elements in randomly mating host populations are reviewed. The data on the distribution of copy numbers between individuals are largely concordant with expectation. The role of regulation of rates of transposition, and of various modes of natural selection, in maintaining an equilibrium in copy numbers in the face of transpositional increase in copy number are discussed. Tests for the role of selection against insertional mutations, and against chromosome rearrangements induced by exchange between homologous elements located at nonhomologous chromosome locations, are described, together with theory and data on the probability distribution of element frequencies at individual chromosomal sites. It is concluded that the available population data are consistent with the notion that element abundances are largely controlled by the interaction of transpositional increase in copy number with opposing forces, although the nature of these forces is not yet clear.

The genomes of most species contain families of repeated DNA that have the property of self-replication and movement to novel locations within the genome (1). This finding has generated a considerable amount of debate concerning their significance in evolution, and the means by which they are maintained within their host populations (2–5). While much of this debate was conducted initially in the absence of data on the distribution of transposable elements between host individuals in natural populations, and of clearly formulated models of their population biology, these deficiencies have, to some extent, been remedied in recent years (6–8).

We are presently far from being able to give conclusive answers to the above questions, but the theoretical and empirical framework for pursuing them has become considerably clearer. Both theoretical and empirical studies are reviewed here. The focus is on models and data relating to randomly mating, diploid populations with elements dispersed over multiple genomic sites, and with relatively high frequencies of recombination between different sites, since these provide the most clear-cut means of discriminating between alternative hypotheses.

I. Population Statics

A. Distribution of Element Numbers between Individuals

Consider a particular family of elements in a given generation. It is assumed that there is a number m of chromosomal sites in a haploid genome, into which members of the family are capable of inserting. At present, we have no reliable information concerning the value of m for any system, but it seems clear that it is usually a large number, probably of the order of several thousand in a multicellular organism such as maize or *Drosophila*. The number of elements belonging to a given family in the diploid genome of a given individual is denoted by n. Studies of populations of yeast, *Drosophila*, and bacteria show wide variation among individuals in the sites at which elements are located (see below), indicating that n is very much less than $2m$, the total number of sites available for occupation in a diploid genome. Such studies also indicate some variation between individuals of a population with respect to the value of n; $\bar{n}$ and V_n denote the mean and variance of n over individuals.

Simple quantitative genetics considerations permit some predictions to be made concerning the properties of the distribution of the number of elements between host individuals. The number of elements per individual can be regarded as a strictly additive, polygenic trait, since it is merely the sum of the number of elements at each site. If there is no linkage disequilibrium, and no variance in element frequencies between different sites, it follows from standard theory that there is a binomial distribution of n across individuals. If $\bar{n} < 2m$, this reduces to a Poisson distribution, such that $V_n \approx \bar{n}$. Variation between sites in element frequencies reduces the variance below binomial expectation; if there is a tendency for the frequency of occupation of one site to be higher if a neighboring site is occupied (positive linkage disequilibrium), there will be a greater-than-binomial variance (9).

Population data to test these predictions for diploid species seem to be available only for *Drosophila melanogaster*. They are mostly from the technique of *in situ* hybridization of labeled element probes to the polytene salivary chromosomes, enabling counts of the numbers of elements in the euchromatin of each chromosome arm to be made (e.g., *10*). The most reliable results are from experiments in which sets of isogenic chromosomes were extracted from natural populations, using the balancer chromosomes uniquely available in *Drosophila*. Overall, agreement of the data with the Poisson expectation is fairly good. For example, Table I shows the results of a survey of the distribution of ten families of elements over 14 X chromosomes isolated from a Maryland population (unpublished data).

At present, there seems little reason to doubt that the nature of the distribution of elements between individuals in natural populations simply reflects the consequences of Mendelian transmission, although more population data for a wider variety of species are needed.

B. Frequencies of Elements at Individual Chromosomal Sites

Identification of the sites of labeling of probes to sets of isolated salivary chromosomes also permits analysis of the frequencies of elements at individual chromosomal sites. The resolution of this technique is somewhat coarse, as the chromosomal locations of elements can only be determined down to the level of a salivary chromosome band. Nonetheless, a useful picture of the general features of the distribution of element frequencies can be obtained (*11, 12*). The results of population surveys show that, most of the time, a site is occupied only once in the sample for each element in the family. Because of the lack of resolution of the method, it is possible

TABLE I
TESTS OF THE POISSON DISTRIBUTION FOR TEN FAMILIES OF *Drosophila melanogaster* ELEMENTS ON A SAMPLE OF 14 X CHROMOSOMES FROM A MARYLAND POPULATION[a]

Element	$\bar{n}$	V_n	Element	$\bar{n}$	V_n
412	2.29	2.53	2217	1.79	1.41
copia	1.21	0.80	2161	4.07	5.46
2181	1.21	0.49	2210	0.64	0.55
2158	0.21	0.18	roo	11.4	4.40
297	4.43	4.88	2156	0.50	0.58

[a] The copy numbers shown omit the values for the base of the X chromosome (18D1–20A4), since many elements show a differential accumulation of elements in this region. Details of the elements studied are given by Brookfield *et al.* [J. F. Y. Brookfield, E. A. Montgomery, and C. H. Langley, *Nature (London)* **310**, 330 (1984)].

that at least some instances of apparent multiple occupancy of the same site are the result of insertions into different sites within the same salivary chromosome band (13). An alternative method is to employ restriction fragment mapping of a defined region of the genome to determine with much greater accuracy the locations of insertions of elements in a sample of genomes from a population. In *D. melanogaster*, this method has been applied to the *Adh* region (14, 15), to the heat-shock region (16) and to the *white* locus (17). Each of these studies indicates that element insertions appear to occur at unique locations within a sample of a few tens of genomes, indicating low frequencies at these locations in the population from which the samples were drawn. Southern restriction studies of complete yeast (18) and *Escherichia coli* (8) genomes similarly indicate low element frequencies at individual sites. Further analysis and interpretation of these data require consideration of possible dynamic processes affecting element frequencies and abundances, which are discussed below.

II. Population Dynamics

A. Population Models

Models of the transmission of a family of elements from generation to generation incorporate the following features.

1. *Change in copy number due to transposition.* The probability per generation that a given element in the germ line of an individual with n elements of the family in question produces a new copy that is inserted elsewhere in the genome is denoted by u_n. The subscript n is used to take into account the possibility of functional dependence of the probability of transposition of an element on the number of elements in the same cell; there is evidence for self-regulation of transposition probability (such that u_n is a decreasing function of n) in a variety of species (19). The value of u for most eukaryote elements seems to be of the order of 10^{-4} or less per generation (9, 20).

2. *Excision of elements.* The probability per generation that an element is excised from its location in the genome, and hence not transmitted to the next generation, is denoted by v (v could obviously also be written as a function of n, but there is little evidence that this is often so). There is experimental evidence for such excision events for bacterial transposable elements (21) and *Drosophila* P elements (22). There is no firm evidence for excision of the abundant retroviral-like elements of *Drosophila* (23). If transposition is regulated, and if the

rate of transposition at low copy number (u_0) exceeds v, there will be an increase in copy number until $\bar{n} \approx \hat{n}$, where $u_{\hat{n}} = v$, or all available sites have filled up with elements (9). In the former case, there will be a stable equilibrium at $\hat{n}$, with the frequency of elements at each site equal to $\hat{n}/(2m)$. If there is no regulation, transposition leads to fixation of elements at each site if $u_0 > v$. Unless regulation of transposition is so strong that there is no transposition at all above a certain threshold copy number, maintenance of an equilibrium with element frequency polymorphism requires a non-zero excision rate.

3. *Natural selection.* There are several ways in which natural selection can affect the abundance of elements within populations; these are outlined below. All of them have the common feature that the fitness of a host individual with copy number n for a given family is assumed to be a decreasing function of n, w_n. If this is the case, and if assumptions similar to those used above are made, the following equation is obtained for the change in mean copy number per generation (6, 9):

$$\Delta \bar{n} \approx \bar{n}(1 - \bar{n}/[2m])(\partial \ln \bar{w}/\partial \bar{n}) + \bar{n}(u_{\bar{n}} - v) \tag{1}$$

where $\bar{w}$ is the mean fitness of the population ($\bar{w} \approx w_{\bar{n}}$).

The approximate equilibrium value of $\bar{n}(\hat{n})$ is given by setting $\Delta \bar{n}$ to 0. It is relatively easy to find functional forms that result in an equilibrium with low element frequencies at each site; for example, with $w_n = \exp - \frac{1}{2}tn^2$, we have $\hat{n} \approx (u_{\hat{n}} + v)/t$. In the absence of regulated transposition, the stability of this equilibrium requires that the logarithm of fitness declines more steeply than linearly with increasing n (6, 9). Given that low frequencies of transposition seem to be the norm, these results imply that even a weak pressure of selection (as measured by t, which is the slope of the relation between the logarithm of fitness and copy number at a copy number of 1) is capable of maintaining a balance with transpositional increase in copy number. For example, a mean copy number of 50 elements per diploid genome would be maintained if t is one fiftieth of the excess of rate of transposition over excision. With transposition rates of the order of 10^{-4} or less, this implies that extremely small selection coefficients would be needed to maintain a balance. The loss in mean fitness to the population would be negligible, since the mean fitness of the population is given by $\bar{w}/ = \exp - \hat{n}(u/_{\hat{n}} - v)$, regardless of the mode of selection (6).

One possible mode of selection on transposable elements is through the deleterious mutational effects of the insertion of elements into or near genes (1). Studies of *Drosophila* mutations affecting viability show that the most frequent class of spontaneous mutations with detectable effects consists of those with relatively small, detrimental effects (24). Recent studies of P-element mutagenesis in *D. melanogaster* show that transposition is frequently accompanied by such detrimental mutations (25–27). The finding that a very weak intensity of selection is sufficient to check the spread of elements casts doubt on the possibility that selection against insertional mutations is the main factor in maintaining element frequencies, in the absence of self-regulated transposition or other forces, since the average effect of a deleterious mutation in *Drosophila* is much larger than 10^{-4} (24). It is hard to account for the low frequencies of elements at most sites, even if the possibility of a wide distribution of fitness effects across sites of insertion is allowed for, since sites where element insertions are nearly neutral would be expected to exhibit high frequencies.

These considerations lead to the conclusion that, although selection against insertional mutations may be a factor in stabilizing element frequencies in natural populations, it may not be the only force involved. One possibility is that regulation of transposition rates may act in addition to selection. This makes it much easier to achieve stabilization of element abundances at reasonable levels. Another possibility is that crossing over between homologous elements located at different chromosomal sites will lead to the production of deleterious chromosome rearrangements (10), and models of this process have been developed (28). If unequal exchange events between pairs occur with a frequency that is dependent only on the rates of crossing over in their neighborhood, the equilibrium density of elements in a genomic region is inversely proportional to the frequency of crossing over in that region. If unequal exchange is confined to elements at nearby locations, as is suggested by data from *Drosophila* (28), densities are inversely proportional to the square root of the frequency of exchange.

4. *Genetic drift*. Since the deterministic forces acting on transposable elements seem to be so weak, the full interpretation of statistics on element frequencies in natural populations requires that effects of genetic drift be taken into account. Stochastic models of element frequencies have been constructed (6, 9, 29–34). The form of the probability distribution of element frequencies for a given family that is attained when the forces of transposition, excision and selection considered above come into statistical equilibrium with random changes in element frequencies—the stationary distribution of element frequencies—has been derived (9, 32).

The following parameters provide a complete approximate description of the form of $\phi(x)$, the stationary probability density for element frequency x: N_e is the effective size of a local population; $\bar{n}$ is the expected value of the mean copy number, which can be approximated by the equilibrium value $\hat{n}$ for a large population; $\alpha = 4N_e u \bar{n}/(2m - \bar{n})$ measures the effect of drift, and the effect of transposition in causing insertions into a given site; $\beta = 4N_e(s + v)$, where s is the value of $-\partial \ln \bar{w}/\partial \bar{n}$ at $\bar{n} = \hat{n}$, and measures the joint effects of drift, excision, and selection.

ϕ is approximated by a β distribution:

$$\phi(x) \approx C\, x^{\alpha-1}(1 - x)^{\beta-1} \qquad (2)$$

where C is a constant of integration, equal to $\Gamma(\alpha + \beta)/\Gamma(\alpha)\,\Gamma(\beta)$ (see 9).

This is the formula for a closed population. If $\bar{n} << 2m$, the effect of migration can be included by adding $4N_e M$ to β, where M is the frequency of immigrants into a local population. If m is sufficiently large compared with $\bar{n}$, α can be neglected in Eq. (2). This yields the formula derived (32) for the expected number of sites per haploid genome with element frequency x, $\Phi(x) = m\, \phi(x)$.

B. Tests of the Models

Given the small probable magnitudes of the forces involved in determining element frequencies, apart from exceptional situations such as hybrid dysgenesis (22), it seems that attempts to discriminate among the various possibilities for the maintenance of stable element frequencies in natural populations by direct measurement of the quantities concerned are doomed to failure. Instead, tests of the predictions of the models for the statistical properties of populations are needed.

1. *Tests of the role of insertional mutations.* One such test has been devised and applied by Montgomery et al. (10). They noted that the hypothesis that frequencies are stabilized by the deleterious effects of insertional mutations predicts that element frequencies at equilibrium will be lower for X-chromosomal sites than for autosomal sites, due to the expression of X-chromosomal mutational effects in the hemizygous state in males, compared with the predominantly heterozygous state of rare autosomal genes. The partial recessivity of mutational effects on fitness (24) thus means that there will be a greater selective impact of X-linked transposable elements relative to autosomal elements.

The results of scoring the numbers of copies of the retroviral elements *copia*, *roo* and *412* on sets of 20 X, 2nd, and 3rd chromosomes from a natural population of *D. melanogaster* show no significant difference between the observed abundance of elements on the X and the expectation on the null hypothesis of no selection for *copia* and *roo,* whereas *412* shows a significant ($p < 0.05$) deviation from the null expectation, but agrees with the expectation on the insertional mutation hypothesis. It is unclear why these differences between families should exist; further data on a wider variety of families are clearly desirable. On the whole, these results suggest that, while there may be some effect of insertional mutations on copy number, it is not necessarily the chief mechanism involved in controlling copy number.

2. *Tests for effects of unequal exchange.* A test for the role of selection against chromosome rearrangements produced by unequal exchange between elements is provided by asking whether or not the abundance of elements tends to be higher in regions where exchange is reduced in frequency. *D. melanogaster* again provides useful material for carrying out such a test, as there is pronounced suppression of meiotic crossing over in the telomeric and centromeric regions of the euchromatin of the chromosome arms (35). It is straightforward to use *in situ* hybridization to determine the distribution of the numbers of elements over the salivary chromosome maps in chromosomes sampled from natural populations, and to compare this with the distribution expected on a null hypothesis of no variation in element abundance with respect to the rate of crossing over. A study of *roo* (28) shows clear evidence for an excess of elements at the base of the X chromosome, of the order of threefold over the number of elements that would be expected if elements were inserted in proportion to the physical size of the region. The picture is less clear for the autosomes. A survey of a sample of 14 X chromosomes from a Maryland population for ten families of elements shows an excess of elements near the base in six families; two of these also show an excess at the tip (Table II). DNA cloned from band 19E8, which is located close to the base of the X, also shows an unusual abundance of elements in this region (36). The data thus suggest that elements do tend to accumulate in regions where exchange is limited, although the effect seems to be much stronger for the base than the tip for reasons which are unclear.

3. *Data on the distribution of element frequencies.* Equation (2) can be applied to the analysis of *in situ* hybridization data on element frequencies, using the sampling theory previously developed (9, 13, 32).

Table III displays estimates of α and β obtained with the method of Charlesworth and Charlesworth (9) for a number of elements scored on chromosomes sampled from natural populations of *D. melano-*

TABLE II
DISTRIBUTION OF TEN FAMILIES OF ELEMENTS ACROSS THE X CHROMOSOME OF
Drosophila melanogaster[a]

Element	Tip	Mid	Base	Element	Tip	Mid	Base
roo	0.108[b]	0.708	0.185[c]	2158	0.000	0.150	0.850[d]
2156	0.000	0.259	0.741[d]	2210	0.000	0.750	0.250
2181	0.125[b]	0.406	0.469[d]	2161	0.088	0.750	0.162
copia	0.111	0.833	0.056	2217	0.066	0.767	0.167
297	0.085	0.671	0.244[d]	412	0.061	0.592	0.347[d]
Expected	0.07	0.810	0.120	Expected	0.07	0.810	0.120

[a] The data are for the same 14 X chromosomes sampled from a Maryland population shown in Table I. The base and the tip of the X chromosome are here defined as salivary chromosome sections 18D1–20A4 and 1A1–3A4, respectively.
[b] $p < 0.05$.
[c] $p < 0.01$.
[d] $p < 0.001$.

gaster. The generally large values of β obtained are highly significant when compared with small values, reflecting the low frequencies of multiple occupation. This effectively rules out the possibility that element frequencies are controlled by regulated transposition in the absence of excision, such that transposition rates are 0 in an equilib-

TABLE III
ESTIMATES OF THE PARAMETERS OF THE PROBABILITY DISTRIBUTION OF ELEMENT
FREQUENCIES FOR *Drosophila melanogaster* ELEMENTS FROM NATURAL POPULATIONS[a]

Element	Chromosome	α	β	Sample size
297	X	0.05	16.7	20
412	X	0	30.0	20
copia	X	0	40.0	20
roo	X	0.8	12.5	14
I	X	∞	∞	20
roo	3L	3.4	32.0	12
roo	3R	2.5	28.0	11

[a] $4N_eM$ is equal to 8.8 for autosomal loci and 6.5 for X-linked loci of East Coast *D. melanogaster* (37). The contributions of selection and excision to β can be estimated by deducting these values from the appropriate entries in the table. Data for 297, 412, and copia are from Montgomery and Langley (11); data for roo X chromosomes are original data of B. Charlesworth; data for roo 3rd chromosomes are from a re-examination by B. Charlesworth of material of Montgomery et al. (10); data for I are from Leigh Brown and Moss (12).

rium population (see Section II,B,2). An excellent fit to the observed occupancy profiles is obtained with these estimates of the parameters. For natural population data such as these, the effects of immigration must be included in β. Estimates of the value of $4N_e M$ in this species are available from studies of the geographic distribution of electrophoretic alleles (37), and these can be deducted from the estimated values of β in order to obtain the contribution from excision and selection. Except for *roo* on the X chromosome, these values are all of the order of 10 or more (note that N_e for the X chromosome is three fourths that for the autosomes, so that the difference between the X chromosome and autosome data for *roo* probably partly reflects the difference in N_e). If the expected mean number of elements per individual is unchanging over time, then from Eq. (1), $\beta \approx 4N_e u$. Estimates of N_e for East Coast *D. melanogaster* populations from the frequencies of allelism between recessive lethals suggest values of the order of 2×10^4 (38). With a β value of 20, this yields an estimate of 2.5×10^{-4} for u. The estimates of N_e are subject to considerable uncertainty, so that this value should not be taken too seriously.

III. Conclusions

The main conclusion from the models and data presented above is that there is nothing that we know at present that is strikingly inconsistent with the view that transposable elements are maintained in populations as a result of transpositional increase in copy number. This view also has the merit of providing detailed explanations for a range of phenomena associated with transposable elements. In particular, it convincingly explains the fact that elements in bacteria, yeast, and *Drosophila* are usually present at low frequencies at individual chromosomal sites into which they can insert; this fact is almost impossible to accommodate on the hypothesis that elements persist as a result of favorable mutations associated with their transpositional activities (5). Indeed, this observation suggests that elements in these organisms rarely or never have a positive role in directly promoting evolutionary change in association with an increase in their own frequency. It is possible, however, that certain classes of mutation, such as chromosome rearrangements (39–43), could be generated as a result of excision of elements or exchange between elements at different locations, and could be fixed in populations by drift or selection, without leading to an increase in element frequency at the sites in question. Thus, transposable elements could act indirectly as a source of mutational variation for evolutionary change, but this would not have much influence on their distribution within populations.

It should be emphasized that low frequencies of elements at individual sites are not necessarily predicted by the models; it is relatively easy to generate situations in which elements rise to high frequencies or even fixation at many chromosomal sites. There are, indeed, examples of such elements, the classic one being the very abundant mammalian *Alu* sequences (*44*), although the status of *Alu* as a genuine transposable element has been questioned (*45*).

The evidence on the distribution of elements within populations suggests that transpositional increase in copy number is generally balanced by some force or forces; it is at present unclear whether this involves self-regulation of transposition, selection against mutational effects of insertions, selection against chromosome rearrangements produced by exchange between elements at nonhomologous sites, or some combination of these.

Acknowledgments

This work was supported by National Science Foundation Grant BSR-8-16629 and U.S. Public Health Service Grant GM 36405-01, and by a grant from the Louis Block fund of the University of Chicago. I am grateful to Charles Langley for several years of discussions and collaboration on this topic.

References

1. J. A. Shapiro, "Mobile Genetic Elements." Academic Press, New York, 1983.
2. W. F. Doolittle and C. Sapienza, *Nature (London)* **284**, 601 (1980).
3. L. E. Orgel and F. H. C. Crick, *Nature (London)* **284**, 604 (1980).
4. A. Campbell, *in* "Evolution of Genes and Proteins" (M. Nei and R. K. Koehn, eds.), p. 258. Sinauer, Sunderland, Massachusetts, 1983.
5. M. Syvanen, *ARGen* **18**, 271 (1984).
6. B. Charlesworth, in "Population Genetics and Molecular Evolution" (T. Ohta and K. Aoki, eds.), p. 213. Springer-Verlag, Berlin, 1985.
7. J. F. Y. Brookfield, *Philos. Trans. R. Soc. London, Ser. B* **312**, 217 (1986).
8. D. L. Hartl, M. Medhora, L. Green, and D. E. Dykhuizen, *Philos. Trans. R. Soc. London, Ser. B* **312**, 191 (1986).
9. B. Charlesworth and D. Charlesworth, *Genet. Res.* **42**, 1 (1983).
10. E. A. Montgomery, B. Charlesworth, and C. H. Langley, *Genet. Res.* **49**, 31 (1987).
11. E. A. Montgomery and C. H. Langley, *Genetics* **104**, 473 (1983).
12. A. J. Leigh Brown and J. E. Moss, *Genet. Res.* **49**, 121 (1987).
13. N. L. Kaplan and J. F. Y. Brookfield, *Genetics* **104**, 485 (1983).
14. C. H. Langley, E. A. Montgomery, and W. F. Quattlebaum, *PNAS* **79**, 5631 (1982).
15. C. F. Aquadro, S. F. Deese M. M. Bland, C. H. Langley, and C. C. Laurie-Ahlberg, *Genetics* **114**, 1165 (1986).
16. A. J. Leigh Brown, *PNAS* **80**, 5350 (1983).
17. C. H. Langley and C. F. Aquadro, *Mol. Biol. Evol.* **4**, 651 (1987).

18. J. R. Cameron, E. Y. Loh, and R. W. Davis, *Cell* **16**, 739 (1979).
19. B. Charlesworth and C. H. Langley, *Genetics* **112**, 359 (1986).
20. W. B. Eggleston, D. Johnson-Schlitz, and W. R. Engels, *Nature (London)* **331**, 368 (1988).
21. N. Kleckner, *ARGen* **15**, 341 (1981).
22. W. R. Engels, *Philos. Trans. R. Soc. London, Ser. B* **312**, 205 (1986).
23. D. J. Finnegan and D. H. Fawcett, *Oxford Surv. Eukaryotic Genes* **3**, 1 (1986).
24. M. J. Simmons and J. F. Crow, *ARGen* **11**, 49 (1977).
25. K. Yukuhiro, K. Harada, and T. Mukai, *Jpn. J. Genet.* **60**, 531 (1985).
26. T. F. C. Mackay, *Genet. Res.* **48**, 77 (1986).
27. B. J. Fitzpatrick and J. A. Sved, *Genet. Res.* **48**, 89 (1986).
28. C. H. Langley, E. A. Montgomery, R. H. Hudson, N. L. Kaplan, and B. Charlesworth, *Genet. Res* (in press) (1988).
29. T. Ohta, *Nature (London)* **292**, 648 (1981).
30. T. Ohta, *Genet. Res.* **41**, 1 (1983).
31. T. Ohta and M. Kimura, *PNAS* **78**, 1129 (1981).
32. C. H. Langley, J. F. Y. Brookfield, and N. L. Kaplan, *Genetics* **104**, 457 (1983).
33. N. L. Kaplan, T. Darden, and C. H. Langley, *Genetics* **109**, 459 (1985).
34. N. L. Kaplan and J. F. Y. Brookfield, *Theor. Pop. Biol.* **23**, 273 (1983).
35. D. L. Lindsley and L. Sandler, *Philos. Trans. R. Soc. London, B* **277**, 295 (1977).
36. G. L. G. Miklos, M. J. Healy, P. Pain, A. J. Howells, and R. J. Russell, *Chromosoma* **89**, 218 (1984).
37. R. S. Singh and L. R. Rhomberg, *Genetics* **115**, 313 (1987).
38. T. Mukai and O. Yamaguchi, *Genetics* **82**, 63 (1974).
39. W. R. Engels and C. R. Preston, *Genetics* **107**, 657 (1984).
40. M. D. Mikus and T. D. Petes, *Genetics* **101**, 369 (1982).
41. G. S. Roeder, *MGG* **190**, 117 (1983).
42. M. L. Goldberg, J.-Y. Shen, W. J. Gehring, and M. M. Green, *PNAS* **80**, 5017 (1983).
43. P. S. Davis, M. W. Shen, and B. H. Judd, *PNAS* **84**, 174 (1987).
44. C. W. Schmid and C.-K. J. Shen, *in* "Molecular Evolutionary Genetics" (R. J. MacIntyre, ed.), p. 323. Plenum, New York, 1985.
45. J. Rogers, *Nature (London)* **317**, 765 (1985).

The *hobo* Element of *Drosophila melanogaster*

WILLIAM M. GELBART[1] AND
RONALD K. BLACKMAN

*Department of Cellular and
Developmental Biology, Harvard
University, Cambridge,
Massachusetts 02138*

Of the many families of mobile elements in *Drosophila melanogaster*, the P element has been the most useful to geneticists and molecular biologists. It can be purposefully mobilized by appropriate outcrosses of strains containing P elements (P-strains) to strains lacking them (M-strains) (*1–4*). While it is not clear if P-element insertion is truly random in the genome, certainly many genes are prone to inactivation by P-element integration. This has permitted the use of P elements for "transposon tagging" of genes of interest (*5*). The hypermobilization of P elements when introduced into the ooplasm of M-strain oocytes (said to have an M cytotype) has also been exploited to develop P-element vector systems for the stable germline transformation of *Drosophila* (*6, 7*).

In the last 2 years, another mobile element family termed *hobo*[2] (*8, 9*), with a molecular structure reminiscent of P (*10*), has been encountered by several laboratories, including our own, in independent examples of *hobo*-associated genetic instability (*11–16;* J. Lim, personal communication). While work on *hobo* is in its early stages, it is clear that the genetics of this transposable element system is quite similar to that of the P element. This report may serve as an introduction to the *hobo* system and summarizes our current state of information on the system.

I. The Molecular Organization of the *hobo* Element

Strains of *Drosophila* contain anywhere from zero to over 50 copies of *hobo* per haploid genome. Typically, strains bearing *hobo*

---

[1] Speaker.
[2] "*hobo*" because it is a wanderer, or transient element [Eds.]

elements have two to ten copies of a 3.0-kb element, and a larger, variable number of internally deleted elements (8, 9, 11). The sizes of the internally deleted elements differ among strains, but there is usually only one or a very few size classes of deleted elements in a given strain.

Based on its larger size, and on the frequency with which it was found in different strains, Streck *et al.* (9) originally suggested that the 3.0-kb version of *hobo* represents the intact transposase-producing element. One isolate of the element (called $hobo_{108}$) has been sequenced by them. This element is 3016 base-pairs (bp) long and has 12-bp inverted terminal repeats. [By comparison, the P element is 2907 bp long with 31-bp inverted terminal repeats (10). However, there is no evidence of homology between P and *hobo* sequences.] Within $hobo_{108}$, there are three large open reading frames along one strand; however, no direct information is available concerning the nature of *hobo* transcripts or protein products. The evidence we present below (Section III) demonstrates that at least one member of the $hobo_{108}$-size class is indeed capable of inducing integration of other *hobo* elements.

Deleted *hobo* elements generally retain both terminal repeats, but lack from 25 to 2000 bp of the internal sequences present in full-length elements. As stated above, the deleted elements represent the preponderant form of *hobo* in most strains, suggesting either that these elements transpose more frequently or that, when present in large number, they are less deleterious to the fly than are the full-length elements.

Strains have been classified as H (*hobo*-containing) or E (empty of *hobo*) based purely on molecular criteria (9). With reference to the P-M and H-E systems, virtually all strains encountered are P-H, M-H, or M-E (*10–11*). We have not encountered any P-E strains in our initial surveys. Unlike P, *hobo* elements can frequently be found in established laboratory strains.

II. The Genetics of *hobo* Element Mobilization

A. Introductory Comments

Several independent examples of *hobo*-associated instability have recently been encountered. We have encountered two completely independent *hobo* insertions into the decapentaplegic[3] gene that can

---

[3] "Decapentaplegic" indicated 15 independent defects in the appendages derived from the imaginal disks in *Drosophila*. [Eds.]

be mobilized at high rates (11, 17). Yannopoulos, Louis, and coworkers (12, 15) independently discovered that *hobo* is associated with a high frequency of chromosomal rearrangements involving a strain called 23.5 MRF. J. Lim (personal communication) found *hobo* to be responsible for at least some of the hypermutability of several X-linked loci associated with the *Uc* (unstable chromosome) system (13, 14). Quite recently, V. Corces and M. M. Green (personal communication) identified unstable mutations of the *yellow* locus associated with *hobo*. As with the history of P, now that *hobo* has been shown to be unstable, it appears likely that *hobo* will prove responsible for a number of unexplained instances of hypermutability in *Drosophila*.

In general, a consistent picture has emerged from observations of these different hypermutable systems. In the case of the other published system (23.5 MRF) (12, 15, 16), the studies are complicated by the presence of P elements in the hypermutable strains. This is not so for our unstable decapentaplegic strains, which arose in MH strains, and so we focus on our results in the following sections.

B. The *decapentaplegic* Gene: An Assay System for *hobo* Mobilization

The *decapentaplegic* (*dpp*) gene (2-4.0, 22F1,2), located near the tip of the left arm of chromosome 2, is a >55 kb gene controlling a variety of morphogenetic events (18–20). This gene appears to encode a single protein product, a member of the transforming growth factor-β family, which is translated from a series of alternatively initiated and spliced transcripts (18, 21). A notable feature of *dpp* is that it contains an extended (>25 kb) 3' *cis*-regulatory region (termed the disk region) devoted to controlling expression of the *dpp* product in specific regions of each of the imaginal disks of the developing larva (17, 18, 20). As the imaginal disks are the anlage (primordia) of the adult epidermis of the head, thorax, and terminalia, it is quite easy to identify recessive dpp^{disk} mutations on the basis of adult morphological defects (20).

Two independent wild-type isoalleles of *dpp* have spontaneously given rise to such morphological mutations. In characterizing these isoalleles and their mutant derivatives, we have found that each contains a *hobo* insertion embedded in the disk region (11, 17). In one strain, the insertion is of a nearly full-length 3.0-kb element (called dpp^{IR+}) and in the other, it is an internally deleted 1.9-kb element (called dpp^{+ORS}). These insertions are located about 6.5 kb apart on the *dpp* molecular map.

In each strain, the *hobo* element is inserted between the *dpp* transcription unit and a 3' *cis*-regulatory element that is required for the proper construction of the wing of the fly. Mutations of *dpp* that remove this *cis*-regulatory element engender the production of flies lacking a portion of the dorsal base of the wing and in which the wings are held out 90° laterally instead of being at rest along the longitudinal body axis. Because of the location of the *hobo* elements in dpp^{IR+} and dpp^{+ORS} between the *dpp* transcription unit and the held-out *cis*-regulatory element, two kinds of mobilization events could be detected as held-out alleles. (In one, the *hobo* element deletes the held-out *cis*-regulatory element and in the other, *hobo* is associated with a rearrangement, such as a translocation or inversion, which severs the *dpp* transcription unit from the held-out element.)

C. Properties of *hobo* Mobilization at the *decapentaplegic* Locus

1. Mating Schemes for Recovering Held-Out Mutations

Our best quantitative measures of *hobo* mobilization have come from outcrosses of strains carrying the dpp^{+ORS} insertion. We have examined the induction of held-out wing mutations in outcrosses of dpp^{+ORS}-bearing strains to other strains, some of which possessed full-length and deleted *hobo* elements (H strains) while others lacked all such elements (E strains) (11, 17). From either type of cross, F_1 individuals were crossed to flies homozygous for a tester chromosome bearing a deletion of the held-out *cis*-regulatory element. F_2 flies in which mobilization of the dpp^{+ORS} hobo element had occurred were detected by virtue of their held-out phenotypes. Thus, the events detected are occurring in the germ line of the F_1 individuals.

2. Rates of Mobilization Events

Using the held-out assay system, we have observed *hobo* mobilization in the germ lines of all of the dpp^{+ORS} outcrosses we have performed. These include crosses of (*a*) dpp^{+ORS} H-strain males by E-strain females, (*b*) dpp^{+ORS} H-strain females by the E-strain males, and (*c*) dpp^{+ORS} H-strain males by H-strain females (11, 17). In each case, the frequency of F_1 individuals carrying mosaic germ lines are approximately equal (in the range of 1–4% of the F_1 generation carry *dpp* mutations in their germ lines). However, in Cross (*a*), the mobilization events frequently occur in large clusters, whereas in Crosses (*b*) and (*c*), the clusters are one third to one quarter this size. This observation suggests that in the direction of H-strain males by

E-strain females, mobilization events might be occurring earlier in the development of the germ line than happens in the reverse direction. In addition, from some (but not all) crosses of type a, we have observed considerable frequencies of infertility of the progeny (11, 17). This infertility is associated with rudimentary development of the gonads, reminiscent of the gonadal dysgenesis (GD) sterility characteristic of crosses of P-strain males by M-strain females (3). Overall, we have some indications that reciprocal cross effects on mutability and fertility do occur, as is true for the P-M system. On the other hand, it is clear that these effects are not nearly as dramatic for H-E as they are for P-M.

We have seen one instance in which a strain bearing a derivative of the dpp^{+ORS} hobo insertion element is not mobilized upon outcross (17). By appropriate matings, the two full-length hobo elements present in this strain (called $dpp^{d\text{-}blk}$) were segregated away from the hobo-bearing $dpp^{d\text{-}blk}$ gene. The resulting strain lacked all full-length elements and retained 10 to 20 deleted elements in its genome. Upon outcross of males of this strain to E-strain females and subsequent mating of the F_1 to appropriate held-out testers, no germ-line mobilization events were observed. This experiment suggests that full-length hobo elements are required in trans for hobo mobilization. As we discuss below (Section III), this suggestion is corroborated by our demonstration of a hobo-mediated germ-line transformation experiment.

3. The Cytogenetics of hobo-Associated Held-Out Mutations

As expected, hobo-associated held-out mutations fall into two major groups. Approximately half are gross chromosomal rearrangements, while the bulk of the remainder are deletions that begin within the dpp gene and extend at least far enough to remove the held-out cis-regulatory element. Our held-out assay system is a very poor detector of mutations caused by the insertions of hobo into new sites within dpp; however, one such insertion of a hobo into the held-out regulatory element was detected in our studies (11). We briefly describe our observations on the two major groups of mutations.

a. Gross Chromosomal Rearrangements. In this category, we include mutations associated with translocations, inversions and more complex rearrangements of chromosomal sequences (11, 17). Three generalizations can be made. First, the sites of preexisting hobo elements are "hot spots" for production of rearrangements. Many of our rearrangements are between polytene regions already occupied by *hobo* elements. Second, the nearer two *hobo* elements are to one another, the more likely they are to be involved in a rearrangement. Third, and most surprisingly, about 95% of all rearrangements we detect are paracentric in nature (i.e., confined to a single chromosome arm). We have recovered rearrangements involving as many as nine breakpoints, all of which are confined to chromosome arm 2L, where the *dpp* locus resides.

This last generalization cannot be explained by the distribution of *hobo* elements in the dpp^{+ORS} strains. Approximately 10 to 15 elements are present in each of the five major chromosome arms. Rather, we think it likely that paracentric rearrangements abound because of (*a*) constraints imposed by the architecture of the gonadal nucleus on the distribution of chromosome arms (see 22 for a description of nuclear organization in *Drosophila* polytene chromosomes) and (*b*) the possible localization of mobilization signals to discrete regions of the nucleus, such that the elements in one chromosome arm might be preferentially rearranged. Similar observations of preferential paracentric rearrangements have been observed by others for *hobo*- (12, 14, 15, 23) and P-associated (24) events.

b. *Deletions.* In our dpp^{+ORS} crosses, approximately one half of all held-out mutations are associated with deletions of portions of the *dpp* gene. Many of these deletions extend beyond *dpp* into adjacent complementation groups, and the largest of these are associated with the removal of several polytene chromosome bands and presumably several hundred kilobases of DNA. More typically, however, our *hobo* deletions are much smaller. Among the cytologically normal *hobo* deletions we have characterized molecularly, the smallest removes approximately 1.1 kb of *dpp* DNA, whereas the largest extends for some 60 kb (R. K. Blackman, H. A. Irick, and W. M. Gelbart, unpublished observations). Most *hobo* deletions occur adjacent to the preexisting site of the *hobo* element in *dpp*, and most retain the *hobo* element intact at the deletion endpoint (as determined by detailed restriction mapping) (11, 17). These properties of *hobo* deletions make it feasible to accumulate a nested array of *hobo*-mediated deletions with one common deletion end point.

III. *hobo*-Mediated Germ-Line Transformation

The detailed characterization of the *hobo* system, as well as the refinement of mutagenesis tools based on *hobo*, would be facilitated by the advent of a method for introduction of exogenous *hobo* element into the genome. Given the precedent of the P-mediated germ-line transformation system (6, 7), our approach was quite straightforward. A plasmid containing a marked *hobo* element, in which the long open reading frame of *hobo* is interrupted by the insertion of a wild-type allele of the *rosy* (*ry*) gene, was constructed and designated H[har1]. A second plasmid, containing one of the two putative full-length *hobo* elements from the $dpp^{d\text{-}blk}$ strain, was also constructed, and is termed pHFL1. We have made several observations from a series of transformation experiments in which appropriate plasmids were injected into recipient embryos homozygous for the mutant *rosy* gene (ry^-). Adults developing from injected embryos were then individually mated to appropriate ry^- testers. Thus, in the progeny of injected individuals, transformants can be identified and recovered as ry^+ flies. The results of these experiments are reported elsewhere in detail (25).

First, from control experiments, in which H[har1] was injected into recipient ry^- E-strain embryos, no transformants were recovered. We presumed that this failure was due to the absence of full-length *hobo* elements. Such full-length elements were introduced in two ways: (1) the H[har1] vector was injected into embryos resulting from a cross of ry^- H-strain males by ry^- E-strain females, and (2) H[har1] and pHFL1 were co-injected into embryos of a ry^- E strain. Both injection series were successful, generating transformants in the germ lines of 25–30% of the fertile adults arising from the injected embryos. The transposons in approximately 40 transformed lines have been mapped to a chromosome and localized by polytene chromosome *in situ* hybridization using a *rosy* DNA probe. Twenty-eight of these have been examined molecularly by means of whole-genome Southern blots.

From the mapping and the molecular analyses, it is clear that the properties of H[har1] integration are the same for both injection series. First, each insertion event involves integration of a single copy of the marked *hobo* element at a given site in the genome. Second, each integration event has occurred at the *hobo* termini (± 100 base pairs). Third, each *hobo* integration is at a different molecular site in the genome, suggesting that a very large number of sites for *hobo* integration are available.

IV. A Comparison of the *hobo* and P Mobilization Systems

All of our evidence to date points to very similar phenomenology in the mobilization of P and *hobo* elements. Deleted elements of each type can be mobilized by the presence of functional full-length elements. Full-length elements can induce mobilization in *trans* if they are already present in the genome or if they are introduced by injection into early embryos. Mobilization generally occurs gonially, leading to the production of clusters of mutant offspring from a mutant germ line. Mobilization involves insertion events exactly at the ends of the short inverted terminal repeats, generating 8-bp direct repeats of host sequences (8, 10). These observations, together with the structural similarities of the elements, suggest that this family of elements operates through mechanisms common to both. However, thus far, it appears that cross-mobilization of *hobo* elements by full-length P elements does not occur (26) although this issue needs to be investigated further. It is intriguing that P and *hobo*, which have short inverted terminal repeats, have proven so tractable as insertion mutagens and transformation vectors, unlike the more common families of mobile elements, such as the retroviral-like elements possessing long terminal repeats (27).

V. Prospects for *hobo* as a Mutagen and a Transformation Vector

It is already clear that *hobo* is responsible for some instances of spontaneous mutation. Although it is difficult to make rigorous comparisons, it is our impression that the rates of *hobo* mobilization in our strains are comparable to rates observed for the P element in similar mutational systems. Indeed, transformation rates for *hobo* and P are quite similar (25, 28). Thus, we anticipate that *hobo* will provide another mutational tool that can be used in parallel or in conjunction with P for various purposes.

Minimally, *hobo* may have a different set of sequence preferences for integration, enabling it to be used as an insertional mutagen for loci refractory to P element insertion. More exciting is the possibility of the combined use of *hobo* and P for transformation and mobilization. For this to be feasible, a critical examination of the independence of the P and *hobo* mobilization systems must first be undertaken. If, as appears likely (26), P and *hobo* cannot cross-mobilize, we can foresee constructs in which, for example, transposition-incompetent P elements can be introduced into the genome using *hobo* termini and *hobo*-mediated germ-line transformation. This would provide us with a refinement to the very useful "jumpstarter" (29 and this volume, Cooley *et al.*, p. 99) and Δ-2,3 (30 and this volume, Preston and Engels, p. 71) elements, in which the sources of transposase can be integrated into any chromosomal location most convenient for particular purposes. In addition, we can envision constructs that will aid in the dissection of *cis*-regulatory sequences. Currently, *cis*-regulatory regions of a gene are dissected by engineering and transforming individual constructs in which a given *cis*-regulatory region placed next to a reporter gene and integrated via P element ends. We envision integrating *hobo* within the *cis*-regulatory region such that, after a stable transformed stock is generated using the P element transformation system, *hobo* can be mobilized, generating a nested set of deletions of the *cis*-regulatory region. These deletions would reside at the same position in the genome, eliminating concerns over position effects of different sites of integration.

While these constructs are fanciful at present, they represent some of the most promising applications of the *hobo* system to *Drosophila* molecular genetics, and work in progress in several laboratories should soon determine if these types of constructs are feasible.

References

1. P. M. Bingham, M. G. Kidwell, and G. M. Rubin, *Cell* **29**, 995 (1982).
2. W. R. Engels, *Genet. Res.* **33**, 137 (1979).
3. M. G. Kidwell, J. F. Kidwell, and J. A. Sved, *Genetics* **86**, 813 (1977).
4. M. J. Simmons, N. A. Johnson, T. M. Fahey, S. M. Nellett, and J. D. Raymond, *Genetics* **96**, 479 (1980).
5. G. M. Rubin, M. G. Kidwell, and P. M. Bingham, *Cell* **29**, 987 (1982).
6. G. M. Rubin and A. C. Spradling, *Science* **218**, 348 (1982).
7. A. C. Spradling and G. M. Rubin, *Science* **218**, 341 (1982).
8. W. McGinnis, A. W. Shermoen, and S. K. Beckendorf, *Cell* **34**, 75 (1983).
9. R. D. Streck, J. E. MacGaffey, and S. K. Beckendorf, *EMBO J.* **5**, 3615 (1986).
10. K. O'Hare and G. M. Rubin, *Cell* **34**, 25 (1983).
11. R. K. Blackman, R. Grimaila, M. M. D. Koehler, and W. M. Gelbart, *Cell* **49**, 497 (1987).
12. P. Hatzopoulos, M. Monastirioti, G. Yannopoulos, and C. Louis, *EMBO J.* **6**, 3091 (1987).
13. D. Johnson-Schlitz, D. Lim, and J. K. Lim, *Genetics* **115**, 701 (1987).
14. J. K. Lim, *Genetics* **93**, 681 (1979).
15. G. Yannopoulos, N. Stamatis, M. Monastirioti, P. Hatzopoulos, and C. Louis, *Cell* **49**, 487 (1987).
16. G. Yannopoulos, N. Stamatis, A. Zacharopoulou, and M. Pelecanos, *Mutat. Res.* **108**, 185 (1983).
17. W. M. Gelbart and R. K. Blackman, unpublished observations.
18. W. M. Gelbart, V. F. Irish, R. D. St. Johnston, F. M. Hoffmann, R. K. Blackman, D. Segal, L. M. Posakony, and R. Grimaila, *CSHSQB* **50**, 119 (1985).
19. D. Segal and W. M. Gelbart, *Genetics* **109**, 119 (1985).
20. F. A. Spencer, F. M. Hoffmann, and W. Gelbart, *Cell* **28**, 451 (1982).
21. R. W. Padgett, R. D. St. Johnston, and W. M. Gelbart, *Nature (London)* **325**, 81 (1987).

22. D. Mathog, M. Hochstrasse, Y. Gruenbaum, H. Saumweber, and J. Sedat, *Nature (London)* **308**, 414 (1984).
23. J. K. Lim, *CSHSQB* **45**, 553 (1981).
24. W. R. Engels and C. R. Preston, *Genetics* **107**, 657 (1984).
25. R. K. Blackman, M. M. D. Koehler, R. Grimaila, and W. M. Gelbart, *EMBO J.* (in press) 1989.
26. W. B. Eggleston, D. M. Johnson-Schlitz, and W. R. Engels, *Nature (London)* **331**, 368 (1988).
27. D. J. Finnegan and D. H. Fawcett *Oxford Surv. Eukaryotic Genes* **3**, 1 (1986).
28. A. C. Spradling, *in* "Drosophila: A Practical Approach" (D. B. Roberts, ed.), p. 175. IRL Press, Oxford, England, 1986.
29. L. Cooley, R. Kelley, and A. Spradling, *Science* **239**, 1121 (1988).
30. H. M. Robertson, C. R. Preston, R. W. Phillis, D. M. Johnson-Schlitz, W. K. Benz, and W. R. Engels, *Genetics* **118**, 461 (1988).

Molecular Biology of *Drosophila* P-Element Transposition

> RHONDA F. DOLL
> PAUL D. KAUFMAN*
> SIMA MISRA* AND
> DONALD C. RIO[1],*
>
> *Whitehead Institute for Biomedical Research, Cambridge, Massachusetts 02142, and*
> ** Department of Biology, Massachusetts Institute of Technology, Cambridge, Massachusetts 02139*

Transposable P elements are a family of mobile genetic elements found in *Drosophila*. They are responsible for the syndrome of genetic traits known as hybrid dysgenesis. Their transposition is genetically controlled and tissue-specific, occurring only in germ lines of progeny from a cross of wild (P-strain) males to laboratory (M-strain) females. We are attempting to understand the molecular basis of the genetic control and tissue specificity of P-element transposition, using a variety of approaches that include molecular biology and biochemistry as well as genetics and P-element-mediated transformation. This paper gives a current view of the molecular biology of *Drosophila* P-element transposition.

P elements are a family of transposable elements found in the fruit fly, *Drosophila melanogaster*. They are the causative agents of a syndrome of correlated genetic traits, known as hybrid dysgenesis, that occurs in the progeny of a cross between males carrying P elements (P-strains) and females that lack them (M-strains). It is known that P-element transposition is controlled genetically and only occurs when a P-strain male is mated to an M-strain female but not in the reciprocal M-male × P-female cross nor in a P × P cross (1, 2). P-element transposition also exhibits tissue specificity, occurring only

[1] Speaker.

in the germ line. P elements have been extensively analyzed at the molecular level and are known to encode two proteins: an 87-kDa protein that appears to be the transposase required for the high rates of P-element transposition as well as for the precise and imprecise excision of P elements, and a second, smaller protein of 66-kDa postulated to be a negative regulator of transposition (3). Furthermore, the tissue specificity of P-element transposition is not regulated at the level of transcription initiation, but at the level of mRNA splicing (4). This germ-line-specific mRNA splicing event allows functional transposase to be synthesized only in the germ line, thereby limiting P-element transposition to those tissues.

The molecular analysis of P elements identified two types of elements: complete autonomous elements of 2907 bp in length, capable of encoding transposase, and smaller, nonautonomous elements 0.5–2.5 kb in length that cannot encode transposase themselves but can be activated to move when transposase is provided in *trans*. All P elements carry perfect 31-bp inverted repeats at their termini and create an 8-bp duplication of DNA at the site of insertion (5). These molecular studies and a number of genetic observations led to the development of P elements as efficient vectors for the transfer of cloned DNA segments back into the *Drosophila* genome (6). Furthermore, P elements are useful in transposon-tagging experiments, as cell autonomous markers, and as mutagenic agents. It is conceivable that the use of P elements as genetic tools could be extended to organisms other than *Drosophila*.

I. Results and Discussion

A. P Elements and Hybrid Dysgenesis

Hybrid dysgenesis is a syndrome of correlated genetic traits that occurs during certain interstrain matings of *Drosophila* (7–9). The phenomenon was discovered by population geneticists who found that when certain *Drosophila* strains were interbred, a series of abnormalities ensued in the germ lines of the progeny from these matings. These defects included high rates of mutation, male recombination, chromosomal rearrangements, sterility, and abnormal germline development. These dysgenic traits were only observed when males isolated from wild populations (termed P, or paternally contributing, strains) were mated to females derived from laboratory populations (termed M, or maternally contributing, strains). No dysgenesis occurred when the reciprocal cross of mating laboratory males to wild females was performed or when wild males and females were mated to each other. Furthermore, the ability of P-strains to induce or repress dysgenic traits (termed "cytotype control") exhibited an interesting pattern of inheritance that was dependent not only on genotype but also on whether the female egg cytoplasm was derived from a P- or M-strain mother (10, 11). Also, a number of dygenesis-induced mutations were unstable and exhibited high reversion frequencies.

These and other genetic observations led to the hypothesis that P-strains carry a family of transposable elements, called P factors, that are the causative agents of hybrid dysgenesis, and that M-strains do not contain P factors. It is now thought that P factors were introduced into *Drosophila* in the wild within the last 30 years.

An excellent review of hybrid dygenesis and the pattern of inheritance of dysgenic traits has appeared recently (12). In addition to the P-M system, there is a second independent system of hybrid dysgenesis in *Drosophila* known as I-R (12). In fact, a third system of hybrid dygenesis in *Drosophila* involving the *hobo* family of transposable elements has recently been described (13).

B. Molecular Analysis of P Elements

The hypothesis that P-M hybrid dysgenesis was caused by a P-strain-specific family of transposable elements led Rubin et al. (14) to test this idea at a molecular level. A number of dygenesis-induced mutations at the *white* (*w*) locus were isolated in a dysgenic screen. The *white* locus had previously been cloned, thus allowing a direct molecular analysis of the mutants isolated in this screen. Analysis of DNA from the dysgenesis-induced w^- alleles indicated that they carried insertions of foreign DNA segments at the *white* locus. Furthermore, reversion of these alleles occurred at high frequency (i.e., they were unstable) and was accompanied by the loss of the foreign DNA insertion. Molecular cloning of DNA from the w^- alleles identified these foreign DNA segments as being middle repetitive DNA sequences present at about 50 copies per genome in P-strains but absent from M-strains. These studies clearly demonstrated that P-M hybrid dysgenesis is caused by a P-strain-specific transposon family, termed P elements.

The isolation of P-element insertions in the *white* locus facilitated a detailed molecular analysis of several independent P elements (5). These studies showed that there are two classes of P elements: a complete element, 2907 bp in length, and a class of elements heterogeneous in size, ranging from about 0.5 to 2.5 kb. The smaller elements appear to be derived from the 2.9-kb element by heterogenous internal deletions. It is known from genetic analysis (15) and embryo microinjection experiments (16) that complete 2.9-kb P elements encode a function required to catalyze transposition and excision (referred to as transposase). The smaller elements are incapable of encoding transposase themselves, but carry all of the *cis*-acting DNA sequences required for transposition and can be activated to move by providing transposase activity in *trans*. There are about 10–15 complete 2.9-kb elements and 30–40 smaller, nonautonomous elements in a typical P-strain (14) DNA sequence analysis indicated that all P elements carry perfect 31-bp inverted repeats at their termini. In addition, there is an 11-bp inverted repeat located 120–140 bp from each end (K. Moses, unpublished observations).

Analysis of many different sites of insertion showed that an 8-bp duplication of target-site DNA is created upon insertion (5). A very weak "consensus" sequence of the target sites was compiled and one common feature was their general (G-C)-rich nature. In addition, analysis of several w^+ revertants of the alleles induced by dygenesis indicated that P elements can also undergo precise excision, leaving only a single copy of the target-site duplication. P elements are also known to be excised imprecisely, leaving a portion of the element behind or taking adjacent non-P-element DNA sequences with them (17). In fact, imprecise excision appears to occur at a higher frequency than precise excision. DNA sequence analysis of the complete 2.9-kb P element indicated the presence of four extended open reading frames (ORFs) presumed to encode P-element functions responsible for transposition of the element and also perhaps for a regulator of transposition.

The molecular characterization of P elements facilitated their development as vectors for the transformation of *Drosophila*. Briefly, this technique involves the microinjection of cloned DNA into preblastoderm *Drosophila* embryos and subsequent P-element transposition from the injected plasmid onto the germ-line chromosomes. The ability to reintroduce altered P elements into the *Drosophila* genome allowed the analysis of the functions of P-transposable elements to be extended. In a series of experiments (18), frame-shift mutations were made in each of the four P-element ORFs and these mutations were introduced separately into a complete 2.9-kb P element. The resulting modified P elements were then transferred into the *Drosophila* germ line, using a helper plasmid that could provide transposase functions in *trans* but was itself unable to transpose. This helper plasmid was important, as the frame-shift mutations could have inactivated transposase functions. The mutant P elements were then examined for their ability to encode transposase, using a sensitive genetic assay for transposase activity (15). This assay relies on *singed-weak* (sn^w), a hypermutable allele of the *signed bristle* (*sn*) locus (20). sn^w is a dygenesis-induced allele that carries two tandem, nonautonomous P elements in a head-to-head orientation (19). In the presence of transposase, sn^w mutates at high frequency (up to 50%) to either a wild-type allele (sn^+) or to an allele with a more extreme phenotype (sn^e), depending on which one of the two tandem P elements is excised. This simple genetic test provides a sensitive assay for transposase function *in vivo*. The wild-type Pc[ry] element gave high levels of sn^w mutability. However, all of the frame-shift mutations in the four P element ORFs failed to give sn^w mutational activity, indicating that each ORF was needed to encode transposase. Furthermore, each pairwise combination of the frame-shift mutants failed to complement one another; this suggested that the four ORFs were spliced together in the P-element transcript to encode one transposase polypeptide. Karess and Rubin (18) also determined that a complete 2.9-kb P element encodes a 2.6-kb mRNA that starts at nucleotide 87 on the P-element sequence (5). In addition, RNA from dysgenic and nondysgenic embryos was examined by RNA blotting methods. The results indicated no change in the P-element transcriptional pattern or level in P-male × M-female (dysgenic) versus M-male × P-female (nondysgenic) embryos. Transcriptional regulation could occur in the germ cells, yet any change in RNA from these cells would be masked by the large proportion (95%) of somatic cells in the embryonic cell population from which the RNA samples were derived.

C. Genetic Control of P-Element Transposition

Genetic studies of hybrid dygenesis, the unique inheritance patterns of P factors, and the nonreciprocity of P-element transposition have led to the definition of two states (1). P-strain eggs are said to possess the P "cytotype" or "the cytoplasmic environment capable of preventing P-element transposition," e.g., in the case in which a P-element-carrying sperm enters an egg cytoplasm derived from a P-strain mother and no transposition occurs. Alternatively, M-strain eggs possess the M "cytotype" or "the cytoplasmic environment capable of allowing P-element transposition," e.g., when a P-element-carrying sperm enters an egg whose cytoplasm was derived from an M-strain mother and transposition occurs. A variety of genetic data following the inheritance and activity of P factors and cytotype suggest that something about the P elements themselves, either the

DNA itself or an encoded product is responsible for P-cytotype regulation (10, 20, 21).

Several models have been proposed to account for cytotype regulation. One, the "transposase titration" model, states that in P-strain eggs the presence of the endogenous P elements (perhaps as episomal circles generated by the action of transposase) titrate the transposase made by the elements introduced by a P-strain sperm; thus, transposase becomes distributed at a wide variety of sites but never reaches an active concentration at any individual P element (22). An alternative model hypothesizes the existence of two P-element-encoded functions, one being the transposase required for P-element transposition and excision, and a second product that functions as a negative regulator of transposition in P-strains (5). This model predicts that the regulator must not only prevent transposase synthesis or activity in P-strains, but must also activate its own synthesis in some way. It should be pointed out that neither of these models is exclusive and cytotype regulation might be brought about by a combination of both mechanisms. There are other scenarios involving either transcriptional or posttranscriptional negative control of transposition in the germ line that might account for cytotype control, but none has been tested experimentally in a definitive way. For example, it is possible that a negative regulator could function in the germ line to reduce transcription or proper splicing of transposase mRNA. In addition, it might be advantageous for the regulator to be present not only in the germ line but also in the soma to prevent "somatic dysgenesis."

Biochemical analyses have been carried out using P-element-specific antibodies to identify P-element-encoded proteins expressed in cultured Drosophila cells (3). These studies identified two proteins: a protein of 87 kDa made from an mRNA in which all three introns are removed, and a smaller protein of 66 kDa made from the somatic mRNA that retains the ORF-2–3 intron (Fig. 1). The 87-kDa protein encoded by ORF-0, -1, -2, and -3 was predicted from genetic experiments (18) to be the transposase, and a transient assay for P-element excision showed it to be the biologically active transposase polypeptide (3). The existence of the 66-kDa protein encoded by ORF-0, -1, and -2 has led to the postulate that this polypeptide is responsible for negatively regulating transposition. In fact, genetic data from W. R. Engels and H. Robertson (personal communication) suggest that truncated forms of the 87-kDa transposase, as well as the 66-kDa protein that carries 15 C-terminal amino acids not found in transposase (S. Misra and D. Rio, unpublished) can function as "repressors" in genetic tests.

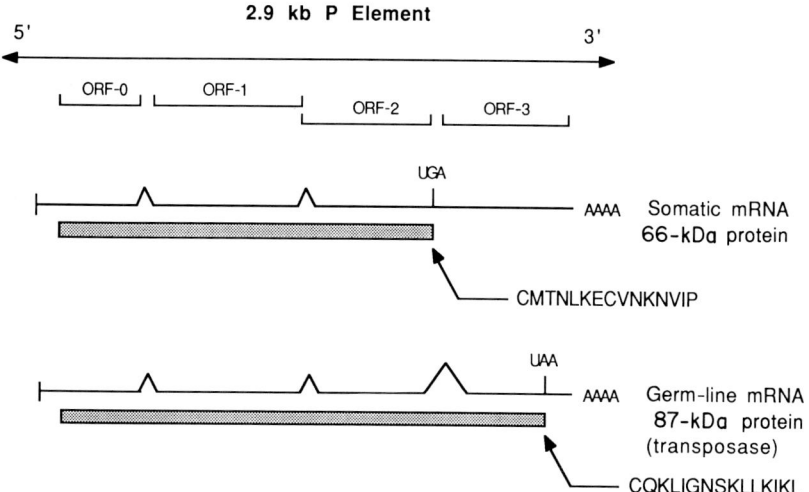

FIG. 1. Structure of the 2.9-kb P element, shown with four open reading frames (ORF-0, -1, -2, and -3). The 31-bp inverted terminal repeats are shown by arrows. Somatic and germ-line mRNA structures are shown with three introns (IVS-1, -2, and -3) denoted by lines joining the open reading frames. C-terminal peptides sequences of the 66-kDa and 87-kDa proteins are shown in the one-letter amino-acid code. [*J. Biol. Chem.* 243, 3557 (1968); Eds.]

Current studies are directed toward the purification and biochemical characterization of the 66-kDa and 87-kDa proteins in order to understand how one protein might negatively regulate transposition while the other one catalyzes it. It is conceivable that if both proteins bind to identical or similar DNA sites, the 66-kDa protein could compete for binding sites with transposase. Alternatively, since both proteins are very homologous except for the extended length of the 87-kDa protein and the 15 C-terminal amino acids of the 66-kDa protein, it is possible that the 66-kDa protein could interact with the 87-kDa transposase to form inactive heteroprotomers. Hopefully, these possibilities can be addressed with monoclonal antibodies recognizing either one or both proteins. These biochemical probes should allow detection of protein–protein or protein–DNA interactions as well as determination of the relative stoichiometries of the two proteins in the *Drosophila* germ line.

The P-element termini are required for P elements to undergo transposition and excision. It is known that, in addition to the 31-bp inverted repeats absolutely required for transposition (29, 18), adjacent unique DNA sequences are also required (29). Furthermore, P-element termini carry 11-bp inverted repeats that differ slightly (by 9 bp) in their location at either end (K. Moses, unpublished observations). M. C. Mullins, D. C. Rio, and G. M. Rubin (unpublished observations) have analyzed the *cis*-acting DNA sequences required for P-element transposition at the 3' P-element end. Using an *in vivo* assay for transposition involving a transposon with duplicated 3' ends, it was shown that over 150 bp of DNA are required at the 3' end for wild-type levels of transposition. This includes the 11-bp inverted repeat and unique DNA sequences between the 11-bp and 31-bp inverted repeats, although it has not been determined what role, if any, these unique DNA segments play in transposition. In addition, DNA sequences upstream from the 11-bp inverted repeat, although not essential for transposition, appear to enhance significantly the levels of transposition. Thus, the picture that emerges is one in which a large region of DNA is required at the 3' end for wild-type levels of transposition. The fact that different unique DNA sequences are located between the 31-bp and 11-bp inverted repeats, as well as the different locations of the 11-bp inverted repeats at either end, suggested that the 5' and 3' ends are nonequivalent. This idea was tested directly (M. C. Mullins *et al.*, unpublished observations) and it was shown that a 5' end will not substitute for a 3' end in transposition. This result may not be surprising, because the 5' and 3' ends are necessarily different; the 5' end also functions as the promoter sequence to initiate transcription of P-element mRNA and so much contain DNA-sequence signals for this process. Current studies are directed toward determining whether the 66-kDa and 87-kDa P-element proteins might bind to the P-element termini, and if other non-P-element-encoded proteins, presumably accessory proteins involved in P-element transposition, might recognize the terminal sequences. These studies should provide insight into the molecular mechanisms involved in P-element transposition and its regulation.

D. Tissue Specificity of P-Element Transposition

Normally, P-element transposition occurs only in the germ line of *Drosophila* (1). This tissue specificity has been studied with modified P-element derivatives, reintroducing them into the *Drosophila* genome and genetically determining if transposition occurs in the germ line, somatically, or both (4). Transcription of P-element coding sequences was placed under the control of the inducible hsp70 (70-kDa heat-shock-protein) promoter, known to be active in somatic tissues. This hsp70 P-element fusion gene was cloned into a transformation vector, introduced into *Drosophila*, and then assayed for its ability to produce P-element transcripts and to undergo transposition either somatically or in the germ line (see below). Although transcription of this gene fusion occurred at high levels in somatic tissues upon induction, no somatic transposition occurred, suggesting a posttranscriptional block to transposase activity.

The detailed structure of the somatic P-element mRNA was determined using RNase protection mapping, cDNA cloning, and direct RNA sequencing (4). The results indicated that, in the soma, a transcript is synthesized that has two intervening sequences removed such that the first three open ORFs (ORF-0, -1, and -2) were joined in the same translational frame (Fig. 1). There was no splicing event that joined ORF-2 to ORF-3. This observation was inconsistent with previous genetic observations (18) that all four ORFs are needed to encode transposase and should be linked in frame to encode a single polypeptide. This led to the hypothesis that there was a third intervening sequence that joined ORF-2 to ORF-3, but that this intron was removed only in the germ line. Thus, the somatic mRNA would encode a truncated, inactive form of transposase. This idea was tested using oligonucleotide-directed mutagenesis (4). Putative 5' and 3' consensus splice-site sequences (23) were located in ORF-2 and ORF-3, respectively, such that when this putative intron was removed, the two ORFs would be joined translationally. Single-base-pair changes were made in each of the conserved dinucleotides at the 5' (GT) and 3' (AG) splice sites. These mutations were known from other work to inactivate splice-site utilization (24, 25). In addition, a deletion was made that precisely removed the putative intron, thus joining ORF-2 to ORF-3 in the same translational frame.

These mutations were introduced into *Drosophila* and then assayed for their ability to produce transposase activity in genetic tests. Both the 5' and 3' splice-site mutations failed to give transposase activity in the sn^w test, suggesting that these single-base changes prevented splicing of the putative third P-element intron. Moreover, using the sn^w test, the element carrying the deletion of the third intron (called Δ2-3) exhibited transposase activity, not only in the germ line, but also in somatic tissues, as observed by bristle mosaics in which sn^w, sn^+, sn^e bristles were observed in the same individual. This somatic transposase activity was visualized independently using P-element transformants carrying the cell autonomous w^+ eye-color gene (26), thus allowing detection of somatic transposase-mediated events as patches of wild-type or mutant-type tissue. Genetic tests using $P[w^+]$ transposons indicated that both somatic excision and transposition could occur in strains carrying the Δ2-3 transposon (4).

These experiments illustrate three important points. First, the fact that single-base changes in the 5' and 3' splice-site consensus sequences abolished the ability of those elements to encode transposase suggests that there is an intron joining ORF-2 and ORF-3. Second, removal of the ORF-2-3 intron appears to be the sole determinant of transposase tissue specificity, since the Δ2-3 P element is capable of producing somatic transposase activity. Third, no germ-line-limited *Drosophila* proteins are required for P-element excision and transposition, since both activities are observed somatically with Δ2-3. However, the fact that transposase expression in somatic tissues allows transposition does not mean that there are not *Drosophila*-specific accessory proteins required for P-element transposition. Thus, expression of transposase in other organisms may not be sufficient to allow transposition.

The regulation of P-element germ-line-specific splicing seems likely to involve a positive factor present in the germ line, since even with the hsp70 P-element fusion construct, where P-element RNA is expressed at about 10 to 20-fold higher levels, no somatic transposase activity is observed. However, it is possible that a negative factor in the soma could be present in vast excess over the P-element mRNA. Candidates for a positive factor in the germ line involved in splicing might be germ-line-specific small nuclear ribonucleoprotein particles (snRNPs), known to participate in mRNA splicing (27), or perhaps an RNA binding protein (specific or nonspecific) that alters the P-element RNA structure such that it can now be recognized and spliced by the normal cellular splicing machinery. These two models make specific and testable predictions. For instance, a germ-line-specific snRNP or specific RNA binding protein would function in *trans* and act at defined sites at or near the intron, whereas a specific RNA structure might be perturbed by *cis*-acting mutations (not necessarily near the intron) that change the RNA–RNA base pairing or tertiary structural interactions. Genetic and biochemical studies aimed at addressing these possibilities are currently under way. Removal of the third P-element intron is unique among examples of alternative mRNA splicing in higher eukaryotes because, rather than using an alternate combination of donor and/or acceptor splice sites, the intron is either removed in the germ line or not removed in the soma. This fact should provide a powerful biochemical assay for experiments to define germ-line-specific factors involved in splicing the ORF-2–3 intron. In order to address this, we have recently developed an *in vitro* system to begin the study of *Drosophila* P-element mRNA splicing (28).

References

1. W. R. Engels, *ARGen* **17**, 315 (1983).
2. G. M. Rubin, *in* "Mobile Genetic Elements" (J. A. Shapiro, ed.), New York, p. 329. Academic Press, 1983.
3. D. C. Rio, F. A. Laski, and G. M. Rubin, *Cell* **44**, 21 (1986).
4. F. A. Laski, D. C. Rio, and G. M. Rubin, *Cell* **44**, 7 (1986).
5. K. O'Hare and G. M. Rubin, *Cell* **34**, 25 (1983).
6. G. M. Rubin and A. C. Spradling, *Science* **218**, 348 (1982).
7. J. A. Sved, *BioScience* **29**, 659 (1979).
8. J. C. Bregliano, G. Picard, A. Bucheton, A. Pelisson, J. M. Lavige, and P. L'Heritier, *Science* **207**, 606 (1980).
9. M. G. Kidwell, J. F. Kidwell, and J. A. Sved, *Genetics* **86**, 813 (1977).
10. W. R. Engels, *Genet. Res.* **33**, 219 (1979).
11. W. R. Engels, *CSHSQB* **46**, 561 (1981).
12. J. C. Bregliano and M. G. Kidwell, *in* "Mobile Genetic Elements" (J. A. Shapiro, ed.), p. 363. Academic Press, New York, 1983.
13. R. K. Blackman, R. Grimaila, M. M. D. Koehler, and W. M. Gelbert, *Cell* **49**, 497 (1987).
14. G. M. Rubin, M. G. Kidwell, and P. M. Bingham, *Cell* **29**, 987 (1982).
15. W. R. Engels, *Science* **226**, 1194 (1984).
16. A. C. Spradling and G. M. Rubin, *Science* **218**, 341 (1982).
17. S. B. Daniels, M. McCarren, C. Lowe, and A. Chovnick, *Genetics* **109**, 95 (1985).
18. R. E. Karess and G. M. Rubin, *Cell* **38**, 135 (1984).
19. H. Roiha, G. M. Rubin, and K. O'Hare, *Genetics* **119**, 75 (1988).
20. W. R. Engels, *PNAS* **76**, 4011 (1979).
21. M. G. Kidwell, *Genetics* **98**, 275 (1981).
22. M. J. Simmons and L. M. Bucholz, *PNAS* **82**, 8119 (1985).
23. S. M. Mount, *NARes* **10**, 459 (1982).
24. R. A. Padgett, P. J. Grabowski, M. M. Konarska, S. Seiler, and P. A. Sharp, *ARB* **55**, 1119 (1986).
25. M. R. Green, *ARGen* **20**, 671 (1986).
26. T. Hazelrigg, R. Levis, and G. M. Rubin, *Cell* **36**, 469 (1984).
27. T. Maniatis and R. Reed, *Nature (London)* **325**, 673 (1987).
28. D. C. Rio, *PNAS* **85**, 2904 (1988).
29. P. M. Bingham, M. G. Kidwell, and G. M. Rubin, *Cell* **29**, 995 (1982).

The Use of Molecularly Tagged P Elements to Monitor Spontaneous and Induced Frequencies of Transposon Excision and Transposition

CHRISTOPHER OSGOOD[1]
AND SONYA SEWARD

Department of Biological Sciences,
Old Dominion University,
Norfolk, Virginia 23529

Transposable elements (TEs) are ubiquitous in the genomes of higher eukaryotes and contribute substantially to the load of spontaneous mutations (1). McClintock (2) has suggested that TEs may be mobilized by so-called genomic stresses, and recent studies with sensitive prokaryotic and eukaryotic test systems support this contention. For example, a variety of chemicals (including, but not limited to, conventional mutagens) induce the mobilization of bacterial TEs (3, 4), *Drosophila* TEs in somatic tissues (5), and mammalian retroviruses (6). There is also suggestive evidence that the well-characterized yeast Ty1 elements may be similarly responsive to physical damage of the yeast genome (7).

The development of genetic test systems to monitor the spontaneous and induced mobilization of eukaryotic TEs is complicated by two factors: (*a*) that there are typically multiple copies of each class of TE in the genome, and (*b*) that TEs have no inherent phenotype. Both of these limitations can be largely overcome through the use of molecularly tagged TEs constructed *in vitro* and then transformed back into the host genome. TEs tagged with a gene conferring a visible or biochemical marker will, in most cases, lose the ability to catalyze their own mobilization; however, if these retain the critical terminal repeat sequences, they may be mobilized in *trans* by other functional TEs in the host genome (8). Thus, tagged TEs in conjunction with functional "helper" elements provide a promising system for identifying those environmental stresses that may promote transposition.

[1] Speaker.

The P elements of *Drosophila* provide a promising system for development of transposition assays. P elements are flanked by inverted repeat sequences and encode a tranposase, and both natural and tagged elements are mobilized by specific genetic crosses (9, 10). Clark and Chovnick first suggested that marked P elements might be used to monitor induced rates of mobilization (11), and in this paper we present the results of efforts in our laboratory to develop sensitive and reproducible assay systems permitting the identification of chemicals ("transposagens") promoting P-element transposition.

I. Genetic Systems Promoting Low Levels of Tagged P-Element Mobilization

The mobilization frequency of tagged P elements is known to be a function, in part, of the size of the inserted tag, as well as the chromosomal integration site (12). In the experiments described here, we have employed a P element marked with the wild-type *white*$^+$ (w^+) gene (designated here $P[w^+]$), which is inserted on the X chromosome (for a full description, see 13). This element is mobilized in *trans* by functional helper elements, but at a reduced rate relative to that of a native P element (13, and unpublished observations). Using the X-linked $P[w^+]$ element, a simple genetic system was devised that permits detection of both losses of $P[w^+]$, recoverable as phenotypically white-eyed males, as well as transpositions of $P[w^+]$ from its resident site on the X chromosome to an autosome, recoverable as w^+ females (see Fig. 1A).

In the absence of genetically introduced helper elements, $P[w^+]$ is stable (see Table I, Cross Type 0). Functional helper elements were introduced in two types of crosses illustrated in Fig. 1B. Cross Type I is a nondysgenic cross in which helper elements are derived from P-cytotype females; the test males derived from this type of cross are also expected to be of P cytotype. In this background, P elements are relatively stable (14). The summary of the control results with Cross Type I are presented in Table I. As can be seen, losses of $P[w^+]$ were recovered at a rate of approximately one w male per 750 offspring, while transpositions occur at a lower rate of one w^+ female per 2400 offspring. Both frequencies are significantly higher than those from Type 0 crosses. Not shown in Table I is the distribution of exceptional offspring; transpositional females, and particularly loss males, were recovered in clusters, suggesting that mobilization events were occurring largely in immature, premeiotic male germ cells. This result is entirely consistent with previous results showing that P elements are most active in immature germ cells (9, 10).

MOLECULARLY TAGGED P ELEMENTS

A. Monitoring mobilization of activated, molecularly tagged P elements

Cross Type I or II:

$$X, P[w^+]*/Y \ \male\male \quad \times \quad X\char`\^X, y\ w/Y\ \female\female$$

$$\downarrow$$

Regular offspring
X, P[w$^+$]/Y ♂
w$^+$

X$\char`\^$X, y w/Y ♀
w

Exceptional offspring
X, w/Y ♂
loss of P[w$^+$] – w

X$\char`\^$X, y w/Y; P[w$^+$]/+ ♀
transposition of P[w$^+$] – w$^+$

B. Genetic schemes for the *trans*-activation of a molecularly tagged P element

Cross Type I (nondysgenic):

$$X, P[w^+]/Y\ \male\male \quad \times \quad X\char`\^X/Y; P/P\ \female\female$$

$$\downarrow$$

X, P[w$^+$]*/Y; P/+ ♂♂ — test as in part A

Cross Type II (weakly dysgenic):

$$X/Y; MR^{h12}/Cy\ \male\male \quad \times \quad X, P[w^+]/X, P[w^+]\ \female\female$$

$$\downarrow$$

X, P[w$^+$]*/Y; MR^{h12}/+
or
X, P[w$^+$]*/Y; Cy/+   ♂♂ – test as in part A

FIG. 1. (A) Genetic strategy for monitoring mobilization of a tagged P element, P[w$^+$]. *Drosophila* males bearing a w$^-$ P[w$^+$] X chromosome, and carrying autosomal helper P elements introduced via either Cross Type I or II (see B), are mated with attached-X females C(1)DX, y w Y (see *17* for definition of symbols. Regular sons from this cross inherit the paternal X chromosome and are w$^+$, while regular daughters inherit the maternal attached-X and are h/w. Excisional losses of P[w$^+$] from the paternal X chromosome are recovered as w$^-$ sons; transpositions of P[w$^+$] to an autosome can be detected by the recovery of y/w females. (B) Two genetic schemes for *trans* activation of the X-linked P[w$^+$] reporter. In Cross Type I, P[w$^+$] males are crossed with attached-X, P-strain females and their nondysgenic sons are recovered for testing. In Cross Type II, helper P elements are derived from a Q-strain male (MR^{h12}/Cy) and their weakly dysgenic sons, carrying either the MR^{h12} or Cy homolog, are tested. Test males recovered from either cross carry autosomal copies of functional P elements that activate P[w$^+$] in *trans*. A modification of Cross Type II (Type II'), in which the female parent supplies autosomes which also bear internally deleted P elements, is described in the text. *, Activation in *trans* of P[w$^+$] by introduced helper elements (see B).

TABLE I
SPONTANEOUS MOBILIZATION FREQUENCIES OF TAGGED P ELEMENTS[a]

Cross (Description)	Frequency of $P[w^+]$	
	Loss	Transposition
1. Type 0 (no helper P elements)	1/21,830 (0.005%)	0/21,830
2. Type I (nondysgenic)	39/29,211 (0.13%)	12/29,211 (0.04%)
3. Type II, MR^{h12} (dysgenic)	136/25,374 (0.54%)	10/25,374 (0.04%)
4. Type II, Cy (dysgenic)	14/22,995 (0.06%)	0/22,995
5. Type II', MR^{h12} (dysgenic)	83/37,660 (0.22%)	22/37,660 (0.06%)
6. Type II', Cy (dysgenic)	20/39,348 (0.05%)	2/39,348 (0.005%)

[a] Shown are the spontaneous frequencies of $P[w^+]$ mobilization in the absence of *trans*-activating helper P elements (Cross Type 0), and in the presence of helper elements introduced via Cross Type I, II, or II' (see Fig. 1). Presented are the frequencies of $P[w^+]$ loss (w^- males) and transposition (w^+ females).

An alternative type of genetic system for introducing helper P elements is illustrated by Cross Type II (see Fig. 1B). In this cross, P elements are derived from the male parent, and the test offspring are therefore expected to be dysgenic (9, 10). In order to minimize the magnitude of *trans* activation of the reporter element, $P[w^+]$, we first screened a series of P element-bearing stocks to identify a candidate that promoted low levels of $P[w^+]$ activation. Q-strains are weak P-strain stocks that carry only a few copies of the functional P element and many copies of internally deleted, nonfunctional elements (15). We therefore surveyed several Q-strain stocks and found that one of these, MR^{h12}, consistently exhibited low levels of *trans*-activation of $P[w^+]$. MR^{h12} is a second chromosome isolated from a natural population; its genetic characterization suggests that one or more P-elements are located on this chromosome, apparently in the centromeric region of the chromosome (16). A molecular analysis of this stock was performed to assess the number of full-length and deleted P elements. The MR^{h12} homolog is maintained over an inverted chromosome II which is marked with Cy (*curly wings*) (see 17 for a full description of symbols). The chromosome II homologs were isolated by outcrossing with an M-strain, and DNA was prepared from animals bearing both homologs and analyzed as described in Fig. 2. It can be seen that there are two to four copies of the intact element; moreover, at least one of these must reside at a site other than on the MR^{h12} homolog (compare lanes 3 and 4). Similar types of analyses have been performed over the past two years and the MR^{h12} stock appears to be stable over this period (data not shown).

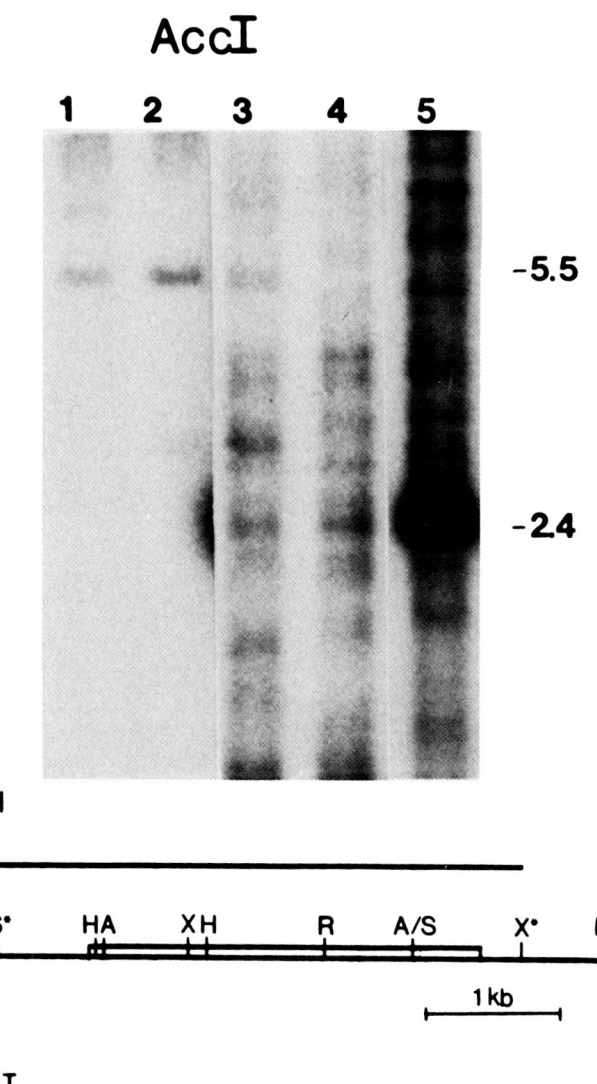

FIG. 2. Shown is P-element content in genomic DNA from $P[w^+]$ and related strains. Total genomic DNA isolated from the indicated stocks was digested to completion with AccI, fractionated on a 1% agarose gel, and blotted to nitrocellulose. The blot was hybridized with [^{32}P]P-element probe (pπ25.1, shown below the autoradiogram). AccI digestion of intact P elements generates a 2.4-kb fragment, the position of which is indicated to the right of the autoradiogram. The 5.5-kb band present in lanes 1 and 2 is derived from $P[w^+]$ internal sequences. (Lanes 1 and 2) DNA from two $P[w^+]$ strains; (lane 3) DNA from $Cy/+$ flies derived from an outcross of MR^{h12}/Cy; (lane 4) DNA from $MR^{h12}/+$ flies from the same outcross; (lane 5) DNA from a strong P-strain, Kerbinou.

The results from the Cross Type II series are also summarized in Table I, where we have indicated the observed rates when trans-activation is promoted by the MR^{h12} chromosome versus its Cy-marked homolog. We note the following observations of interest: (1) losses of $P[w^+]$ are substantially higher in the MR^{h12} dysgenic series as compared to the nondysgenic Cross Type I, while transposition frequencies are nearly equivalent (cf. lines 3 and 2), and (2) the rates of both apparent losses and transpositions of $P[w^+]$ are substantially higher in the presence of the MR^{h12} chromosome than its Cy homolog (compare lines 3 and 4). The latter finding is of interest in view of the apparent equality in the number of full-length P elements observed in stocks carrying one, but not both, of these homologs (see Fig. 2). A possible explanation is that the Cy homolog carries more copies of nonfunctional P elements and that these reduce the effective level of transposase in the test males.

Shown in lines 5 and 6 of Table I are the results from a modified version of Cross Type II, designated II', in which deleted, nonfunctional copies of the P element have been introduced into the genomes of the test animals. The rationale for this modification is that internally deleted P elements that retain their terminal tandem repeats are thought to compete for transposase generated by functional P elements (18). Nonfunctional elements were introduced by replacing the autosomes of the $P[w^+]$ females used in the initial generation of Cross Type II (see Fig. 1B) with autosomes derived from an M'-strain designated Muller-5, Birmingham (19). M'-strains have been shown by molecular analyses to carry multiple copies of deleted, nonfunctional P elements (19) and, hence, provide an ideal source of competing elements. As in the conventional Cross Type II, data from test males bearing the MR^{h12} and Cy homologs are presented separately. Note that in comparing the loss frequencies in the MR^{h12} series (see lines 3 and 5 for Crosses II and II' respectively) that losses of $P[w^+]$ are depressed by approximately 50% in the M'-strain autosomal background, while transpositions occur at approximately equivalent rates. For the Cy series (lines 4 and 6) there is no apparent effect of added nonfunctional elements; however, the rates of loss and, especially, transposition are so low that a small inhibitory effect of the introduced elements cannot be excluded.

In summarizing these control experiments, we have found that either dysgenic or nondysgenic crosses can be used to mobilize a tagged P element at reproducible low rates. Of the various genetic schemes we have employed, Crosses II and II', in the presence of the Cy homolog, appear to promote the lowest levels of $P[w^+]$ transposition and are thus best suited to tests for induction of transposition. It is of interest that while transposition rates can be varied 10-fold (from 0.04% in Cross Type I to 0.005% in Cross Type II'), spontaneous rates of P-element loss are less easily controlled. These results are consistent with the notion that transposition and excision of P elements may occur by different biochemical pathways. In agreement with previously published results (20), we have found that only a fraction of apparent losses of $P[w^+]$ are associated with complete excisions of the element; the remainder of cases are derived from internal deletions and from mutational events involving no gross structural change within the element. We have also observed, however, that this distribution of mutational events is similar among the various crosses tested here. We therefore think it unlikely that, for example, bona fide excisions are depressed in one of the crosses as compared to the others.

II. Tests for Mutagen-Induced Mobilization of Tagged P Elements

As noted earlier, evidence from a variety of prokaryotic and eukaryotic systems suggests that transposable elements are mobilized by mutagenesis, consistent with McClintock's genomic stress hypothesis (2). The question of whether specific classes of chemical agents preferentially induce transposition, or whether transposons in some way respond generally to chromosomal damage, remains to be resolved. Extensive testing with provirus (4) and retrovirus (6) activation assays has identified a number of model chemicals that can be tested for transposition induction in *Drosophila*.

Preliminary studies have begun in our laboratory to identify chemicals that induce mobilization of tagged P elements. Crosses II and II', described above, have been modified by treating test males during late larval development. Surviving adult males are then tested in crosses with attached-X, w females, as illustrated in Fig. 1, and frequencies of loss and transposition of $P[w^+]$ are compared between treated groups and concurrent controls. We have found that a variety of potent, direct-acting mutagens, including ethyl methanesulfonate (EMS), substantially increase the frequency of recovered loss males (i.e., those with w eye color) while having no detectable effect on the frequency of transposition. Molecular analysis of the w males recovered from treated fathers reveals that these mutants are as structurally complex as those recovered from controls, and are qualitatively similar (i.e., comprised of apparent complete excisional losses of the element, as well as internal deletions and apparent point mutations not associated with gross structural change). At present, it is unclear whether these loss mutants are derived from aborted transpositions, or whether they simply reflect conventional lesions induced within the w^+ marker carried by the P element. Until additional molecular diagnostics are completed, we cannot conclude that agents that only promote loss are indeed transposagens.

In contrast to the situation with apparent losses of $P[w^+]$, transpositions of the tagged element, recovered in daughters of test males, are a diagnostic end point. For this reason, we are particularly interested in identifying chemicals that promote the recovery of w^+ females. While our data are only preliminary, we have suggestive evidence that at least three chemicals act as transposagens in *Drosophila*. Chemicals that promote transposition frequencies over five times the concurrent control rates include: aflatoxin B_1, a fungal metabolite which has been shown to promote the addition of bulky adducts to DNA (21); mitomycin C, a potent alkylating agent and crosslinker of DNA (21); and 5-azacytidine, a nucleoside analog with potent effects in higher eukaryotes (22). It is of interest that each of these chemicals is active in retrovirus activation assays (6). Experimentation is continuing to confirm the classification of these three chemicals as P-element transposagens, as well as to extend the range of chemicals tested. It will be of interest, for example, to test the effects of compounds that inhibit DNA synthesis, since current models envision P-element transposition occurring via DNA synthesis rather than through an RNA intermediate (23).

III. Conclusions

The genetic test systems presented here provide good control of spontaneous P-element transposition; losses are less amenable to genetic regulation and, as we have argued above, may be ambiguous end points for assays designed to identify transposagenic chemicals. Preliminary testing with model chemicals suggests that *Drosophila* P elements may be mobilized by mutagenesis. In the absence of a detailed understanding of the mechanism(s) of P-element transposition, it is impossible to rationally select chemicals for testing; however, we anticipate that experiments with chemicals that disrupt DNA synthesis may be informative.

An obvious area for improvement in the test systems described here would be the replacement of the Q-strain currently used as a source of helper P elements (MR^{h12}/Cy) with a genetically better defined source of helper. Ideally, one would like to employ a single immobile but fully functional P element as the source of transposase. Two promising systems that may meet this requirement have recently been described (24, 25); these are being evaluated in our laboratory. Efforts are also under way to develop analogous test systems for other classes of *Drosophila* transposons, including the retrotransposons.

Acknowledgment

The work performed was supported by an award from the NIEHS (NO1ES75181). Dr. Laura Lacy ably performed the molecular analyses described here.

References

1. J. A. Shapiro, "Mobile Genetic Elements." Academic Press, New York, 1983.
2. B. McClintock, *Science* **226**, 792 (1984).
3. A. R. Datta, B. W. Randolph, and J. L. Rosner, *MGG* **189**, 245 (1983).
4. R. K. Elespuru, in "Chemical Mutagens" (F. de Serres, ed.), Vol. 9, p. 213. Plenum, New York, 1984.
5. B. Rasmuson, A. Rasmuson, and J. Nygren, in "Handbook of Mutagenicity Test Procedures" (B. J. Kilbey, ed.), p. 603. Elsevier, New York, 1984.
6. R. W. Tennant, J. A. Otter, F. E. Meyer, and R. J. Rascati, *Cancer Res.* **42**, 3050 (1982).
7. M. Rolfe, A. Spanos, and G. Banks, *Nature (London)* **319**, 339 (1986).
8. T. Hazelrigg, R. Levis, and G. M. Rubin, *Cell* **36**, 469 (1984).
9. W. R. Engels, *ARGen* **17**, 315 (1983).
10. M. G. Kidwell, in "*Drosophila*: A Practical Approach" (D. B. Roberts, ed.), p. 59. IRL Press, Oxford, England, 1986.
11. S. H. Clark and A. Chovnick, *Environ. Mutagen.* **7**, 439 (1985).
12. A. C. Spradling, in "*Drosophila*: A Practical Approach" (D. B. Roberts, ed.), p. 175. IRL Press, Oxford, England, 1986.
13. R. Levis, T. Hazelrigg, and G. M. Rubin, *Science* **229** 558 (1985).
14. C. R. Preston and W. R. Engels, *Drosophila Inf. Serv.* **60**, 169 (1984).
15. M. J. Simmons and M. Bucholz, *PNAS* **82**, 8119 (1985).
16. B. Slatko and M. M. Green, *Biol. Zentral.* **99**, 149 (1980).
17. D. L. Lindsley and E. H. Grell, *Carnegie Inst. Washington Publ.* **627**, (1968).
18. M. J. Simmons, J. D. Raymond, M. J. Boedigheimer, and J. R. Zunt, *Genetics* **117**, 671 (1988).
19. P. M. Bingham, M. G. Kidwell, and G. M. Rubin, *Cell* **29**, 995 (1982).
20. L. L. Searles, R. S. Jokerst, P. M. Bingham, R. A. Voelker, and A. L. Greenleaf, *Cell* **31**, 585 (1983).
21. P. A. Jones, *Cell* **40**, 485 (1985).
22. E. Vogel, A. Schalet, W. R. Lee, and F. Wurgler, in "Comparative Mutagenesis" (F. de Serres and M. Shelby, eds.), p. 175. Plenum, New York, 1981.
23. M. J. Simmons and R. E. Karess, *Drosophila Inf. Serv.* **61**, 2 (1985).
24. H. M. Roberston, C. R. Preston, R. W. Phillis, D. M. Johnson-Schlitz, W. R. Benz, and W. R. Engels, *Genetics* **118**, 461 (1988).
25. L. Cooley, R. Kelly, and A. Spradling, *Science* **239**, 1121 (1988).

II. Transposable Elements in *Drosophila* (Continued)

Asymmetrical Exchanges and Chromosomal Rearrangements in *Drosophila* (Abstract only, see p. 325)
B. H. Judd, E. A. Montgomery, S.-M. Huang, and C. H. Langley

Spread of P Transposable Elements in Inbred Lines of *Drosophila melanogaster*   71
Christine R. Preston and William R. Engels

Suppressible Insertion-Induced Mutations in *Drosophila*   87
Zuzana Zachar and Paul M. Bingham

Identifying and Cloning *Drosophila* Genes by Single P Element Insertional Mutagenesis   99
Lynn Cooley, Celeste Berg, Richard Kelley, Dennis McKearin, and Allan Spradling

Molecular Lesions Induced by I-R Hybrid Dysgenesis in *Drosophila melanogaster*   111
Isabelle Busseau, Alain Pelisson, Michèle Crozatier, Chantal Vaury, and Alain Bucheton

Spread of P Transposable Elements in Inbred Lines of *Drosophila melanogaster*

CHRISTINE R. PRESTON AND
WILLIAM R. ENGELS[1]

Laboratory of Genetics, University of Wisconsin, Madison, Wisconsin 53706

The invasion of the *Drosophila melanogaster* genome by members of the P family of transposable elements was monitored by *in situ* hybridization to polytene chromosomes. Populations consisted of inbred lines starting with a single element. There were several cases of very rapid proliferation of element copy numbers, going from a single copy to more than 20 in the span of a few generations. These episodes, which occurred during periods of intense inbreeding and usually led to extinction of the stock, were preceded by 6–20 generations, during which the copy number remained at 1. In other cases, the founding P element was lost before any proliferation could occur, and in still others, the stock gradually developed the "P cytotype," in which P-element mobility is suppressed. Finally, there was one transposase-producing P element that appeared to be stable, or nearly so. Possible explanations for this behavior and implications for natural populations are discussed.

There are many families of transposable elements in *Drosophila melanogaster* (1). One of these, called the P family, has been extensively studied by molecular and genetic approaches (2). P elements are common in natural populations, where they typically occupy 30–50 positions scattered throughout the genome. However, many of the old laboratory stocks of this species are entirely devoid of P elements. Strains with and without P elements are called P and M, respectively.

P elements are relatively quiescent within P-strains, but when such strains are crossed with M-strain females, the P elements transpose and excise at high frequencies in the germ line of the F_1

---

[1] Speaker.

hybrid progeny. This high mobility and its accompanying syndrome of abnormal traits, such as elevated mutability and chromosome rearrangements, are known collectively as hybrid dysgenesis. The relative inactivity of P elements within P-strains is associated with a state called the P cytotype (3–7). A hypothetical element-encoded repressor product is thought to be responsible for this regulation (H. Robertson and W. Engels, unpublished results, described in 2).

The P cytotype only appears in the presence of many P elements (4–7). Therefore, when only a single P element is present in the genome, as might be the case in the early stages of "invading" a population of species, the element is expected to remain in its active state until the copy number is built up. This process has been observed following the introduction of P elements via transformation into M-strains (8, 9) or into the genome of the closely related species, *D. simulans*, which normally lacks P elements (10, 11). In each case, the resulting stock was maintained by mass transfers with uncertain population numbers. The data consisted primarily of genomic Southern blots using mass-extracted DNA from the population. This procedure has the advantage of revealing all P elements, including those in heterochromatin. However, it provides only a rough idea of the state of the P elements, since such stocks are expected to be highly polymorphic.

We describe experiments in which single P elements were observed to spread in M strain genomes. Some of the uncertainties inherent in the previous experimental designs were avoided through the use of *in situ* hybridization of P-element sequences to polytene chromosomes, as opposed to genomic Southern blots of mass-extracted DNA. In addition, to follow the chromosomes as closely as possible, the populations were maintained by brother–sister matings, as opposed to mass transfers, during the generations when P-element sites were being monitored. We consider control over the population size and structure to be a critical feature in this experiment, since the data indicate that these parameters are important factors in determining the fate of invading P elements.

I. Materials and Methods

A. *Drosophila* Crosses

All flies were raised on standard cornmeal–molasses–agar medium at room temperature (~22°), or 28° in the case of GD sterility tests (see below).

B. GD Sterility

When dysgenic hybrids are raised at elevated temperatures, their gonads frequently fail to develop, resulting in a condition known as GD sterility (12). Females were tested for GD sterility by squashing them between two glass plates and scoring for the presence of oocytes. A female was scored as fertile if one or more oocytes were present.

C. sn^w Hypermutability

A P element-induced mutation at the X chromosomal *singed* bristle locus, known as sn^w, provides a convenient and sensitive assay for hybrid dysgenesis (6, 13–15). When P transposase is present and the P cytotype is absent, sn^w mutates to two alternative forms designated sn^e and $sn^{(+)}$, which have extreme *singed* and nearly wild-type bristles, respectively. These events occur frequently enough to allow detection of a single transposase-producing P element. To test sn^w males, we crossed them to compound-X females and scored the sons as sn^w, sn^e, or $sn^{(+)}$. Females of the genotype $sn^w/+$ were tested by scoring sons as either sn^w or sn^e, ignoring the $sn^{(+)}$ males, which cannot be distinguished from those bearing the wild-type homolog.

D. Tests for Transposase and Cytotype

To test for P-element transposase in the genome of a particular strain (U), we performed crosses of the type M♀ × U♂ and tested for GD sterility or sn^w hypermutability in the germ line of the progeny. The test for cytotype was U♀ × P♂, where the absence of dysgenic traits (GD sterility or sn^w hypermutability) in the germ line of the progeny indicated the P cytotype in strain U. The M strains used in these crosses were $bw; st$ or $y\ sn^w; bw; st$ (M) (14) The P strains were π_2 (12) or sn^w (ii); π_2 (13, 14). Sons from the cross $y\ sn^w; bw; st$ (M) ♀ × U♂ were test-crossed in order to detect transposase-making elements on the autosomes or the Y chromosome. To test the X chromosomes, we used daughters from the same cross; these carry the X chromosome and a complement of autosomes from strain U.

E. *In situ* Hybridization to Polytene Chromosomes

Preparation of slides, treatment with tritiated or biotinylated probes, and detection of label were as described previously (16, 17). In most cases, the probes were plasmids pπ25.1 (18) or py5.4 (W.

Eggleston, personal communication), which contain complete P elements plus genomic flanking sequences from cytological positions 17C and 1B, respectively. For chromosomes suspected of having P elements near one of these locations, we used subclones obtained from K. O'Hare (personal communication) containing nearly complete P-element sequences but no flanking DNA.

II. Results

A. Generation of Single P-Element Stocks

The first step was to obtain stocks with only one or a small number of complete P elements. One set of lines, designated 1039, was initiated with a single P-element site at cytological position 17B. This element had been placed in the genome previously by Spradling and Rubin (19) via germ line transformation using the plasmid pπ25.1. We determined by *in situ* hybridization that the progenitor stock had several P-element sites. Therefore, we eliminated all but one conveniently located site (cytological position 17B) by replacing most of the genome with that of an M strain bearing the visible mutations *y cho cv v f* on the X chromosome and *bw* and *st* on the two major autosomes (symbols as in 20). The resulting stocks, designated 1039-3 and 1039-5, had all of these markers except *f*, which is near the P element at 17B. Each stock was initiated with a brother–sister mating and maintained in two parallel lines, one by repeated pair matings and the other by mass transfers of approximately 20 adults. The presence of the 17B element and the absence of any other P-element sites was confirmed by two *in situ* hybridization slides from each stock following the initial pair mating.

Another set of lines was obtained by isolating naturally occurring P elements on the Y chromosome, which is completely heterochromatic in this species. Since P elements are relatively rare in the heterochromatin (2) and cannot be detected there by *in situ* hybridization, it was necessary to screen a collection of P strains using sn^w hypermutability as an assay for transposase. The Y chromosomes from each of 26 inbred P strains, derived independently from a local wild population as described previously (21), were isolated and tested as shown in Fig. 1. Two of these Y chromosomes, designated Y85 and Y86, proved to carry transposase-making P elements. Given the relatively low frequency of P elements in heterochromatic regions, it is likely that these Y chromosomes had only a single transposase-making P element. However, the presence of nonautonomous ele-

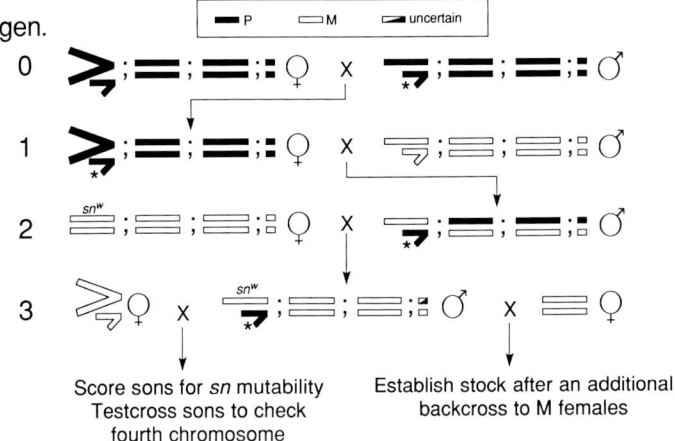

FIG. 1. Shown is the mating scheme to recover Y chromosomes bearing transposase-making P elements. P chromosomes are shown as black bars and M chromosomes are shown as shaded bars. The Y chromosome to be tested is marked with an asterisk. The male in generation 0 is derived from an inbred P strain, and the female is from the compound-X stock C(1)DX, yf; π_2 (3). The M-derived second and third chromosomes are marked with bw and st, but the fourth chromosome is unmarked. Note that the males in generation 2 are nondysgenic, thus minimizing the chance of transposition before generation 3. The presence of sn^e and $sn^{(+)}$ in generation 4 indicates a transposase-making P element on either the Y chromosome or the fourth chromosome. To distinguish between these possibilities, we mated 10–20 of the sn^w males from generation 3 individually to compound-X females. Half of these males are expected to carry the P-derived fourth chromosome. The absence of sn^e and $sn^{(+)}$ in all crosses implies the transposase source is on the Y chromosome. Among the 26 P-strains tested, nine had transposase associated with the fourth chromosome, and two had transposase-producing Y chromosomes.

ments or possibly some additional autonomous P elements is not ruled out. *In situ* hybridization confirmed that no euchromatic sites were present in two of the three Y85 sublines and the single Y86 line that was made. The remaining Y85 subline did have a small number of euchromatic sites observed in the second generation of inbreeding, ranging from one to three sites in the larvae tested. These might have originated as transpositions from the Y chromosome in generation 3 of Fig. 1 or in the initial generation of sib mating. As was the case for the 1039 lines, each subline was kept as two parallel stocks, one maintained by mass matings and the other by brother–sister matings. The brother–sister mating lines were used to monitor the P-element sites by *in situ* hybridization.

B. Evolution of P-Element Sites

Approximately six brother-sister pair matings were initiated per line in each generation, only one of which would be used to establish the next generation. The purpose of the extra matings was to minimize the likelihood of extinction. When a stock became weak, we made more than six pair matings whenever possible. Despite these precautions, however, several of the lines died out, presumably due to deleterious P-element-induced mutations, discussed in more detail below. When this happened, we continued the experiment by replacing the extinct line with a new pair-mating line derived from the corresponding mass-mated stock. In three cases, fertile flies of one sex survived, allowing us to salvage the stock by outcrossing to an M-strain.

Female larvae from each sib-mating line were collected each generation to prepare *in situ* hybridization slides. One or two slides were analyzed per line per generation, with the larvae taken, in most cases, from the same cross as was used for perpetuation of the line. In some cases, however, the line was so weak that it was necessary to utilize larvae from the other sib-matings as well. A few line-generations were missed entirely. Approximately 400 *in situ* hybridization slides were analyzed for this experiment. All lines surviving to generation 30 were tested for transposase and cytotype at approximately that time (generations 24–42), using both sn^w mutability and GD sterility as assays. Following these tests, all lines were maintained as small mass matings, and *in situ* hybridization slides were prepared less frequently. Tests for cytotype and transposase were repeated at approximately generation 90.

The results are summarized in Figs. 2 and 3. Two of the lines (Y85a and Y85d) have lost their transposase-making P element(s). The *in situ* hybridization data do not allow us to determine whether the loss of the transposase activity was due to complete or partial excision. A third case (1039-5a) lost its *in situ* sites and therefore, presumably, its transposase. The loss of some P elements, or at least their transposase activity, is expected from previous data indicating that P elements have a high rate of internal excision when they are in the dysgenic state (2).

More suprising was the fate of lines Y85b1, Y85c, 1039-3a, and 1039-3c. In these cases, the P-element sites remained similar to the initial state for a variable number of generations, ranging from 6 to 20. Then the number of sites increased very rapidly to more than 20 in the span of only a few generations. This period of rapid proliferation was terminated in each case by extinction of the stock. Figure 4 shows the

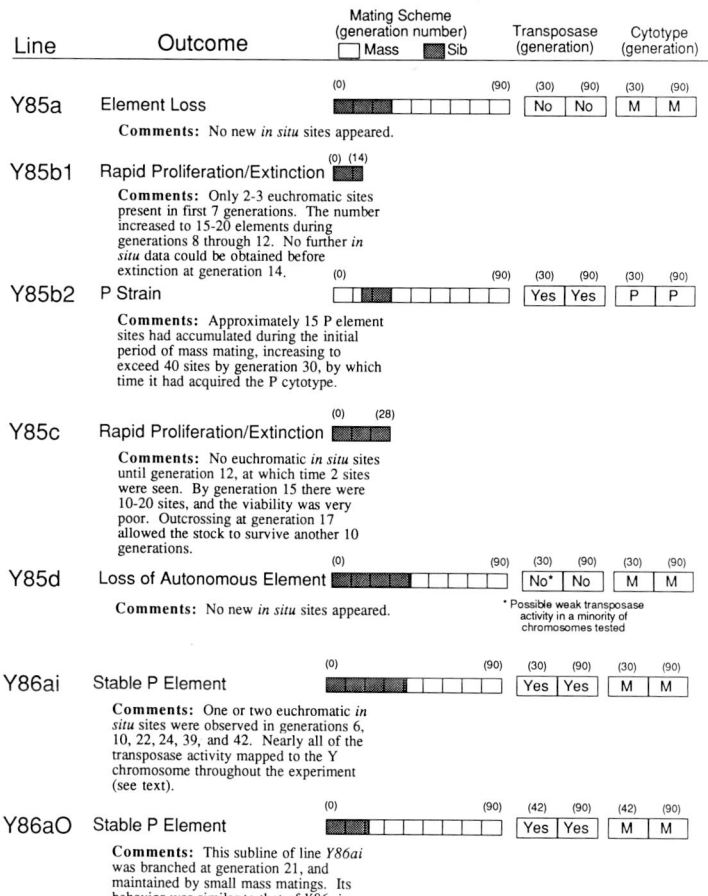

FIG. 2. The evolution of P elements in lines Y85 and Y86 is shown through generation 90. All lines started with transposase only on the Y chromosome, unless otherwise stated.

in situ hybridization results for one of these cases. The reasons for extinction were varied. In one line (1039-3a) most of the inviability occurred at the pupal stage. Usually, however, there was both sterility and inviability, the latter occurring at all stages of development. The close correlation between the rapid proliferation of P elements and stock extinction strongly suggests that the apparent losses in fitness were caused by P-element activity. Note that only one line (1039-3b) was lost without a period of rapid P-element proliferation immediately preceding extinction.

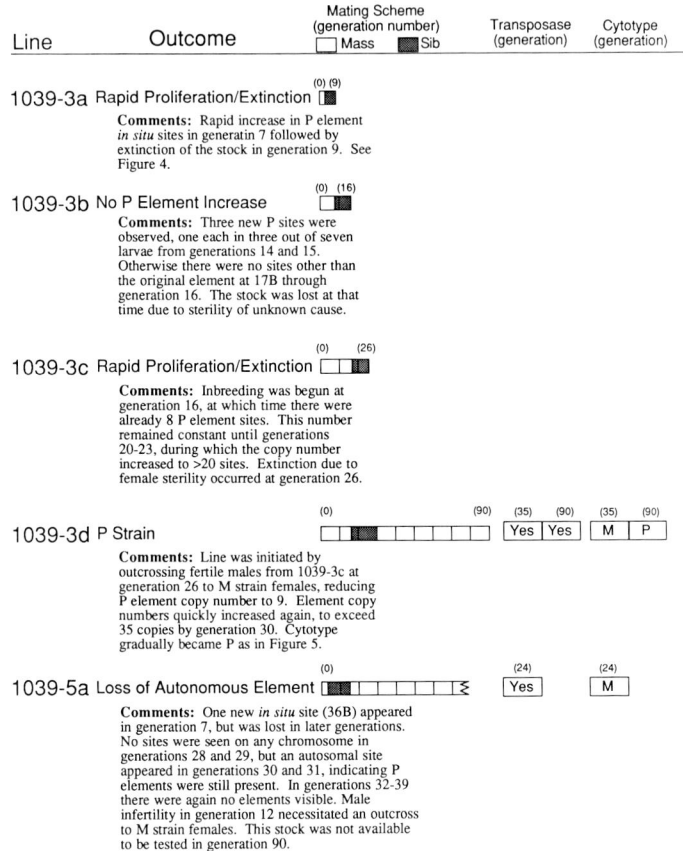

FIG. 3. The evolution of P elements in lines 1039 is shown through generation 90. All lines started with a single transposase-making P element at cytological position 17B.

Only two of the lines (Y85b2 and 1039-3d) evolved to become stable P-strains in which the cytotype eventually switched to P, thus suppressing further P-element mobility. This switch was monitored for line 1039-3d by additional cytotype and transposase tests during generations 35–47. The transposase level was high throughout this period, at least equivalent to that of a strong P-strain such as π_2. Surprisingly, however, the cytotype change occurred only gradually over many generations and was not apparent until 10–30 generations after P elements had spread to a large number of chromosomal sites. (Note that line 1039-3d was started by outcrossing line 1039-3c.

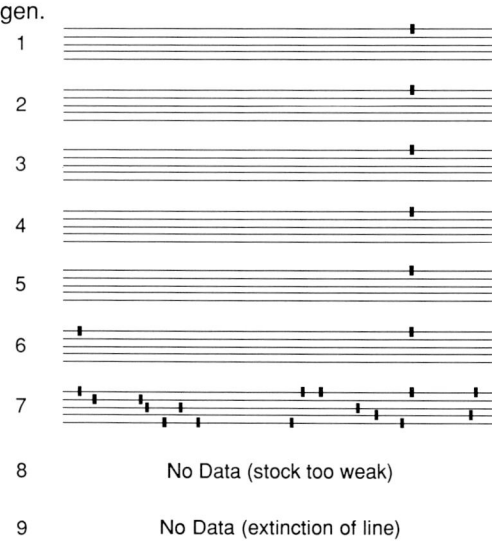

FIG. 4. Shown are P-element *in situ* hybridization sites in line 1039-3a. The five lines represent the arms of the major chromosomes. Starting from the top they are: X, 2L, 2R, 3L, and 3R. P-element sites are indicated by black boxes. The width of the box represents the range of uncertainty in placement of the site. Generations are counted from the first generation following isolation of the P element at position 17B (see text). Data from each generation came from one larva which was a full sibling of the parents of the next generation.

Therefore, more than 20 P elements were already present by generation 23.) Figure 5 shows the change in cytotype as measured by decreases in both GD sterility and sn^w hypermutability following crosses to a P-strain male. Even after 90 generations, the switch to P cytotype, measured by lack of GD sterility, was not yet complete. Less extensive data are available for line Y85b2. However, since P-element proliferation had occurred by generation 14, it is possible that this line also maintained the M cytotype for many generations prior to switching to the P cytotype, which had taken place by generation 30. Persistence of the M cytotype for many generations in the presence of a dense population of P elements is in sharp contrast to previous experiments (4–7) indicating that P-strain-derived chromosomes change the cytotype to P within a few generations. A possible explanation for this difference will be discussed below.

One difference between these two lines and the four cases in which there was rapid proliferation of P elements followed by extinction of the stock is that, in the lines evolving to P-strains, much

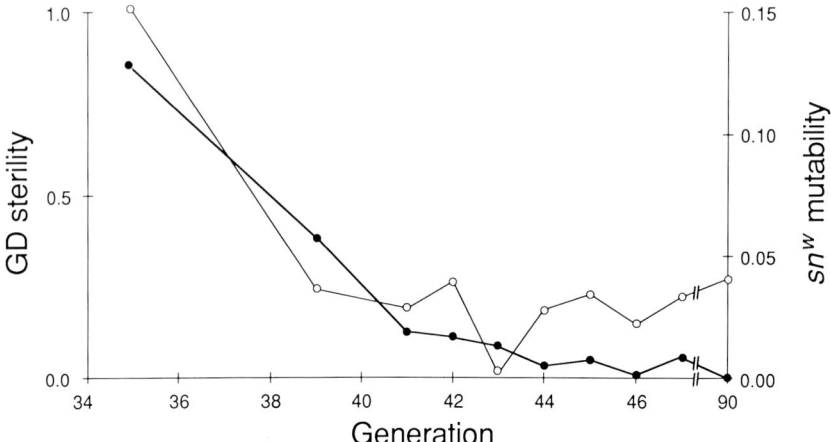

FIG. 5. The change from M to P cytotype for line 1039-3d is indicated by decreasing GD sterility (○) and sn^w mutability (●). GD sterility was measured as the proportion of daughters from the cross 1039-3d ♀ × π_2 ♂ with no oocytes present. A similar cross utilizing sn^w(ii); π_2 males was used to generate heterozygous sn^w females. The proportion of sn^e sons among sn^e and sn^w is plotted. Each point represents an average of 80 females tested for GD sterility or 1719 flies scored for *singed* phenotype.

of the increase in P-element copy number occurred during periods of mass matings, as opposed to brother–sister crosses. This difference suggests that extinctions were due to the fixation of deleterious mutations produced by P-element activity. Such mutations are much more likely to become fixed by random drift in small inbred populations than in large ones, where natural selection is likely to eliminate the most harmful alleles.

Finally, there was one case (the related lines Y86ai and Y86aO) in which the autonomous P element appeared to persist for at least 90 generations with no loss of transposase activity and relatively few cases of new P elements produced. To determine the source of transposase in this line, we used the sn^w assay in tests of individual chromosomes from the stock. Isolation of the chromosomes was achieved by crosses to unmarked M-strain females (recall that Y86a carries *bw* and *st*) with or without a compound-X chromosome, then crossing the sons to sn^w; *bw*; *st* females to produce sn^w-bearing progeny with each combination of Y86a chromosomes. The Y86a fourth chromosome was assumed to be present in half the bw^+; st^+ progeny. These sn^w males (or heterozygous females for testing the X chromosomes) were then test-crossed to determine the mutability of

sn^w in their germ lines. Approximately 1000 progeny of sn^w males or females were scored per chromosome tested. No transposase was detected from three X chromosomes, six second chromosomes, and six fourth chromosomes extracted from the stock. Only one of six third chromosomes showed transposase activity. In contrast, all four of the Y chromosomes tested had strong transposase activity. Thus, the transposase-making P element(s) on the Y chromosome of Y86 appears to be much less active in transposition and excision than its counterparts in the Y85 and 1039 lines, despite producing comparable levels of transposase as indicated by sn^w hypermutability. This Y chromosome may carry one or more elements with properties similar to the element P[ry^+Δ2-3](99B), described elsewhere (22), which is highly active in transposase production but cannot mobilize itself.

III. Discussion

A. Rapid Proliferation of P Elements

The precipitous increases in P-element copy numbers, as exemplified by generations 5–7 in Fig. 4, were not observed in previous experiments of this kind (8–11). However, it is unlikely that the earlier studies would have detected such events if they had occurred. One reason is that those experiments relied exclusively on relatively large, randomly mating populations rather than on controlled crossing schemes. Thus, any rapid increase of P elements in a particular germ line would be diluted by the rest of the population within a few generations, and would appear as part of a much more gradual invasion of the genome.

Another relevant difference between the present work and previous studies of the spread of transformed P elements (8–11) is our use of *in situ* hybridization to monitor the elements, as opposed to genomic Southern blots on DNA extracted from many individuals in the population. Our approach does not allow detection of heterochromatic P elements. However, it affords much higher resolution, and is less likely to obscure any events that require observation of individual genomes.

The very rapid proliferation of P elements we observed in four lines seems at odds with transposition rate measurements (reviewed in 2), indicating that the expected number of new P elements produced each generation in P-M dysgenic hybrids is only about 0.25 per existing element. One possible explanation comes from a difference in the ratio of complete to defective (i.e., autonomous to nonautonomous) P elements in stocks generated from a single autonomous P element, as compared to ordinary P strains. Whereas the latter usually contain a preponderance of nonautonomous elements, the strains used here begin with only autonomous ones, and generate nonautonomous elements only gradually through internal deletion events. There is some evidence that transposase is rate-limiting, such that the presence of many nonautonomous P elements results in less mobility per element. This "titration" effect was proposed as an explanation for the decrease in sn^w hypermutability in the presence of chromosomes with many nonautonomous P elements (23). Similarly, M. Fortini, W. Eggleston, and W. Engels (unpublished results) have shown that defective P elements injected into dysgenic embryos decrease sn^w mutability. Therefore, the transposition rates measured previously for M × P-strain hybrids probably reflect considerable transposase "titration" which would not occur in the early stages of this study.

A second explanation for the relatively high transposition rates observed here compared with M × P-strain hybrids is that a P-element repressor is thought to be produced by some of the P elements in most P strains. For reasons discussed below, this repressor is probably reduced or absent in our lines.

The above arguments might explain the high transposition rate during periods of rapid P-element proliferation. They do not, however, account for the lag time of 6–20 generations, during which no proliferation was observed. For line Y86a, which appears to be in a perpetual lag phase, we postulated that the starting P element might be deficient in transposition and excision, perhaps as a result of its genomic position, while still retaining full ability to produce transposase. Several previous studies have implicated genomic position effects in determining P-element stability (reviewed in 2). The lag period of other lines (Y85b1, Y85c, 1039-3a, and 1039-3c) might be explained if the P elements initially present were intermediate in this respect. That is, they had a lower-than-average transposition rate as a consequence of their genomic position. Although we have no *a priori* reason for expecting the 17B element in the 1039 lines to be relatively stable, heterochromatic effects might play a role in the case of Y85. According to this interpretation, such elements occasionally produce a new copy in a different genomic position, where it is likely to have greater mobility. When that happens, the P elements can begin a period of exponential growth. Since the initiation of this phase depends on a single, relatively rare transposition event, the number of generations required will be variable and unpredictable, as was observed.

B. Gradual Appearance of the P Cytotype

Stocks that survived proliferation of P elements (Y85b2 and 1039-3d) eventually developed the P cytotype and became stable P-strains. The development of P cytotype was predictable from previous work (4–7) indicating that P-strain chromosomes tend to bring about the P cytotype. However, in the earlier studies, the switch from M to P cytotype took place within a few generations following the introduction of a large proportion of the P-strain genome. This contrasts with the present experiments in which the M cytotype persisted for 10–30 generations following the spread of P elements. Furthermore, comparable delays in cytotype switching have been observed by Anxolabéhère et al. (8), Daniels et al. (11), and Scavarda and Hartl (10) in strains containing complete P elements introduced by transformation.

We suggest that this delay can be explained by the rare occurrence of repressor-making P elements, and subsequent natural selection for these elements. Recent work by H. Robertson, R. Phillis, D. Johnson-Schlitz, and ourselves (unpublished results) indicates that certain P elements produce a gene product capable of suppressing P-element activity in *trans*. This repressor is thought to be responsible for the P cytotype. Only a small fraction (approximately one sixth in preliminary experiments) of the insertions of a given P element produced significant repressor activity, suggesting that the ability to make this repressor is highly dependent upon an element's genomic position. Furthermore, all repressor-making P elements observed to date are defective in the fourth open reading frame of the transposase gene, but it is not known whether such a defect is necessary for repressor. If it is, this would make the occurrence of repressor-making P elements even more rare.

Therefore, when a newly introduced complete P element proliferates in a population, none of the new copies is likely to be a repressor maker. However, after many generations during which the P-element positions are being reshuffled in the genome and defective P elements are being produced at random by internal deletions, it is probable that one or more repressor-making P elements will occur. Natural selection can then act to increase the frequency of these repressor-makers, leading to the P cytotype and a stable P-strain. Accordingly, ordinary P-strain chromosomes are expected to have already undergone these steps of reshuffling and selection. They should be enriched for repressor-making P elements, thus explaining why they are much more effective than our "synthetic" P-strain chromosomes in switching the cytotype to P.

C. Implications for Natural Populations

What do these experiments tell us about the process of P-element invasion of M populations? The intense inbreeding makes our results a poor model for wild *Drosophila* populations, which are expected to have a much larger effective population number. The observed rapid increases in P-element copy number, followed by extinction of the line, would therefore not be expected in nature. Instead, a gradual increase in copy number, as seen in previous experiments (*8–11*), with natural selection eliminating the most deleterious insertions, is more probable. The result is still expected to be a loss of average fitness prior to the cytotype switch, but not as drastic as was seen in some of our lines.

On the other hand, the long delay we observed between the time of P-element copy number increase and development of the P cytotype might occur in natural populations, depending on the nature of the invasion. Simple migration of individuals from a P to an M population would introduce a mixture of autonomous, nonautonomous, and repressor-making P elements. The result, according to our interpretation and some previous data from experimental populations (*24–26*), would be a relatively rapid development of the P cytotype or other type of suppression of P element activity. However, *de novo* introduction of P elements into a species, presumably by some form of horizontal transfer, would result in a population similar to those studied here. The expected outcome would be many generations during which autonomous P elements would coincide with the M cytotype. Throughout this period, the population would be in a continuous state of hybrid dysgenesis. The frequency of mutations and chromosome rearrangements would be drastically elevated, and GD sterility would occur in areas where the developmental temperature is sufficiently high. These effects would result in a considerable genetic load.

This period of dysgenesis is short by evolutionary standards, but it might have lasting consequences in some cases. If the decrement in fitness were sufficient to prevent effective competition with other species, it might lead to extinction of the dysgenic species. Alternatively, some of the mutations or chromosome rearrangements produced by the P-element activity could be beneficial, and eventually become fixed in the population by natural selection.

Acknowledgments

We thank Douglas Kellogg for technical help, and Wendy Benz and Randy Phillis for comments on the manuscript. This work was supported by Grants GM-30948 and

GM-35099 from the National Institutes of Health. This is paper number 2996 from the Laboratory of Genetics, University of Wisconsin–Madison.

References

1. D. J. Finnegan and D. H. Fawcett, *Oxford Surv. Eukaryotic Genes* **3**, 1 (1986).
2. W. R. Engels, in "Mobile DNA" (D. Berg and M. Howe, eds.), in press. Am. Soc. Microbiol., Washington, D.C., 1989.
3. C. R. Preston and W. R. Engels, *Drosophila Inf. Serv.* **60**, 169 (1984).
4. W. R. Engels, *Genet. Res.* **33**, 219 (1979).
5. J. A. Sved, *Genetics* **115**, 121 (1987).
6. W. R. Engels, *Genetics* **98**, 565 (1981).
7. M. G. Kidwell, *Genetics* **98**, 275 (1981).
8. D. Anxolabéhère, H. Benes, D. Nouaud, and G. Périquet, *Evolution* **41**, 846 (1987).
9. S. B. Daniels, S. H. Clark, M. G. Kidwell, and A. Chovnick, *Genetics* **115**, 711 (1987).
10. N. J. Scavarda and D. L. Hartl, *J. Genet.* **66**, 1 (1987).
11. S. B. Daniels, A. Chovnick, and M. G. Kidwell, *Genetics* 1989 (in press).
12. W. R. Engels and C. R. Preston *Genetics* **92**, 161 (1979).
13. W. R. Engels, *PNAS* **76**, 4011 (1979).
14. W. R. Engels, *Science* **226**, 1194 (1984).
15. H. Roiha, G. M. Rubin, and K. O'Hare, *Genetics* **119**, 75 (1988).
16. W. R. Engels and C. R. Preston, *Genetics* **107**, 657 (1984).
17. W. R. Engels, C. R. Preston, P. Thompsons and W. B. Eggleston, *Focus* **8**, 6 (1986).
18. K. O'Hare and G. M. Rubin, *Cell* **34**, 25 (1983).
19. A. C. Spradling and G. M. Rubin, *Science* **218**, 341 (1982).
20. D. L. Lindsley and E. H. Grell, *Carnegie Inst. Washington Publ.* **6**. (1968)
21. W. R. Engels and C. R. Preston, *Genetics* **95**, 111 (1980).
22. H. M. Robertson, C. R. Preston, R. W. Phillis, D. Johnson-Schlitz, W. K. Benz, and W. R. Engels, *Genetics* **118**, 461 (1988).
23. M. J. Simmons and L. M. Bucholz, *PNAS* **82**, 8119 (1985).
24. D. Anxolabéhère, D. Nouaud, G. Périquet, S. Ronsseray, *Genetica* **69**, 81 (1986).
25. D. M. Black, M. S. Jackson, M. G. Kidwell, G. A. Dover, *EMBO J.* **6**, 4125 (1987).
26. M. G. Kidwell, J. B. Novy, and S. M. Feeley, *J. Hered.* **72**, 32 (1981).

Suppressible Insertion-Induced Mutations in *Drosphila*

Zuzana Zachar and
Paul M. Bingham[1]

Department of Biochemistry, State
University of New York, Stony
Brook, New York 11794

Several genes (loci) have been identified in *Drosophila* that modify the phenotypes of specific subsets of alleles at various other loci. The specific subset of mutant alleles suppressed (rendered more nearly wild-type) or enhanced (rendered more extremely mutant) by these loci are idiosyncratic to the suppressor (or enhancer) locus in question (1). Molecular analysis of *Drosophila* mutant alleles, begun in the early 1980s, immediately revealed that the mutant alleles acted on (suppressed or enhanced) by these suppressor/enhancer loci inevitably resulted from insertion of transposable elements (2–6). Moreover, alleles suppressed by the one suppressor locus for which a large sample was available (7) almost always resulted from insertion of a specific transposon (5, 6). It was concurrently recognized that many *Drosophila* transposons were themselves developmentally programmed transcription units (8). Collectively, these various observations suggested that allele-specific suppressor loci might be regulatory genes responsible for developmental programming of transposon expression, and led several laboratories, including our own, to investigate them in detail.

This essay has two objectives. First, in the next section, we review the basic elements of our current picture of the origin and behavior of transposons and their relationship to their cellular hosts. Second, we focus our attention, in the final two sections, on one specific aspect of the analysis of transposons in *Drosophila*: the molecular genetic analyses of suppressor/enhancer loci. In this case, our objective will be to draw attention to details we consider important and to develop a new, detailed proposal for the role of suppressor/enhancer loci.

---

[1] Speaker.

I. Genomic Economics and Transposon Parasites

Genomes of cellular organisms are generally present in only a few copies per cell and, in most cases, genomic replication and maintenance require a minute fraction of a cell's resources. As a result, many cellular organisms can carry large amounts of DNA in addition to that required for cellular genetic function. Genomes of cellular organisms thus represent a potentially rich niche for parasitic DNA molecules. Moreover, the entire gene pool of a sexually reproducing species is ultimately accessible to a genomic parasite after it initially gains access to the genome of a single individual. (Equivalently, every currently extant member of sexual populations such as fruit flies or humans has numerous common ancestors in the relatively recent past.) The existence of this rich, accessible niche has engendered a class of parasites adapted to invasion of and persistence in the genomes of cellular organisms. These parasites are transposons.

Since the recognition by most transposon investigators by around 1977 that transposons were likely to be parasitic organisms—rather than, for example, intrinsic genomic components functioning in the day-to-day lives of their hosts—there has been occasional confusion as to the probable role of these parasites in the biology of their hosts. It now seems certain that essentially all transposons in prokaryotic and eukaryotic organisms persist on a day-to-day basis because they are infectious, invasive parasites, and for no other reason. (The relatively unusual exceptions to the simplest form of this rule are the bacterial drug-resistance transposons.) This conclusion is reached for the same reasons the analogous conclusion is reached regarding, for example, human cold viruses or canine fleas: the infectiousness of the presumptive parasite is apparently sufficient to account for its presence, and no evidence can be found or inferred for significant short-term benefit to individual host animals conferred by the presumptive parasite. Thus, transposons probably do not, in general, persist in populations because of any advantage they confer on their hosts.

It is important to notice that the fundamentally parasitic nature of transposons does not preclude an adventitious, or even crucial role for them in the evolution of their hosts. In other words, the fundamental parasitism of transposons is not inconsistent with the hypothesis that currently surviving lines of descent come exclusively or preferentially from previously existing lines of descent that carried transposon parasites because of some inadvertent role of these parasites in the evolution of the host genome.

Transposons are a diverse group of organisms. This group is almost certainly polyphyletic: there are at least several transposon classes of entirely independent origin (see 9 for a collection of reviews). However, transposons can reasonably and productively be included in a single group because of properties they share as a result of occupying the same niche. There are at least two such general properties.

1. *Transposons are invasive.* Transposon insertion often causes mutational alteration or inactivation of functional host gene loci. Mutation is a fundamentally random process and will be actively destructive to a system of high order (like a cellular organism), as required by the second law of thermodynamics. Thus, transposons add to the mutational loads of their hosts. Moreover, transposon copies are usually indissolubly linked to any deleterious mutations they cause. Therefore, transposon parasites must be sufficiently invasive to offset elimination resulting from these selective consequences to the host. However quantitatively modest these consequences might be in any particular case, they are utterly inevitable and persistent. High invasiveness is likely to reduce the probability of stochastic loss of the few transposon copies present in a population during the very early stages of invasion. Invasiveness is achieved by replication occurring at frequencies adequate to replenish copies eliminated by selective or stochastic processes. This replication is achieved in most or all cases by duplicative transposition: generation of a new transposon copy templated on a previously existing copy and insertion of this new copy in a new location in the host genome while the original template copy remains intact and in its original location.

2. *Transposon invasiveness is subject to moderation.* Duplicative transposition, however, is intrinsically exponential and would inevitably overwhelm the host unless moderated. This leads to a second fundamental property shared by most or all transposons: duplicative transposition is moderated in a copy-number-dependent fashion, leading to an effective upper limit to the number of copies of a transposon in an organism. Precisely how this moderation is achieved is very poorly understood at present.

These two properties allow transposons to be sufficiently invasive to persist without compromising the fitness of the host to the point of host extinction. These properties also create a dynamic, steady-state condition in which transposons are continuously functioning generators of mutation in their hosts.

In addition to the two properties listed above that all transposons probably must share, there is a third property that some transposons clearly have and that is likely to be shared by many (though probably not all) transposons inhabiting multicellular sexual populations. This third property is recognition of and response to the germ line–soma distinction. This property was first clearly demonstrated experimentally during the initial genetic (10) and molecular (11) studies of the P element in *Drosophila*. These studies demonstrated that even in individuals in which P is transposing to new sites at very high rates in germ-line cells, no detectable transposition occurs in somatic tissues in spite of the application of highly sensitive assays to detect somatic transposition.

Our current view of the most economical theory accounting for this phenomenon is as follows. First, the persistence of obligatory germ-line parasites in a sexual population requires, of course, transposition in the germ line. Conversely, transposition in the short-lived, sterile soma is irrelevant at best. Second, though somatic transposition is irrelevant to transmission, it is not necessarily innocuous. For example, germ-line P transposition is associated with chromosome fragmentation (possibly due to excessive cleavage by transposon-encoded activities responsible for cleavage of host chromosomes at potential insertion sites) that can be so severe as to produce sterility (complete obliteration of the germ line) in extreme cases (12). Presumably, somatic transposition of P would be capable of producing similarly extreme effects on the health and survival of somatic cells. Thus, demonstrably for P and probably for many other transposons, somatic transposition is likely to be actively deleterious: it reduces the health of the soma (protector and transmitter of the critical germ line) without offsetting benefit in increasing the representation of the transposon in the host gene pool. It follows from these considerations that transposons will respect the germ line–soma distinction (transposing only in the germ line) in cases where effects on somatic health are sufficiently severe and where the genetic circuitry to make the germline–soma distinction exists.

II. Allele-Specific Suppressor Loci from the Point of View of the Transposon

As reviewed above, transposons are quite sophisticated organisms. They carry on a life-style requiring very complex behaviors and decisions. In spite of this, transposons are very small (typically 3 to 10 kb in size), with limited amounts of genetic information. This paradoxical juxtaposition of genetic simplicity and a complex life-style is achieved by extensive parasitism of information produced by the cellular host. Equivalently, transposons carry only a tiny fraction of the genetic information necessary to their existence. This tiny fraction consists of a few functions idiosyncratic to transposition and whatever information is necessary to organize parasitized host functions and information. This latter "organizational" function is incompletely understood but certainly consists (at least in part) of cis-acting regulatory elements responsive to host regulatory information.

Immediately below, we review currently available information regarding the effects of suppressor loci on transposons and transposon insertion alleles. These studies provide essential background to the new proposal developed in the succeeding section. While a number of "suppressible" transposon insertion alleles have been identified, the molecular basis of suppression has been investigated in only a few cases, as follows.

1. *suppressor-of-white-apricot [su(w^a)]*. $su(w^a)$ acts on the suppressible *copia* insertion allele, *white-apricot [w^a]*. The w^a *copia* insertion is in the same transcriptional orientation as the *white* gene and occurs in the second *white* intron (*13–15*). This *copia* retrotransposon insertion behaves as a complex, slightly leaky transcript terminus formation site. Moreover, it does not entirely prevent the splicing of the intron into which it is inserted (*15*). This results in a situation in which most w^a primary transcripts are cleaved and polyadenylated in the *copia* transposon [in the second or 3′ long terminal repeat (LTR) of the transposon] but in which a small minority (roughly 5%) of transcripts are spliced at the normal second *white* intron splice sites to produce a reduced level of a structurally wild-type *white* transcript (*15*). Mutational inactivation of $su(w^a)$ leads roughly to a doubling of the levels of structurally wild-type mature *white* RNA (the properly spliced transcript form). This increase in the level of the properly spliced RNA is likely to be at the expense of the mature form truncated in *copia* sequences (*15*); however, this has not been conclusively demonstrated to date. In contrast, $su(w^a)$ appears to have no effects on levels of expression of the *white* promoter in the w^a allele. Thus, $su(w^a)$ is very likely influencing processing of w^a primary RNA transcript. Moreover, the strict specificity of $su(w^a)$ for the w^a allele suggests that the sequence target for $su(w^a)$ is within the *copia* transposon.

The allelic state of $su(w^a)$ has no discernible effects on the steady-state levels of the major *copia* transcript class in RNA preparations from whole organisms at a variety of larval, pupal, and adult stages; however, small effects (of the order of twofold) would probably escape notice, as would larger effects on minority *copia* transcript classes (our unpublished results). This indicates that $su(w^a)$ effects on *copia* are either small [possibly because available mutations are leaky, or the participation of $su(w^a)$ in *copia* expression is particularly redundant with some other function such that its absence produces small effects] or involve a minority tissue or transcript class. The notable implication of these observations is that the absence of detectable $su(w^a)$ effects on steady-state *copia* transcript levels indicates that the *copia* promoter is not a likely target for $su(w^a)$ action.

In summary, these observations suggest that $su(w^a)$ influences processing of *copia* and *copia*-containing RNA transcripts. Studies of the $su(w^a)$ locus itself (16, 17; reviewed in 18) strongly support this proposal. These studies indicate that the $su(w^a)$ autoregulates its expression by repressing splicing of its primary transcripts. The studies of w^a reviewed above are consistent with $su(w^a)$ repressing splicing of the *copia*-containing second *white* intron.

2. *suppressor-of-hairy-wing [su(Hw)]*. *su(Hw)* alters the phenotype of insertions of the *gypsy* retrotransposon at a variety of loci (5, 6). Molecular studies of the effects of *gypsy* insertions at the *yellow* [y], *forked* [f], and *hairy-wing* [Hw] loci have been carried out. A *gypsy* insertion upstream from the *y* promoter (presumably in 5' regulatory sequences) reduces pupal *y* transcript levels substantially (19). Mutational inactivation of *su(Hw)* restores pupal *y* transcript to or above the original wild-type level (19). *gypsy* insertions in the *achaete–scute* region producing dominant, gain-of-function mutant *Hw* alleles cause overproduction of transcripts in the region (20). Mutational inactivation of *su(Hw)* returns these elevated transcript levels to nearly wild-type (20). A *gypsy* insertion in an intron of *f* results in substantial reduction in pupal *f* transcription levels (21). Mutational inactivation of *su(Hw)* restores pupal *f* transcript levels to nearly normal (21).

The allelic state of *su(Hw)* has a large effect on steady-state levels of the major *gypsy* transcript (19). Mutational inactivation of *su(Hw)* produces decrease to one fifth to one twentieth in these levels (19).

In summary, the currently available evidence suggests that *su(Hw)* acts by affecting the frequency of initiation of transcription at the *gypsy* promoter. [Molecular analysis of the *su(Hw)* locus product supports this inference (V. Corces, personal communication).] Activation of the *su(Hw)*-dependent *gypsy* promoter presumably competes with activation of the *y* and *f* promoters while causing "sympathetic" activation of promoters in the *achaete–scute* region in the cases of *Hw* alleles.

3. *suppressor-of-forked [su(f)]*. $su(f)$ suppresses the mutant phenotype of *gypsy* insertions at the *forked* locus. As noted above, insertion of *gypsy* into *f* results in reduction of pupal *f* transcript levels. Mutational inactivation of $su(f)$ results in a restoration of pupal *f* transcript levels to nearly wild-type (*21*). However, $su(f)$ mutations do not influence expression of the gypsy insertion allele at *y* (see above).

$su(f)$ enhances (makes more mutant) the eye color phenotype of the w^a *copia* insertion allele at *white* (*22*). $su(f)$ has no obvious effects on w^a transcript levels but apparently results in a reduction in the proportion of w^a transcripts that are spliced to produce structurally wild-type *white* message (see above) (*13*). Thus, the molecular and phenotypic effects of $su(f)$ on w^a are apparently precisely the opposite of the those of $su(w^a)$.

Mutational inactivation of $su(f)$ produces an approximately tenfold activation of *gypsy* transcription as assessed by steady-state transcript levels (*23*).

In summary, some observations suggest that $su(f)$ is a negative effector of expression of the *gypsy* promoter. In contrast, the failure of $su(f)$ to influence steady-state levels of w^a transcripts suggests that it may not act on the *copia* promoter [$su(f)$ effects on *copia* expression itself have not been studied to date] but rather on processing of *copia*-containing transcripts. This apparently paradoxical set of observations remains to be integrated. However, we note that the hypothesis that $su(f)$ acts to regulate other suppressor loci [including $su(w^a)$ and $su(Hw)$] would account economically for these various observations.

In conclusion, these various studies suggest that different suppressors act in mechanistically different ways, ranging from effects on RNA processing or termination to effects (apparently) on initiation of transcription. This, in turn, immediately raises the question as to whether these loci represent a group unified merely by the phenomenological similarity of having effects on phenotypes of insertion alleles rather than by some more fundamental (and, thus, useful and predictive) general property. For example, these studies are consistent with the view that suppressor loci are a subset of all fly genes regulating gene expression with this subset consisting of the sum of willy-nilly choices (that is, as a result of evolutionary accident) by individual transposons. This issue remains to be resolved unambiguously; however, based on several observations, we propose in the succeeding section more fundamental relationship between a subset of suppressor loci.

III. Allele-Specific Suppressor Loci from the Point of View of the Fly

As reviewed above, transposons must be expressed in the germ line. Moreover, at least some, and probably many, transposons will be effectively or fully expressed *only* in the germ line. Thus, if suppressor loci are the fly genes producing information parasitized by transposons, these loci are presumably those necessary for allowing germ-line transposon expression and/or restricting that expression to the germ line.

Some or many gene products necessary for germ-line expression could also be present in the soma (for example, generalized transcription factors). Further, gene products necessary for restriction of transposon expression to the germ line could, formally, consist of germ-line-specific positive effectors and soma-specific negative effectors. Moreover, the transposon targets of the well-characterized suppressor loci (the *copia* and *gypsy* retrotransposons, for example) all produce relatively high steady-state levels of their major transcripts in somatic tissues (8, 23, and our unpublished results). This indicates either that effective expression (and thus transposition) of these elements is relatively promiscuous (occurring in both the soma and the germ line) or that restriction of effective expression to the germ line is not a simple matter of germ-line-specific initiation of transcription. In the simplest version of this second hypothesis (the "combinatorial" hypothesis), germ-line restriction of transposition would result from the combination of promiscuous initiation of transcription and some posttranscriptional mechanism (for example, RNA processing or translation) for restricting effective generation of gene products to the germ line.

Below we develop a detailed proposal for the role(s) of suppressor loci in the basic biology of the fly with these considerations in mind. We will propose that a subset of suppressor loci is responsible for somatic repression of full, effective transposon expression. Equivalently, we propose that a subset of suppressor loci is all or a part of the fly's genetic regulatory machinery determining the germ line–soma distinction. The argument is as follows:

First, we note that the P transposon makes the germ line–soma distinction on the basis of a "combinatorial" mechanism. P is promiscuously transcribed; however, splicing of its transcript to produce a functional mRNA occurs only in the germ line (18, 24). This certainly establishes a precedent for combinatorial recognition of the germ line–soma distinction—it is attractive to speculate that this strategy might be general.

Second, our analysis of $su(w^a)$ indicates that its product is a splicing repressor (though it could also have additional functions) and that its expression might be restricted to the soma. The evidence that the $su(w^a)$ protein is a splicing repressor has been described previously (16, 25; see 18 for a review). The evidence that $su(w^a)$ expression might be soma-limited is as follows. $su(w^a)$ transcripts accumulate in all developmental stages, somatic tissues, and somatic tissue fractions analyzed (17). Moreover, the $su(w^a)$ protein autoregulates its expression by repressing splicing of its primary transcript leading to production of recognizable "blocked" (incompletely processed) transcript forms (16, 25). As a result, accumulation of blocked transcript forms represents an assay for the presence of the $su(w^a)$ protein. These blocked forms accumulate as major products in all postcellular blastoderm stages and somatic tissues examined. In contrast, the very first $su(w^a)$ transcripts present in precellular blastoderm embryos are exclusively fully spliced, indicating that the $su(w^a)$ protein was not present when they were synthesized (16, 25). Collectively, these observations indicate that the $su(w^a)$ protein is *not* present in germ-line cells (at least in the female) as they differentiate into mature gametes, in spite of the apparently universal presence of the protein in zygotic, somatic tissues.

Third, the effects of the allelic state of $su(f)$ on *gypsy* transcript levels indicate that $su(f)$ is necessary for (incomplete) repression of *gypsy* transcription (see above). Moreover, genetic analysis of $su(f)$ suggests that it might be a soma-specific function. Specifically, temperature-sensitive lethal $su(f)$ alleles exist and are autonomous-cell lethals killing mitotically dividing somatic cells of (apparently) all types (26–29). In contrast, these same alleles appear not to be lethals with respect to germ-line cells (30, 31). These observations suggest that $su(f)$ is required in somatic cells and dispensable in germ-line cells, consistent with the hypothesis that $su(f)$ is normally expressed in somatic cells and not in germ-line cells.

In summary, the $su(w^a)$ gene was identified originally based on its effects on somatic processing of germ-line parasite-containing (*copia*-containing) transcripts and it appears to be a soma-specific splicing repressor. On this basis, we speculate that $su(w^a)$ is part of the fly's regulatory machinery for defining the germ-line–soma distinction. On the simplest version of this proposal, various fly genes (and parasitic transposons) whose expression recognizes the germ-line–soma distinction produce transcripts that are processed in one way by default in the germ line. In the soma, the $su(w^a)$ protein binds to these transcripts blocking the default processing choice and leading to some alternative processing mode. In turn, this regulatory behavior could produce several different scenarios. (*a*) Both the default (germ line) and somatic processing patterns could be functional mRNAs coding for distinct, partially overlapping polypeptides. (*b*) The default germ-line processing pattern could be a nonfunctional mRNA and the somatic processing pattern a functional mRNA. (*c*) The default germ-line processing pattern could be a functional mRNA while the somatic processing pattern could be nonfunctional. [P displays pattern *c*, and it is tempting to speculate that P might also be a target for $su(w^a)$ regulation. Moreover, *copia* is presumed to show pattern *c* in our proposal.]

Further, on the basis of available evidence, we speculate that $su(f)$ is a soma-specific activity whose effects include repression of *gypsy* transcription. A simple specific version of this proposal would suppose that *gypsy*'s recognition of the germ line–soma distinction is incomplete or quantitative rather than qualitative. This quantitative decrease in the level of somatic *gypsy* activity could result from the approximately tenfold reduction in *gypsy* transcript levels resulting from the presence of the $su(f)$ gene product. [Theoretical considerations reviewed above require only a level of distinction on the part of transposon parasites adequate to prevent an unacceptable decrease in the fitness of the host; absolute discrimination is not necessarily required.] On this model, $su(f)$ is most simply interpreted as a soma-specific repressor of transcription. It is also noteworthy here that $su(f)$ shows a highly specific interaction with the same allele interacting specifically with $su(w^a)$, w^a. Though the detailed mechanistic basis of $su(f)$ action on w^a is obscure, this relationship between $su(w^a)$ and $su(f)$ suggests that they are part of the same genetic circuitry. The various results reviewed above suggest that the genetic circuitry in which $su(w^a)$ and $su(f)$ participate is that defining the germ line–soma distinction.

Lastly, the question remains as to the probable role of $su(Hw)$ in the life of the fly. Available evidence suggests that the $su(Hw)$ product is an activator of transcription acting (at least) in the various somatic tissues affected by the *forked*, *yellow*, and *achaete–scute* loci (see above). The simplest interpretation of these observations in the context of this discussion is that $su(Hw)$ codes for a generalized transcription factor used by *gypsy* to allow transcription in the germ line but present in the soma leading to (at least limited) somatic *gypsy* transcription.

We have reviewed our current understanding of transposons as parasites occupying the gene pools of sexual organisms. These parasites are genetically simple but lead complex lives. This paradoxical attribute requires that they extensively parasitize the regulatory machinery of their cellular hosts. As a result, transposons represent excellent probes of host regulatory machinery. We review studies of a set of host genes—the allele-specific suppressor loci—that apparently represent those producing regulatory information parasitized by transposons. On the basis of these studies and of various theoretical considerations, we make the novel proposal that a subset of allele-specific suppressor loci are (in part or entirely) the genetic regulatory circuitry of the fly responsible for the germ line–soma distinction.

Acknowledgments

Work from the authors' laboratory was supported by National Institutes of Health Grant GM32003 to P.M.B.

References

1. D. L. Lindsley and E. H. Grell, *Carnegie Inst. Washington Publ.* **627**, (1968).
2. P. M. Bingham, M. G. Kidwell, and G. M. Rubin, *Cell* **25**, 693 (1981).
3. P. M. Bingham and B. H. Judd, *Cell* **25**, 705 (1981).
4. Z. Zachar and P. M. Bingham, *Cell* **30**, 529 (1982).
5. W. Bender, M. Akam, F. Korch, P. A. Beachy, M. Peifer, P. Spierer, E. B. Lewis, and D. Hogness, *Science* **221**, 23 (1983).
6. J. Modollel, W. Bender, and M. Meselson, *PNAS* **80**, 1678 (1983).
7. E. B. Lewis, *Drosophila Inf. Serv.* **23**, 59 (1949).
8. H. E. Schwartz, T. J. Lockett, and M. W. Young, *JMB* **157**, 49 (1982).
9. D. E. Berg and M. M. Howe (eds.), "Mobile DNA," in press. Am. Soc. Microbiol, Washington, D.C., 1988.
10. W. R. Engels, *PNAS* **76**, 4011 (1979).
11. P. M. Bingham, M. G. Kidwell, and G. M. Rubin, *Cell* **29**, 995 (1982).
12. W. R. Engels, *in* "Mobile DNA" (D. E. Berg and M. M. Howe, eds.), in press. Am. Soc. Microbiol., Washington, D.C., 1988.
13. R. Levis, K. O'Hare, and G. M. Rubin, *Cell* **36**, 471 (1984).
14. V. Pirrotta and C. Brockl, *EMBO J.* **3**, 563 (1984).
15. Z. Zachar, D. Davison, D. Garza, and P. M. Bingham, *Genetics* **111**, 495 (1985).
16. T.-B. Chou, Z. Zachar, and P. M. Bingham, *EMBO J.* **7**, 4095 (1987).
17. Z. Zachar, D. Garz, T.-B. Chou, J. Goland, and P. M. Bingham, *MCBiol* **7**, 2498 (1987).
18. P. M. Bingham, T.-B. Chou, I. Mims, and Z. Zachar, *Trends Genet.* **4**, 134 (1988).
19. S. M. Parkhurst and V. C. Corces, *MCBiol* **6**, 47 (1986).
20. S. Campuzano, L. Balcells, R. Villares, L. Carramolino, L. Garcia-Alonso, and J. Modelell, *Cell* **44**, 303 (1986).

21. S. M. Parkhurst and V. G. Corces, *Cell* **41**, 429 (1985).
22. M. M. Green, *Heredity* **13**, 302 (1959).
23. S. M. Parkhurst and V. G. Corces, *MCBiol* **6**, 2271 (1986).
24. F. A. Laski, D. C. Rio, and G. M. Rubin, *Cell* **44**, 7 (1986).
25. Z. Zachar, T.-B. Chou, and P. M. Bingham, *EMBO J.* **7**, 4105 (1987).
26. M. A. Russell, *Dev. Biol.* **40**, 24 (1974).
27. W. C. Clark and M. A. Russell, *Dev. Biol.* **57**, 160 (1977).
28. G. Jurgens and E. Gateff, *Wilhelm Roux's Arch. Dev. Biol* **186**, 1 (1979).
29. E. Fekete and A. Lambertsson, *Hereditas* **93**, 169 (1980).
30. T. G. Wilson, *J. Embryol. Exp. Morphol.* **55**, 247 (1980).
31. K. Lineruth and A. Lambertsson, *Hereditas* **104**, 103 (1986).

Identifying and Cloning *Drosophila* Genes by Single P Element Insertional Mutagenesis

Lynn Cooley[1]
Celeste Berg
Richard Kelley*
Dennis McKearin and
Allan Spradling

Department of Embryology, The Carnegie Institution of Washington and Howard Hughes Medical Institute Research Laboratories, Baltimore, Maryland 21210, and
*Department of Zoology, University of Texas, Austin, Texas 78712

Cloning genes by transposon tagging has been successful in several organisms (1–4). This method often represents a significant savings in time compared to cloning genes by microdissection of chromosome segments and walking/jumping through DNA libraries following precise genetic mapping. Instead, transposon insertion provides a direct molecular tag of the gene, both identifying its location and facilitating its cloning. In *Drosophila,* P elements are the most useful transposable elements for insertional mutagenesis because they transpose at high rates and are subject to genetic controls (reviewed in 5, 6). Both full-length (complete) and internally deleted (defective) P elements are present in P strains. Transposition of P elements in P strains is repressed by a cytoplasmic state known as P cytotype. When a P strain is crossed to a strain lacking P elements (M-strain), transposition is derepressed in the germ line of the hybrid progeny, resulting in a syndrome called hybrid dysgenesis. New insertion alleles of a gene to be cloned can be identified in the subsequent generation by complementation tests.

Unfortunately, newly induced insertion mutations are highly unstable due to the continued presence of transposase-encoding P

[1] Speaker.

elements and may be lost. Mutations stabilized in the P cytotype remain in the presence of dozens of additional P elements, greatly complicating cloning of the locus, since the relevant element must be identified against a background of irrelevant ones. Problems caused by element multiplicity have strongly favored the application of P element insertional mutagenesis to only a subset of genes: those with preexisting alleles, or those lying within small chromosome regions defined by deletions. A large number of uncharacterized genes lying on the major autosomes fall outside these classes.

The introduction of P element-mediated germ-line transformation (7, 8) provided an alternative method for producing insertion mutations. Strains made by germ-line transformation contain only one (or a few) insertion(s) of a defective transposon carrying an easily scored genetic marker. These transposons are stable at their site of integration unless a source of transposase is reintroduced by a genetic cross to a P-strain (7, 9, 10) or by transient expression of microinjected P-element DNA (11). The insertion site in such strains is easily determined by *in situ* hybridization (12, 13), and sequences flanking the insertion site are cloned using transposon or marker sequences to probe a genomic library prepared from DNA of the insertion strain. However, the number of single-insert lines that can be constructed by transformation is severely limited, since 10 to 100 embryos must be microinjected with P element DNA for each independent insertion obtained (reviewed in 14).

Experience with germ-line transformation established some important facts that led to the idea of using genetic manipulations for making single-insert lines efficiently. A single integrated complete P element is sufficient to activate the excision (7) or transposition (11, 15) of defective elements, following a genetic cross to a strain containing defective transposons. A complete P element, once introduced by transformation into the *Drosophila* genome, does not immediately multiply to a high copy number. Such strains, however, are unstable and eventually accumulate large numbers of complete and defective elements (16, and Preston and Engels, this volume).

We have used a strain (termed "jumpstarter") containing a complete P element maintained as a single copy to demonstrate that stable insertions of a defective P element (termed "mutator") can be efficiently produced in genetic crosses, eliminating the need for microinjection (17). The generation of strains containing single P element insertions is sufficiently advantageous as a mutagenesis procedure that, in many cases, it would be more efficient than chemical mutagenesis, particularly when molecular analysis is desired. Mutations caused by the insertion of a single P element can be readily mapped cytologically, and DNA flanking the insertion site can be recovered with ease. We have made several thousand single P element insertion stocks and have examined them for recessive mutations; approximately 10% of the insertions are homozygous lethal, and 2–3% render females sterile. In this paper, we report advances in methods for efficiently creating single-insert lines and discuss the significance of maintaining single-insert lines in a *Drosophila* Stock Center as a permanent resource for *Drosophila* biologists.

I. Mutagenesis with Single P Elements

A mutator P element should contain a marker easily scored in *Drosophila* adults and, optimally, a marker and orgin of replication that function in *Escherichia coli* to allow plasmid rescue of DNA flanking the insertion site. The size and structure of the mutator element should also be compatible with high rates of mobilization in response to transposase provided in *trans* from the jumpstarter element. Two different mutator elements have been used in extensive mutagenesis screens (Fig. 1A and B). The first element, pUChsneo (*18*), is scored by resistance to the antibiotic G418[2] present in the culture medium; it contains pUC8 plasmid sequences allowing plasmid rescue. The second mutator element we have used, ry11(1F) (*19*), contains the *rosy* gene within P element ends; its presence in

[2] G418 is a diglycoside of streptamine (*Chem Abstracts* No. 49863-47-0; TM Geneticin). [Eds.]

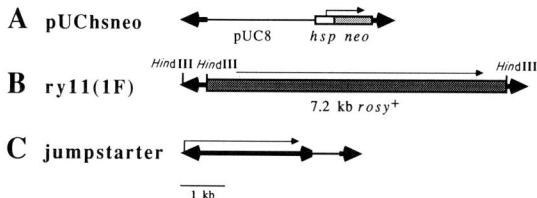

FIG. 1. P elements used in single-element insertional mutagenesis. Two different mutator elements have been used. pUChsneo (A), constructed by Steller and Pirrotta (*18*), contains pUC8 plasmid sequences and a *Drosophila* heat-shock protein (*hsp*) 70 promoter fusion to the *E. coli* neomycin resistance gene, *neo*. The ry11(1F) mutator (B) encodes the *Drosophila rosy* gene (*19*). The jumpstarter element (C) is a complete P element with a duplication of the 3' end (see *17*). Heavy lines and arrows indicate P element sequences.

adults is recognized by a distinctive eye color phenotype. It does not contain plasmid sequences.

Mutagenesis is initiated by crossing a strain containing either of these mutator elements inserted on the X chromosome to a strain containing a single complete P element, jumpstarter (Fig. 2, F_0 generation). The jumpstarter element (Fig. 1C) is a deletion derivative isolated by F. Spencer of a larger transposon made by Laski et al. (20) and produces transposase protein necessary for P-element transposition. F_1 males containing both mutator and jumpstarter elements are chosen and crossed to two or three virgin females. Transposition of the mutator element can occur in the germ-line cells of the F_1 males. In the absence of transposition, all the daughters of an F_1 male will inherit his X chromosome and will express the mutator element marker gene. Sons of an F_1 male will inherit his Y chromosome and therefore will not express the marker gene. If transposition of the mutator element from the X chromosome to an autosome occurs, that autosome can be inherited by an F_2 son, which will then express the marker gene. F_2 males expressing the marker gene are retained if they do not contain the jumpstarter chromosome and are used as founders of insertion strains. The new insertion of the mutator element is mapped to one of the autosomes by genetic segregation and then tested for a mutant phenotype in homozygous animals.

Several important factors contribute to the success of this approach. First, only one F_2 son with an autosomal insertion of the mutator transposon is saved for every F_1 cross. This ensures that each insertion line established represents a unique insertion event, thus avoiding the accumulation of insertions generated premeiotically. Second, the transposition events catalyzed by this jumpstarter element are rare enough (approximately one per 100 germ cells) that the vast majority of the lines established contain a single insert (17). This fact allows rapid correlation of any resulting mutant phenotype with one and only one easily mapped P element insertion. The mutant phenotype and site of insertion can be compared to compilations of previously identified mutations in *Drosophila* to determine if other alleles of the insertion mutation are extant. Third, the low transposition frequency also reduces to negligible the chance that the jumpstarter element, which is capable of catalyzing its own transposition, will transpose in the same germ cell as the mutator element. The majority (we estimate about 98%) of the mutator element insertions will therefore be stable to secondary transpositions unless a source of transposase is reintroduced.

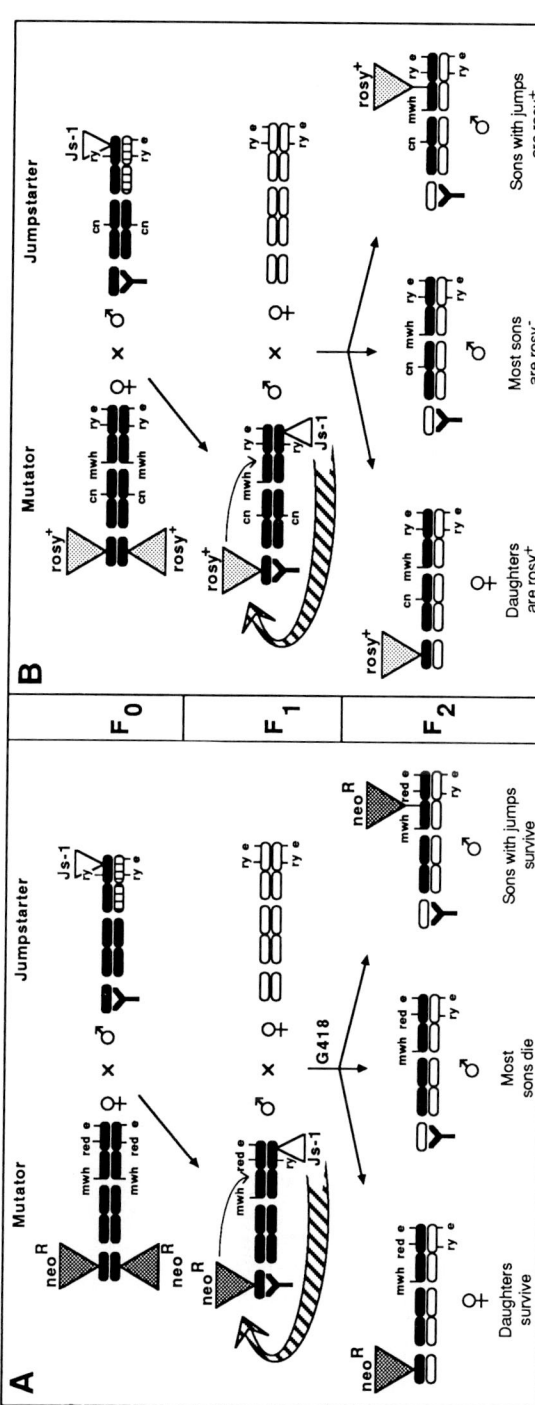

FIG. 2. Single-element insertional mutagenesis of the autosomes. The original X-linked inserts of mutator elements are at 9C and 1F, respectively, for pUChsneo (A) and ry11(1F) (B). See text for details of crosses. Triangles indicate P elements. Chromosomes are diagrammed in the order X, Y; 2nd; 3rd. Crosshatched chromosome symbols represent balancer chromosomes. The large hatched arrow represents transposase.

The prototypical single-element mutagenesis screen using the pUChsneo mutator element (17) produced a jumping frequency of 8%; that is, 8% of F_1 crosses produced at least one fertile son with an autosomal insertion (which did not inherit the chromosome containing the jumpstarter element) (Table I). Substitution of the ry11(1F) mutator element resulted in a dramatic increase in the mobilization frequency. In one mutagenesis experiment, 31% of the crosses yielded a jump (C. Berg, D. McKearin, and A. Spradling, unpublished results) and in a subsequent experiment, a further increase to 44% was obtained (L. Yue, C. Berg, and A. Spradling, unpublished results).

Several factors contribute to overall jumping frequency. Since the total number of transposition events is relatively low in the germ line of F_1 males, the chance of recovering any event is maximized by scoring a large number of progeny for each F_1 cross. The use of G418 in the culture medium to score the presence of the pUChsneo element has deleterious effects on viability and fertility even of resistant individuals (see 17). The ry11(1F) mutator, with its visible eye color marker, eliminates the need for G418 and therefore a greater number of progeny are produced, which, in part, accounts for the increased jumping frequency with the ry11(1F) mutator. The structure of the mutator element also contributes to its mobilization frequency. A mutator element containing a fragment with the *rosy* gene 0.9 kb

TABLE I
SINGLE P ELEMENT INSERTIONS

	Mutator		
	pUChsneo[a]	ry11(1F)[b]	rh11(1F)[c]
Crosses[d]	17,568	3267	5329
Jumps[e]	1317	1019	2364
Frequency[f]	8%	31%	44%

[a] Mutagenesis described by Cooley *et al.* (17).

[b] C. Berg, D. McKearin, and A. Spradling, unpublished results.

[c] L. Yue, C. Berg, and A. Spradling, unpublished results.

[d] F_1 crosses as indicated in Fig. 2.

[e] Jumps are fertile F_2 sons with an autosomal mutator insertion that do not contain jumpstarter. Only one jump for any F_1 cross is counted.

[f] Frequency is the percentage of F_1 crosses that yield at least one autosomal jump.

larger than the one in ry11 that produced the 44% freqency had a mobilization frequency of only 2–8% (*20a*). The jumping frequency of a particular mutator element is largely independent of the original chromosomal site of insertion (*20a*).

The increase in jumping frequency from 31 to 44% can be accounted for as the result of practical experience with single-element mutagenesis. Fertility of F_1 males and females was highest when adults 2–3 days old were mated. Culture conditions for the large number of F_1 crosses were optimized by careful monitoring of humidity and temperature which allowed 150–200 F_2 progeny to be scored for each cross. Some F_2 males from the pUChsneo mutator mutagenesis screen were sterile because they were X/O products of maternal nondisjunction that inherited the paternal X chromosome with the original pUChsneo element (*17*). These individuals could not be distinguished from true autosomal transpositions and thus represent wasted effort in subsequent steps. X/O infertile males were avoided in the ry11(1F) screens by including a y^+ Y chromosome that allows recognition of males containing the marked Y chromosome.

Obtaining the maximum jumping frequency of 100% might be possible using an X chromosome with two or three copies of the ry11(1F) element or by using a jumpstarter element that produces higher levels of transposase such as the somatically active Δ2-3(99B) P element (*20, 21*). Both of these approaches are being tested (J. Tower and A. Spradling, personal communication; C. Goodman, personal communication). When combinations of mutator and jumpstarter elements are found that produce one jump for each F_1 cross, it will be necessary to determine how much the average number of insertions per strain is increased.

II. Characterization of Insertion Lines

The recessive phenotypes of lines homozygous for 1317 pUChsneo (*17*) and 1019 ry11(1F) (*20a*) insertions were examined. For either mutator element, 10% of the insertions resulted in lethal mutations, 2–3% resulted in female sterility, and 1% resulted in male sterility. The distribution of phenotypes is therefore independent of the structure of the mutator element. We are examining whether the two different mutator elements show any insertion site specificity for subsets of genes within these general phenotypic classes.

The site of insertion of the P element in a large number of the pUChsneo insertion lines was determined by *in situ* hybridization to salivary gland polytene chromosomes. Using two different assays, we demonstrated that the majority of the mutations (with the exception of background mutations contaminating the starting strains) were caused by new mutator insertions (17). Further proof that mutations are caused by the P element present in the stock is the growing list of cases in which the cytogenic location of the insert and/or the phenotype of the mutation leads to its identification as an allele of a previously identified gene. This list so far includes: *big brain* (Y. N. Jan, personal communication), *string* (B. Edgar, personal communication), *vin* (22; allelism demonstrated by Y. N. Jan), *kelch* and *chickadee* (loci previously identified and mapped by T. Schüpbach), *flipper* (23), and *rotated abdomen* (24).

Secondary mutations generated at the site of P-element insertions are useful in characterizing gene function. For single-element insertion lines, this is accomplished by reintroducing a jumpstarter element for one generation. Since the ry11(1F) element provides a visible phenotype, it offers the advantage over pUChsneo insertions in that excisions of ry11(1F) can be recovered by screening for loss of $rosy^+$. Thus, excisions can be recognized independently of effects on the mutant gene. Precise excisions will revert the mutant phenotype, and imprecise excisions may generate new alleles (deletions) of the gene.

III. Cloning Insertion Site DNA

One of the powerful advantages of mutagenesis with single P elements is the ease with which DNA flanking the site of insertion can be cloned, thus initiating molecular analysis of the affected gene. Clones of DNA flanking insertions of the pUChsneo element can be recovered by plasmid rescue as follows: DNA is extracted from adults bearing the insertion mutation, then digested to completion with a restriction endonuclease that cleaves in the pUC8 polylinker. The restricted DNA is ligated under dilute conditions to circularize the restriction fragment containing pUC8 and flanking genomic DNA. Clones are recovered by transforming *E. coli* cells to ampicillin resistance. The fragment of genomic DNA from a recovered clone can then be used as a probe to recover clones of genomic DNA from a library of wild-type DNA. The ry11(1F) mutator does not contain plasmid cloning sequences. However, the presence of only one P element in the stock greatly simplifies recovery of clones from a small genomic library constructed with DNA from adults containing the insertion.

IV. An Insertion Library of the Drosophila Genome

Mutagenesis with single P elements generates a large number of lines containing stable P element insertions. While the majority of lines from any one mutagenesis experiment may not be of interest to an individual investigator, collectively they represent a valuable resource to Drosophila biologists. The Drosophila Stock Center at Indiana University will provide a centralized facility to house, characterize, and disperse insertion lines. One part of the insertion library will be composed of lines with an easily identified mutant phenotype, such as recessive lethality or sterility. New contributions to the library will be mapped by in situ hybridization and tested by complementation with any other line having an insertion in the same chromosomal region. Allelic disruptions can be discarded, easing the burden of stock maintenance. It has been estimated that about 50% of the 5000 to 10,000 Drosophila genes are mutable by P element insertion under conditions of hybrid dysgenesis (25, 26). Similarly, we have estimated that the third chromosome contains at least 582 essential loci readily mutable by single P element insertion (17). Based on these estimates, a collection of about 2500 single P element insertions would identify most of the genes easily mutable by P element insertion. Investigators will be able to request copies of strains with a relevant phenotype that contain insertions within a chromosome region of interest.

Another part of the insertion library could be composed of lines with no obvious mutant phenotype. These might include lines with subtle defects in behavior that require detailed anatomical characterization or biochemical screening to identify. Allelic insertions could not be removed initially from this collection. To benefit from this part of the insertion library, an investigator would need to analyze all of the insertion strains for mutants of interest. This requirement would present logistical problems for the Stock Center since mailing large numbers of strains to many investigators is impractical. A format such as a summer course sponsored by the Stock Center would provide a solution to this problem. The researcher could bring equipment for a particular assay to the site, screen a large number of lines during the summer and return home to analyze lines of interest.

An insertion library would also provide an efficient means for cloning genes that are not targets for P element insertion. A collection of 2500 insertions would provide an insertion, on average, every 60 kb throughout the euchromatin. Any mutation could be genetically mapped between the two closest insertions in the library, using the dominant markers encoded by mutator elements, and the insertions could then provide starting points for a chromosomal walk to the gene. Additionally, insertions flanking a gene could provide defined end points for chromosomal deficiencies (L. Cooley and A. Spradling, unpublished results) that would be useful in characterizing the gene. Clearly, an insertion library has the potential to greatly facilitate molecular genetic analysis of the *Drosophila* genome.

ACKNOWLEDGMENTS

This work was supported in part by U.S. Public Health Service Grant GM27875 and by postdoctoral fellowships from the Carnegie Institution of Washington (L.C.), the American Cancer Society (C.B. and D.M.), and the National Institutes of Health (R.K.).

REFERENCES

1. P. M. Bingham, R. Levis, and G. M. Rubin, *Cell* **25**, 693 (1981).
2. N. V. Fedoroff, D. B. Furtek, and O. E. Nelson, *PNAS* **81**, 3825 (1984).
3. I. Greenwald, *Cell* **43**, 583 (1985).
4. E. M. Rinchik, L. B. Russell, N. G. Copeland, and N. A. Jenkins, *Genetics* **112**, 321 (1986).
5. G. M. Rubin, in "Mobile Genetic Elements" (J. Shapiro, ed.), p. 329. Academic Press, Orlando, Florida, 1983.
6. W. R. Engels, in "Mobile DNA" (D. Berg and M. Howe, eds.), in press. ASM Publ. Washington, D.C., 1988.
7. A. C. Spradling and G. M. Rubin, *Science* **218**, 341 (1982).
8. G. M. Rubin and A. C. Spradling, *Science* **218**, 348 (1982).
9. W. R. Engels, *Science* **226**, 1194 (1984).
10. S. B. Daniels, M. McCarron, C. Love, and A. Chovnick, *Genetics* **109**, 95 (1985).
11. R. E. Karess and G. M. Rubin, *Cell* **38**, 135 (1984).
12. M. L. Pardue, in "*Drosophila*: A Practical Approach" (D. B. Roberts, ed.), p. 111. IRL Press, Oxford, England, 1986.
13. W. R. Engels, C. R. Preston, P. Thompson, and W. B. Eggleston, *Focus* **8**, 6 (1986).
14. A. C. Spradling, in "*Drosophila*: A Practical Approach" (D. B. Roberts, ed.), p. 175. IRL Press, Oxford, England, 1986.
15. S. M. Mount, M. M. Green, and G. M. Rubin, *Genetics* **118**, 221 (1988).
16. S. B. Daniels, S. H. Clark, M. G. Kidwell, and A. Chovnick, *Genetics* **115**, 711 (1987).
17. L. Cooley, R. Kelley, and A. Spradling, *Science* **239**, 1121 (1988).
18. H. Steller and V. Pirrotta, *EMBO J.* **4**, 167 (1985).
19. A. Spradling and G. Rubin, *Cell* **34**, 47 (1983).
20. F. A. Laski, D. C. Rio, and G. M. Rubin, *Cell* **44**, 7 (1986).
20a. L. Cooley, C. Berg, and A. Spradling, *Trends Genet.* **4**, 254 (1988).
21. H. M. Robertson, C. R. Preston, R. W. Phillis, D. Johnson-Schlitz, W. K. Benz, and W. R. Engels, *Genetics* **118**, 461 (1988).
22. M. E. Akam, D. B. Roberts, G. P. Richards, and M. Ashburner, *Cell* **13**, 215 (1978).

23. C. B. Bridges and O. L. Mohr, *Genetics* **4**, 304 (1919).
24. T. H. Morgan and C. B. Bridges, *Carnegie Inst. Washington Publ.* **327**, 190 (1923).
25. M. Kidwell, in *"Drosophila:* A Practical Approach" (D. B. Roberts, ed.), p. 59. IRL Press, Oxford, England, 1986.
26. M. Kidwell, *Drosophila Inf. Serv.* **66**, 81 (1987).

Molecular Lesions Induced by I-R Hybrid Dysgenesis in *Drosophila melanogaster*

ISABELLE BUSSEAU[1]
ALAIN PELISSON
MICHÈLE CROZATIER*
CHANTAL VAURY AND
ALAIN BUCHETON

Laboratoire de Génétique,
Université Blaise Pascal, Aubière,
France, and
* Division of Molecular Biology,
Netherland Cancer Institute, 1066
CX Amsterdam, The Netherlands

We have identified molecular lesions associated with six mutations induced by I-R (inducer-reactor) hybrid dysgenesis in the *yellow* gene of *Drosophila melanogaster*. Three of these mutations are due to insertions: one is a complete I element, the other two are shorter I elements, truncated at their 5' ends. The last three mutations are associated with chromosomal rearrangements. In two cases, I elements are found at the rearrangement breakpoints. Their organization suggests that they result from recombination between I elements.

The transposable I element (or I factor) from *Drosophila melanogaster* is of particular interest, as it has been identified as the main genetic factor controlling I-R hybrid dysgenesis (1). All *Drosophila melanogaster* strains can be classified into two categories: inducer (I) strains, which contain active I elements, and reactive (R) strains, which do not (2). I-R hybrid dysgenesis occurs in the germ line of F_1 females resulting from crosses between I males and R females, whereas the progeny of reciprocal crosses are apparently normal. I elements are normally stable when maintained in I strains, but are activated when introduced into the cellular environment of R strains, and transpose with unusually high frequencies in these conditions (3).

[1] Speaker.

This results in various genetic abnormalities chracteristic of this system, such as sterility, due to early embryo death (4), and high rates of mutations (5). Six complete I elements have been cloned from *white* gene mutations produced during I-R hybrid dysgenesis (6–8). They are 5.4 kb long, and are flanked by target-site duplications varying in length from 10 to 14 bp. They lack terminal repeats, and are terminated at their 3′ ends by several repeats of the triplet TAA. One complete I element has been entirely sequenced. It contains two large open reading frames (Fig. 1a). ORF-1 encodes a protein with a putative nucleic-acid-binding domain also found in *gag* polypeptides; ORF-2 encodes a protein showing similarities with retroviral reverse transcriptases and RNase H (8, 9).

The structure of the I element is totally different from that of the P and *hobo* elements (10, 11), which are involved in two other independent systems of hybrid dysgenesis (12–14). However, I elements resemble the "noviral retrotransposons." This category of transposable elements includes LINES (*l*ong *i*nterspersed *n*ucleotide *e*lements, L1) in mammals (15–19), F, G, Doc in D. melanogaster (20–22), ingi in *Trypanosoma brucei* (23), and Cin 4 in *Zea mays* (24).

I. Molecular Lesions Induced by I-R Hybrid Dysgenesis

Very few mutations induced by I-R hybrid dysgenesis have been analyzed at the molecular level (6, 7). From eight mutations of the *white* gene studied so far, six are due to complete I-element inser-

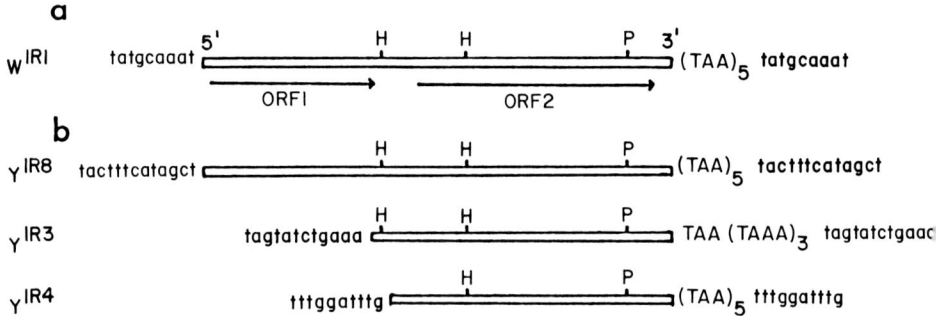

FIG. 1. (a) Structure of the complete I factor inserted in the mutation w^{IRI} (see 8). (b) Structure of y^{IR8}, y^{IR3}, and y^{IR4} insertions. Sequences from the 3′ ends of I elements are indicated by uppercase letters. Sequences from target-site duplications are in lowercase letters. H, *H*3; P, *Pst*I

tions, while the other two are due to deletions removing some coding sequences from the *white* gene (8).

In order to extend this analysis, we have studied six mutations that arose in the *yellow* gene. Only two of these are associated with molecular lesions similar to those found in the *white* gene: y^{IR7} is due to a deletion removing some of the *yellow* coding sequences; y^{IR8} is due to the insertion of a complete I factor. The last four mutations fall into two additional classes: insertions of truncated I elements and chromosomal rearrangements.

A. Mutations Resulting from Insertions of Truncated I Elements

The alleles y^{IR3} and y^{IR4} differ from wild-type by insertions of DNA fragments of 3893 and 3813 bp in size, respectively. These are incomplete I elements, lacking about 1.5 kb from the 5' end compared to the complete 5.4-kb I element (Fig. 1). Both are flanked by target-site duplications, suggesting that they inserted as truncated I elements rather than having been deleted after insertion. Surprisingly, the I element inserted in y^{IR3} terminates at the 3' end by the sequence TAA (TAAA)$_3$ instead of the usual run of TAA.

We have found more truncated I elements as well as complete I elements inserted in the y^{IR5} and y^{IR6} alleles. They are associated with the breakpoints of chromosomal rearrangements.

B. Mutations Associated with Chromosomal Rearrangements

The y^{IR6} rearrangement is too small be be visible cytologically. It is defined by two breakpoints, one of which extends into the *yellow* gene. One is associated with the insertion of a complete I factor, and the other with the insertion of a 5' truncated I element. Each of these insertions is flanked at one side by its own target-site duplication, and at the other side by that of the other element. This organization strongly suggests that the DNA fragment between them has been inverted (Fig. 2).

The simplest hypothesis to explain the y^{IR6} rearrangement is that it is the result of two events: (*a*) insertion of two I elements in opposite orientation on the same chromosome; and (*b*) recombination between the two I elements.

The recombination step has occurred simultaneously, or immediately after transposition, since we have checked that the two I elements involved in this rearrangement were not present on parental chromosomes. This could be the result of homologous pairing be-

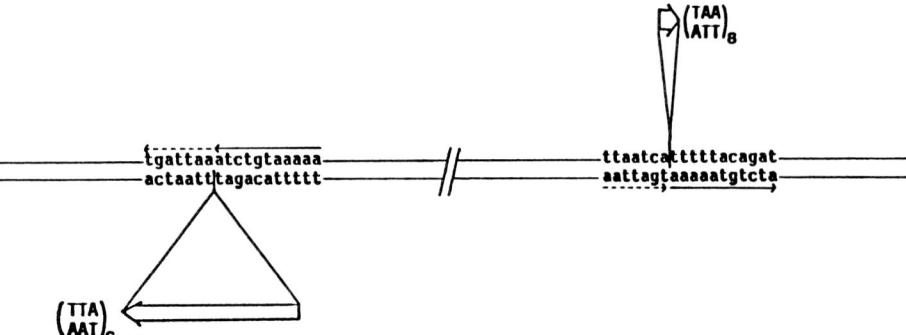

Fig. 2. Organization of sequences at y^{IR6} breakpoints. Open boxes represent I elements, with the 3' end indicated by the arrowhead. Dotted and plain arrows underline target sequences that have been duplicated and exchanged.

tween the two I elements, according to the model proposed by Schnewly et al. (22), or due to another, more specific mechanism, linked to I-element transposition.

The rearrangement associated with the mutation y^{IR5} is more complex. A chromosomal inversion, extending approximately from 1A (*yellow* locus) to 2B, is cytologically visible on polytene chromosomes.

Molecular analysis of this allele is still in progress, but preliminary data indicate that it involves at least three breakpoints, each associated with DNA insertions. One of these insertions is a complete I element. The other two are 5' truncated I elements of 38 and 66 bp, including TAA repeats. Surprisingly, both are associated at their 5' ends with unrelated AT-rich tracks of DNA (Fig. 3). The organization of target-site duplications, although disrupted by the rearrangement, implies that these AT-rich sequences were either associated with I elements before integration or created at the time of insertion.

II. Conclusions

Several I elements truncated at the 5' end are described in this paper. They vary in length from 38 to 3895 bp. They result from recent transposition events.

Fawcett et al. (8) have suggested that elements transpose via reverse transcription of an RNA intermediate transcribed from an internal promoter located near the 5' end. If this is true (5), truncated I elements are unlikely to be exact copies of deleted donor elements,

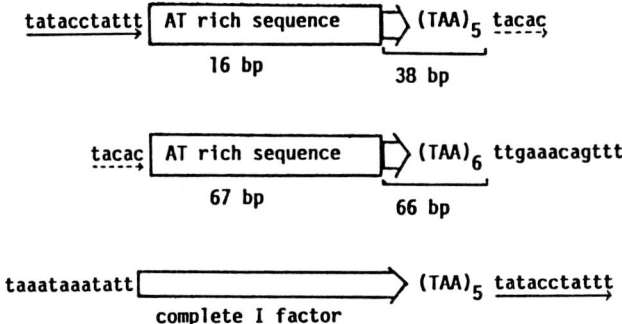

FIG. 3. Organization of sequences at three of the y^{IR5} breakpoints. Symbols are the same as for Fig. 2.

which would obviously be unable to synthesize this RNA. More probably, complete I elements tend to lose their 5' ends during the course of transposition or at the time of integration. This could be due to the mechanism of integration leading to preferential loss of the 5' end, or to early arrest of reverse transcription.

This behavior of the I element has been observed for other nonviral retrotransposons, which are often found truncated at their 5' ends. This probably reflects some similarities in the mechanism of transposition of all these structurally related mobile elements.

Chromosomal rearrangements associated with y^{IR6} and y^{IR5} mutations are not the results of rare events. A high rate of lethal mutations is induced by I-R hybrid dysgenesis (5, 25), and many of these lethal mutations are associated with various chromosomal rearrangements (C. Prudhommeau and J. Proust, personal communication). Their breakpoints are often associated with I-element sequences, as shown by *in situ* hybridization with salivary gland chromosomes (J. Proust, personal communication). The results presented here strongly suggest that rearrangements can be produced by recombination between I elements, and that this phenomenon occurs during the process of transposition. This does not seem to be true, however, for the deletions associated with w^{IR7}, w^{IR8}, and y^{IR7} mutations, as no I-element sequence has been found at their breakpoints (8, and unpublished data).

Homologous recombinations between repeated, dispersed sequences such as transposable elements are believed to play an important role in the evolution of eukaryotic genomes (26–28). I-R hybrid dysgenesis contributes to these phenomena in two ways: (*a*) new insertion sites of I elements are created by transposition favoring subsequent homologous recombination events; and (b) recombinations between I elements occur at unusually high frequencies, stimulated by an unknown mechanism during I-R hybrid dysgenesis.

ACKNOWLEDGMENT

This work has been supported by grants from the Centre National de la Recherche Scientifique (UA360), the Université Blaise Pascal, and the Association pour la Recherche sur le Cancer.

REFERENCES

1. G. Picard and P. l'Heritier, *Drosophila Inf. Serv.* **46**, 54 (1971).
2. A. Bucheton, J. M. Lavige, G. Picard, and P. l'Heritier, *Heredity* **36**, 305 (1976).
3. G. Picard, *Genetics* **83**, 107 (1976).
4. J. M. Lavige, *Biol. Cell* **56**, 207 (1986).
5. G. Picard, J. C. Bregliano, A. Bucheton, J. M. Lavige, A. Pelisson, and M. G. Kidwell, *Genet. Res* **32**, 275 (1978).
6. A. Bucheton, R. Paro, H. M. Sang, A. Pelisson, and D. J. Finnegan, *Cell* **38**, 153 (1984).
7. H. M. Sang, A. Pelisson, A. Bucheton, and D. J. Finnegan, *EMBO J.* **3**, 3079 (1984).
8. D. H. Fawcett, C. K. Lister, E. Kellett, and D. J. Finnegan, *Cell* **47**, 1007 (1986).
9. D. J. Finnegan, in "Transposition" (A. J. Kingsman, F. R. Kingsman, and K. F. Charter, eds.), Soc. Gen. Microbiol. Symp. 43, p. 271. Cambridge Univ. Press, Cambridge, England, 1988.
10. K. O'Hare and G. M. Rubin, *Cell* **34**, 25 (1983).
11. R. D. Streck, J. E. MacGaffey, and S. K. Beckendorf, *EMBO J.* **5**, 3615 (1986).
12. M. G. Kidwell, J. F. Kidwell, and J. A. Sved, *Genetics* **86**, 813 (1977).
13. R. K. Blackman, R. Grimaila, M. M. D. Koehler, and W. M. Gelbart, *Cell* **49**, 497 (1987).
14. G. Yannopoulos, N. Stamatis, M. Monastirioti, P. Hatzopoulos, and C. Louis, *Cell* **49**, 487 (1987).
15. M. F. Singer, *Cell* **28**, 433 (1982).
16. E. D'Ambrosio, S. D. Waitzkin, F. R. Witney, A. Salemme, and A. V. Furano, *MCBiol* **6**, 411 (1986).
17. M. Hattori, S. Kuhara, O. Takenaka, and Y. Sakaki, *Nature (London)* **321**, 625 (1986).
18. D. D. Loeb, R. W. Padgett, S. C. Hardies, W. R. Shehee, M. B. Comer, M. H. Edgell, and C. A. Hutchinson III, *MCBiol* **6**, 168 (1986).
19. H. H. Kazazian, Jr., C. Wong, H. Youssoufian, A. F. Scott, D. G. Phillips, and S. E. Antonarakis, *Nature (London)* **332**, 164 (1988).
20. P. P. Di Nocera and G. Casari, *PNAS* **87**, 5843 (1987).
21. I. B. Dawid, E. O. Long, P. P. di Nocera, and M. L. Pardue, *Cell* **25**, 399 (1981).
22. S. Schnewly, A. Kuroiwa, and W. J. Gehring, *EMBO J.* **6**, 201 (1987).
23. B. E. Kimmel, O. K. Moiyoi, and J. R. Young, *MCBiol* **7**, 1465 (1987).
24. Z. Schwarz-Sommer, L. Leclercq, E. Gobel, and H. Saedler, *EMBO J.* **6**, 3873 (1987).
25. J. Proust and C. Prudhommeau, *Mutat. Res.* **95**, 225 (1982).
26. G. S. Roeder, *MGG* **190**, 117 (1983).
27. M. L. Goldberg, J. Y. Sheen, W. J. Gehring, and M. M. Green, *PNAS* **80**, 5017 (1983).
28. P. S. Davis, M. W. Shen, and B. H. Judd, *PNAS* **84**, 174 (1987).

III. Regulation of Gene Expression

X-Chromosome Inactivation as a System of Gene Dosage Compensation to Regulate Gene Expression — 119
MARY F. LYON

The Developmental Regulation of Albumin and α-Fetoprotein Gene Expression — 131
SALLY A. CAMPER, ROSELINE GODBOUT, AND SHIRLEY M. TILGHMAN

A Methylation Mosaic Model for Mammalian Genome Imprinting — 145
CARMEN SAPIENZA, THU-HANG TRAN, JEAN PAQUETTE, ROSS MCGOWAN, AND ALAN PETERSON

Insertional Mutations in Transgenic Mice — 159
FRANK COSTANTINI, GLENN RADICE, JAMES J. LEE, KIRAN K. CHADA, WILLIAM PERRY, AND HYEUNG JIN SON

X-Chromosome Inactivation as a System of Gene Dosage Compensation to Regulate Gene Expression

MARY F. LYON

Medical Research Council,
Radiobiology Unit, Chilton,
Didcot, Oxon OX11 ORD,
England

The two most extensively studied systems of gene dosage compensation are those in *Drosophila melanogaster* and in mammals, and these systems show considerable differences. The most striking feature of the mammalian system is that two homologous chromosomes, the two X chromosomes of females, behave differently within the same cell, whereas in *Drosophila* homologs within a cell behave similarly, and the compensation is obtained by differential chromosome behavior in the two sexes (1, 2). A further difference is seen when the X chromosome is broken by translocations; in *Drosophila*, the different segments seem to behave autonomously, whereas in mammals the behavior of separated segments is controlled by the X-inactivation center. This paper deals with some aspects of current knowledge of X-inactivation.

I. Initiation and Maintenance of X-Inactivation

As already mentioned, in X chromosome inactivation one X chromosome becomes differentiated in such a way that it is insensitive to the signals that normally control transcription during development. The differential behavior is initiated in early development, first in the trophectoderm and endoderm, and last in the primitive ectoderm, which gives rise to the embryo proper (3, 4).

A. X-Inactivation Center

It is in this initiation of X-inactivation in early development that the X-inactivation center (5–8) is thought to play a critical role, and the evidence for this comes from mouse X-autosome translocations. In such translocations, inactivation spreads from the X chromosome into the attached autosomal material, and can be detected by variegation in expression of the coat-color genes located in the autosomal segment (6, 7). However, Russell and colleagues (6, 7) showed very early that variegation only occurs for genes in one of the two segments. In addition to this, only one of the two segments into which the X chromosome is broken shows the late replication (9) or dark Kanda staining (8) characteristic of the inactive X chromosome. All this evidence suggests that only one of the two X-chromosomal segments can undergo inactivation.

This was tested using the translocations T(X;4)37H and T(X;11)38H, both of which have breakpoints in the proximal region of the X chromosome, near the locus of the *sparse-fur* gene, *spf*, a deficient allele of the gene for ornithine transcarbamoylase (OTC).[2] Animals were bred that were heterozygous for one of the translocations and for *sparse-fur*, with the normal allele of OTC on the translocated chromosome. When the livers of these animals were stained histochemically for OTC, the T38H animals showed a mosaic pattern of positive-negative-staining cells, as expected for heterozygotes for a deficient allele, and as seen also in chromosomally normal +/*spf* heterozygotes. However, in T37H females heterozygous for *spf*, all cells showed uniform positive staining. *In situ* hybridization to the chromosome showed that the translocation breakpoint was distal to the OTC locus in T37H, and proximal to it in T38H. Thus, the interpretation was that in T38H the OTC locus is in physical continuity with the inactivation center, and hence the locus is inactivated normally. In T37H, on the other hand, the OTC locus is separated from the center and so remains active in all cells (*10*).

As well as showing the presence of an inactivation center, these results with translocations also provide evidence for spreading of the inactivation along the X chromosome, and that inactivation is a positive process. X-chromosomal genes do not spontaneously or autonomously become inactive. They remain active unless they receive an inactivation signal.

B. Signals for Inactivation

Obvious questions concern the nature of the inactivation signal that spreads along the X chromosome, and also of the signal to the inactivation center itself that determines whether it becomes active or inactive. Other problems concern the means by which the active or

---

[2] Ornithine carbamoyltransferase, EC 2.1.3.3. [Eds.]

inactive state is maintained and stabilized. However, answers are not known to any of these questions. To consider the inactivation center first, dosage compensation in *Drosophila* is bound up with a mechanism for sensing the X:autosome ratio (*1, 2*), and there is probably an element of X:autosome ratio in determining the number of inactivation centers that remain active in mammals. In diploids, only a single X chromosome remains active no matter how many are present, whereas in triploids in the mouse (*11*), human (*12*), and rabbit (*13, 14*) two X chromosomes may remain active, at least in some cases or some cells. Although the importance and significance of the X:autosome ratio in X-inactivation are far from clear, it is interesting that, in animal groups as widely separated in evolution as insects and mammals, this ratio should be involved in dosage compensation.

1. PARENTAL CHROMOSOME IMPRINTING

An interesting feature of X-inactivation concerns the role of parental chromosome imprinting. In marsupials (reviewed in *15*), in the extraembryonic membranes of mice and rats (*16, 17*), and apparently, but not so clearly, also in extraembryonic membranes of humans (*18*), the paternally derived X chromosome is preferentially inactivated.

In eutherians, the cell lineages that show the imprinting are those in which inactivation occurs first in development. Thus, it may be that the imprint is lost before inactivation occurs in the primitive ectoderm (embryonic) lineage, or that the mechanism loses its sensitivity to the imprint. In parthenogenetic embryos, in which both X chromosomes are maternally derived, X-inactivation can still occur, both in the embryo itself (*19*) and in the extraembryonic membranes (*20, 21*). Similarly, in human diploid moles, in which both X chromosomes are paternally derived, one X chromosome can remain active (*22*). Thus, it appears that the imprint is not an essential part of the inactivation mechanism. Rather, it determines the probability of a particular X chromosome remaining active, and perhaps also the particular developmental stage or cell division at which inactivation occurs. The imprint may affect either the maternal or the paternal chromosome or both (*23*) and its nature is still unknown, but differential methylation has been widely discussed as one possibility (*24, 25*).

2. ALLELIC STATE OF INACTIVATION CENTER

Another factor that affects the probability of a particular X chromosome remaining active is the allelic state of the inactivation center. Johnston and Cattanach (*26*) described three alleles of the X-inactivation center (Xce), differing in strength such that, in heterozygotes, the stronger of the two has a higher probability of becoming the active center, the relative strengths in descending order being Xce^c, Xce^b, Xce^a.

3. Developmental Changes and Reactivation

Whatever the signals may be that are associated with a change in state of the inactivation center from active to inactive or the reverse, they appear to be developmentally fairly specific. Experimentally, inactivation can be induced in cultured embryonic mouse cell lines by allowing them to differentiate (27). Monk and Harper (3) suggested that, in the mouse embryo, inactivation in the various cell lineages, trophectoderm, primitive endoderm, and primitive ectoderm, is associated in time with the differentiation of these lineages. However, in the marsupial mouse *Antechinus*, X-inactivation had already occurred in the protoderm of unilaminar blastocysts at a stage when these cells were thought to be still undifferentiated and totipotent (28). Once inactivation has been established, the state of the inactivation center is typically highly stable. Reactivation occurs normally in female germ cells approaching meiosis (29).

In cultured cells, some experimental procedures, such as treatment with 5-azacytidine, lead to reactivation at either one or a few loci (30), suggesting an interference with the spreading signal, rather than a change in the inactivation center itself. However, complete reactivation has been obtained by treatment of an embryonic cell line with 5-azacytidine (31). This suggests that inactivation in the embryonic cell line might have been less stable, and hence that stabilization is a separate event occurring after the initial inactivation (31). Migeon et al. (32) obtained complete reactivation in a human chorionic villus cell line, i.e., an extraembryonic cell type, by fusing it with a mouse cell line, thus adding to other evidence suggesting that stabilization may be weaker or absent in extraembryonic cells. Interestingly, there is an association between imprinting of the X chromosome and instability of X-inactivation (reviewed in 33).

4. Spreading of Inactivation

As with the nature of the imprint and the signal to the inactivation center, the nature of the spreading signal remains unknown. Again, methylation has been put forward as a possibility (reviewed in 30). However, the observed methylation differences between active and inactive X chromosomes appear later than inactivation (34), suggesting that methylation may be involved in stabilizing the inactive state, but perhaps not in spreading. Further evidence for a role of methylation in stabilization is that in marsupials, where X-inactivation is relatively unstable, the glucose-6-phosphate dehydrogenase (EC-1.1.1.49) (G6PD) gene on the inactive X chromosome lacks some specific methylations present in inactive eutherian G6PD genes (35). Gartler and Riggs (30) have suggested that the spreading signal itself may involve a change in the state of the chromatin fiber, leading to the condensation which results in sex chromatin formation.

As already mentioned, in X-autosome translocations, inactivation can travel into autosomal material, and it can in some cases travel for a considerable distance. In T37H, the coat color locus *brown* (*b*) shows variegation, presumably because of inactivation, and this locus is over 20 centiMorgans (cM) in map distance and several G-bands in physical distance from the translocation breakpoint (36). This makes it likely that there is a true spread into the autosome, rather than a position effect because of proximity to inactive chromatin. However, the signal appears not to travel as readily in autosomal material. This suggests that there is either something about the X chromosome that promotes travel, which Gartler and Riggs (30) call "way stations," or in autosomes that impedes it.

II. Effects in a Segment Homologous with Y

There is the possibility also of stop signals that arrest the spread of inactivation, e.g., into the region of homology with the Y. It has long been thought that because X-inactivation serves the function of dosage compensation, genes in the segment of X homologous with the Y would not be inactivated (37). In the human this appears to be the case, in that the *MIC2* antigen locus is not inactivated (38). However, in the mouse the first indication that genes in the pairing segment might in fact be inactivated came from the sex-reversing factor, *Sxr* (39). *Sxr* involves the transposition and duplication of the testis-determining gene of the Y to the pairing segment (40). Here, it can undergo crossing-over onto the X, and XX animals carrying *Sxr* are sex-reversed to males. In females heterozygous for the X-autosome translocation T(X;16)16H, there is nonrandom X-inactivation with the translocated X being active in all cells. Animals that carry T16H and have *Sxr* on the normal (hence inactive) X chromosome can be females (41, 42). This implies that *Sxr* can be inactivated, at least partially or in some cells. Jones *et al.* (43) then studied steroid sulfatase, *Sts*, the locus for which had previously been shown (44) to be in the mouse homologous segment, with genes on both the X and the Y. The evidence indicated that *Sts* too could be inactivated. Males had a higher STS level than females, and this was not a hormonal effect, because XX sex-reversed males had values comparable with XX females. Moreover, females heterozygous for a null allele had a still lower level. Thus, in this case, there is X-inactivation without dosage compensation. At present, the significance of this is not understood.

III. Possible Role of Dosage Differences in Sex Determination and Differentiation

One possibility is that sex determination and differentiation in mammals are more complex than previously thought and that the animal indeed makes use of dosage differences. This was first suggested by Chandra (45), and more recently by Page *et al.* (46) and German (47). In the marsupial, sexual differentiation is controlled directly by the chromosomes, rather than hormonally as in eutherians (48), and dosage differences are apparently involved, as an XXY animal had female-like and an XO animal had male-like characteristics (49). In the eutherian, one possibility is that dosage differences might be involved in determining the size differences found in male and female mouse embryos before the gonads are differentiated (50), as well as possibly in determining the gonads themselves (46, 47). However, another possibility is that dosage effects in the pairing segment are simply a result of inefficiency in the system. We know that X-inactivation spreads into attached autosomal segments. Perhaps in the X chromosome itself there are in fact no effective stop signals. Inactivation fails to stop accurately at the junction of the differential and pairing segments, but spreads beyond, and the animal tolerates the dosage differences thus created.

IV. Dosage Effects in Germ Cells

Dosage differences are important also in the germ cells. In the male, the X chromosome becomes inactive during the early part of spermatogenesis, as seen by condensation and lack of RNA production. In the female, conversely, the inactive X chromosome is reactivated just before the onset of meiosis, so that in the oocytes both X chromosomes are active. It appears that if the male germ cell has two X chromosomes, as in XX or XXY males, then reactivation occurs also in the male (51). Correct activity level of the X chromosome in germ cells appears to be important in more ways than one.

In both the mouse and human male, there are many different chromosomal causes of sterility, including X-autosome translocations, that share the common feature that they impede meiotic pairing. Miklos (52) and others (e.g., 53) have suggested that the presence of unpaired or unsaturated pairing sites at meiosis is highly detrimental to a male germ cell, and is likely to lead to failure of spermiogenesis. The mechanism of this is still largely unknown. However, it appears likely that one function of X-inactivation during spermatogenesis is the inactivation of unpaired pairing sites on the X. If unpaired sites remain in the active state, some detrimental event occurs, but if these sites are inactivated, they are in some way protected.

This could explain the differing effects of different types of X-autosome translocation on fertility. All known reciprocal X-autosomal translocations in the mouse are male-sterile (7). However, Cattanach's translocation (54), which involves an insertion of autosomal material into the X chromosome (55), is male-fertile. Similarly, a second recently discovered autosomal insertion into the X chromosome is male-fertile, as is a Robertsonian translocation involving the X chromosome and chromosome 2 (56). In the X-insertions, the inserted autosomal segment presumably has inactivated pairing sites. Certainly in Cattanach's translocation it is only rarely seen to pair with the autosomes (55). There are two types of male, with and without a corresponding autosomal deletion. Those without the deletion would thus have no unsaturated pairing sites, and are fertile. Those with the deletion would have unsaturated sites only on the undeleted autosome, and are partially fertile. Similarly, in the Robertsonian translocation, the X-chromosomal arm should have normally inactivated pairing sites, and the autosomal arms should pair normally, giving no impairment of fertility. Conversely, in X-autosome reciprocal translocations, inactivation of pairing sites would probably spread from one X segment into attached autosomal material, while in the other segment of the X chromosome, inactivation might be absent or incomplete. Thus, the total extent of unsaturated pairing sites would lead to male sterility.

In addition to this effect on pairing, there appears to be an entirely separate effect of X-chromosome dosage in male germ cells. In males with two X chromosomes, such as XXY males or XX sex-reversed males, spermatogenesis fails at the spermatogonial stage. This appears to be due to the double X chromosome dosage. In XX males sex-reversed with the *Sxr* factor, small patches of spermatogenesis occur in a few testis tubules. The cells in these patches are chromosomally XO (51), indicating that the loss of one X chromosome from the primitive stem cell in these patches has allowed spermatogenesis to proceed. Similarly, in XO sex-reversed males (XO Sxr), spermatogenesis can proceed up to at least the late spermatid stage (39). This lethal effect of double X-chromosome dosage on the male germ cell is apparently not due to pairing failure, first because the death of the XX cells happens too early in spermatogenesis, before pairing, and second because the effect appears to be in the wrong direction. The pairing failure would appear to be worse in XO Sxr. Thus, in the male germ cell, one appears to need both a correct X dosage and inactivation of unpaired sites at meiosis.

Conversely, in the female germ cell, the presence of only a single X chromosome leads to very early loss of all oocytes in humans and to excess but not total loss of oocytes in mice (57, 58). The interpretation of these effects in female cells is somewhat difficult. It has now been shown that failure of pairing is detrimental to female as well as male germ cells (53), and hence reactivation of the X chromosome in oocytes just before meiosis is probably required to allow normal pairing and recombination of the X chromosomes of the female. XO oocytes obviously have unsaturated pairing sites and may die for this reason. However, it is also possible that a double X-chromosome dosage is required in addition.

V. Multiple Roles of X-Inactivation

Thus, the present roles of X-inactivation in mammals are multiple. There is dosage compensation in the soma, inactivation of unsaturated pairing sites in the male, and reactivation of needed pairing sites in the female germ cell. In addition, there may be a role for dosage compensation in sex determination, and the X chromosome must have a correct but different gene dosage with respect to the autosomes in male and female germ cells.

VI. Resistance and Sensitivity to Inactivation

Further information on the mechanism and possible evolutionary progress of X-inactivation has come from studies of introduced transgenes in the mouse. Krumlauf et al. (59) found that an α-fetoprotein transgene was not inactivated in the extraembryonic membranes, but was inactivated in the embryonic tissues (59), and a chicken transferrin gene was not inactivated at all (60). This suggests that these inserted sequences somehow resisted inactivation. There is also evidence that genes in the human X-Y pairing segment can resist inactivation. Mohandas et al. (61) found a case in which the distal part of the long arm of the X chromosome, including the G6PD gene, had been duplicated and translocated to the distal tip of the short arm. The G6PD gene then underwent inactivation, although genes in the distal short arm, which normally escaped inactivation, still did so. Thus, the spreading signal for inactivation must have passed through the noninactivated region and the genes there must have resisted its effect. Since many autosomal genes do become inactivated when involved in translocations with the X chromosome, it seems likely that most genes are sensitive to inactivation and that the resistant genes have some special feature that withstands either the inactivating signal itself, or the stabilizing signal. However, this is an open question at present.

Until recently, it was thought that the activity of X-linked genes in eutherian mammals was all-or-none, either fully active or fully inactive. However, in humans, where the steroid sulfatase gene is regarded as noninactivated, Migeon et al. (62) showed that the activity of the gene on the inactive X chromosome was less than that on the active one. In the mouse, the data so far do not show whether the inactivation of *Sts* is complete or not. However, as mentioned earlier, in animals with the *Sxr* sex-reversal gene on the tip of the X chromosome, the inactivation of this appears to be incomplete, in that some animals with this X chromosome inactive in all cells can indeed be sex-reversed to males. Migeon et al. (63) raised the possibility that the activity of genes on the active X chromosome might be enhanced. If so, the mammal would be showing an effect like that in *Drosophila*, in which the transcription of genes on the single X chromosome of males is doubled, to give an activity equal to that of two X chromosomes in the female (1). In mammals, one would have a single X chromosome active in both sexes, but with enhanced activity.

If there were such an enhancement, it might be mediated through the X-inactivation center, as is the inactivation. If this were so, genes on segments separated from the center by an X-autosome translocation should show the unenhanced activity. To test this, further work was done on OTC activity in the translocation T37H, mentioned earlier. Animals heterozygous for the OTC-deficient allele, *spf*, and also those homozygous for the wild type were studied. In both cases, animals carrying the translocation had higher OTC activity than corresponding chromosomally normal animals. By contrast, animals carrying the translocation T38H, in which the OTC locus is carried on the inactivated segment, had OTC levels like the normal controls (64). These results are in accord with expectation if the locus on the segment of T37H separated from the inactivation center had full activity. Thus, the conclusion from this is that, if there is any enhancement of activity of genes on the active X chromosome, it is not mediated through the inactivation center. The possibility that genes are acting autonomously in this respect cannot, of course, be ruled out. As a point of interest, the data on STS in the mouse indicate that the activity of this locus is the same when on the X chromosome as on the Y chromosome (43). The most probable explanation of the partial activity of STS on the inactive human X chromosome is that there can be partial inactivation of genes outside the region of complete inactivation.

VII. Evolution of X-Inactivation

The inactivation of the mouse *Sts* locus raises the possible paradox that, in the evolution of vertebrates, X-inactivation may originally have begun not as a system of dosage compensation, but rather as a mechanism to create dosage differences involved in sex determination and differentiation (45–47, 65). At an early stage of evolution, there may have been an inactivation center with only a short spread of effect (66). As the sex chromosomes became heteromorphic, there may then have been a gradual increase in spreading, by selection in the differential segment for sensitivity to inactivation in the control sequences of genes, and for promotion of travel in the intervening sequences. By contrast, in the homologous segment, there would have been selection for resistance to inactivation. At the same time, an intermediate region would have been retained, where dosage differences remain important in sex determination or differentiation.

Acknowledgments

I am very grateful to many colleagues for their discussions and contributions to this work. These colleagues include Debbie Badger, Bruce Cattanach, Ted Evans, Peter Glenister, Janet Jones, Jo Peters, Kevin Whitehill, Jo Zenthon, and many others.

References

1. B. S. Baker and J. M. Belote, *ARGen* **17**, 345 (1983).
2. E. Jaffe and C. Laird, *Trends Genet.* **2**, 316 (1986).
3. M. Monk and M. J. Harper, *Nature (London)* **281**, 311 (1979).
4. J. D. West, in "Genetic Control of Gamete Production and Function" (P. G. Crosignani, B. L. Rubin, and M. Fraccaro, eds.), p. 49. Academic Press, London, 1982.
5. M. F. Lyon, in "Second International Conference on Congenital Malformations," p. 67. Int. Med. Congr., New York, 1963.
6. L. B. Russell, *Science* **140**, 976 (1963).

7. L. B. Russell, in "Cytogenetics of the Mammalian X Chromosome. Part A: Basic Mechanisms of X Chromosome Behaviour" (A. A. Sandberg, ed.), p. 205. Liss, New York, 1983.
8. S. Rastan, *J. Embryol. Exp. Morphol.* **78**, 1 (1983).
9. L. B. Russell and N. L. A. Cacheiro, in "Genetic Mosaics and Chimeras in Mammals" (L. B. Russell, ed.), p. 393. Plenum, New York, 1978.
10. M. F. Lyon, J. Zenthon, E. P. Evans, M. D. Burtenshaw, K. A. Wareham, and E. D. Williams, *J. Embryol. Exp. Morphol.* **97**, 75 (1986).
11. S. Endo, N. Takagi, and M. Sasaki, *Dev. Genet.* **3**, 165 (1982).
12. P. A. Jacobs, A. M. Matsuyama, I. M. Buchanan, and C. Wilson, *Am. J. Hum. Genet.* **31**, 446 (1979).
13. H. Jerome, L. Cathelineau, and O. Bomsel-Helmreich. *Proc. Congr. Int. Reprod. Anim. Insem. Artif.*, 6th **2**, 893 (1968).
14. O. Bomsel-Helmreich, *Adv. Biosci.* **6**, 381 (1971).
15. J. L. Van de Berg, E. S. Robinson, P. B. Samollow, and P. G. Johnston, *Curr. Top. Biol. Med. Res.* **15**, 225 (1987).
16. N. Takagi and M. Sasaki, *Nature (London)* **256**, 640 (1975).
17. N. Takagi, in "Genetic Mosaics and Chimeras in Mammals" (L. B. Russell, ed.), p. 341. Plenum, New York, 1978.
18. K. B. Harrison and D. Warburton, *Cytogenet. Cell Genet.* **41**, 163 (1986).
19. M. H. Kaufman, M. Guc-Cubrilo, and M. F. Lyon, *Nature (London)* **271**, 547 (1978).
20. S. Rastan, M. H. Kaufman, A. Handyside, and M. F. Lyon, *Nature (London)* **288**, 172 (1980).
21. S. Endo and N. Takagi, *Jpn. J. Genet.* **56**, 349 (1981).
22. M. Tsukahara and T. Kajii, *Hum. Genet.* **71**, 7 (1985).
23. M. F. Lyon and S. Rastan, *Differentiation* **26**, 63 (1984).
24. M. Monk, *Nature (London)* **328**, 203 (1987).
25. D. Solter, *ARGen* (in press).
26. P. G. Johnston and B. M. Cattanach, *Genet. Res.* **37**, 151 (1981).
27. G. R. Martin, C. J. Epstein, B. Travis, G. Tucker, S. Yatsiv, D. W. Martin, Jr., S. Clift, and S. Cohen, *Nature (London)* **271**, 329 (1978).
28. P. G. Johnston and E. S. Robinson, *Chromosoma* **95**, 419 (1987).
29. P. G. Johnston, *Genet. Res.* **37**, 317 (1981).
30. S. M. Gartler and A. D. Riggs, *ARGen* **17**, 155 (1983).
31. G. D. Paterno, C. N. Adra, and M. W. McBurney, *MCBiol* **5**, 2705 (1985).
32. B. R. Migeon, M. Schmidt, J. Axelman, and C. R. Cullen, *PNAS* **83**, 2182 (1986).
33. M. F. Lyon, *Am. J. Hum. Genet.* **42**, 8 (1988).
34. L. F. Lock, N. Takagi, and G. R. Martin, *Cell* **48**, 39 (1987).
35. D. C. Kaslow and B. R. Migeon, *PNAS* **84**, 6210 (1987).
36. A. G. Searle, C. V. Beechey, E. P. Evans, and M. Kirk, *Cytogenet. Cell Genet.* **35**, 279 (1983).
37. M. F. Lyon, *Am. J. Hum. Genet.* **14**, 135 (1962).
38. P. N. Goodfellow and S. M. Darling, *Development* **102**, 251 (1988).
39. B. M. Cattanach, C. E. Pollard, and S. G. Hawkes, *Cytogenetics* **10**, 318 (1971).
40. E. P. Evans, M. Burtenshaw, and B. M. Cattanach, *Nature (London)* **300**, 443 (1982).
41. B. M. Cattanach, E. P. Evans, M. Burtenshaw, and J. Barlow, *Nature (London)* **300**, 445 (1982).
42. A. McLaren and M. Monk, *Nature (London)* **300**, 446 (1982).
43. J. Jones, B. M. Cattanach, and C. Rasberry, *Genet. Res.* **52**, 59 (1988).
44. E. Keitges, M. Rivest, M. Siniscalco, and S. M. Gartler, *Nature (London)* **315**, 226 (1985).

45. H. S. Chandra, *PNAS* **82**, 6947 (1985).
46. D. C. Page, R. Mosher, E. M. Simpson, E. M. C. Fisher, G. Mardon, J. Pollack, B. McGillivray, A. de la Chapelle, and L. G. Brown, *Cell* **51**, 1091 (1987).
47. J. German, *Am. J. Hum. Genet.* **42**, 414 (1988).
48. W.-S. O, R. V. Short, M. B. Renfree, and G. Shaw, *Nature (London)* **331**, 716 (1988).
49. G. B. Sharman, E. S. Robinson, S. M. Walton, and P. J. Berger, *J. Reprod. Fertil.* **21**, 57 (1970).
50. M. Seller and K. J. Perkins-Cole, *J. Reprod. Fertil.* **79**, 159 (1987).
51. M. F. Lyon, *in* "Physiology and Genetics of Reproduction" (E. M. Coutinho and F. Fuchs, eds.), Part A, p. 63. Plenum, New York, 1974.
52. G. L. G. Miklos, *Cytogenet. Cell Genet.* **13**, 558 (1974).
53. P. S. Burgoyne and T. Baker, *in* "Controlling Events in Meiosis" (C. W. Evans and H. G. Dickinson, eds.), p. 349. Co. Biol., Cambridge, England, 1984.
54. B. M. Cattanach, Z. *Vererbungsl.* **92**, 165 (1961).
55. S. Ohno and B. M. Cattanach, *Cytogenetics* **1**, 129 (1962).
56. I. D. Adler and A. Neuhauser-Klaus, *Mouse News Lett.* **77**, 139 (1987).
57. M. F. Lyon and S. G. Hawker, *Genet. Res.* **21**, 185 (1973).
58. P. S. Burgoyne and T. G. Baker, *J. Reprod. Fertil.* **75**, 633 (1985).
59. R. Krumlauf, V. M. Chapman, R. E. Hammer, R. L. Brinster, and S. M. Tilghman, *Nature (London)* **319**, 224 (1986).
60. M. A. Goldman, K. R. Stokes, R. L. Idzerda, G. S. McKnight, R. E. Hammer, R. L. Brinster, and S. M. Gartler, *Science* **236**, 593 (1987).
61. T. Mohandas, R. L. Geller, P. H. Yen, J. Rosendorff, R. Bernstein, A. Yoshida, and L. J. Shapiro, *PNAS* **84**, 4954 (1987).
62. B. R. Migeon, L. J. Shapiro, R. A. Norum, T. Mohandas, J. Axelman, and R. L. Dabora, *Nature (London)* **299**, 838 (1982).
63. B. R. Migeon, S. F. Wolf, C. Mareni, and J. Axelman, *Cell* **29**, 595 (1982).
64. J. Peters, S. T. Ball, J. Zenthou, and M. F. Lyon *Genet. Res.* **52**, 62 (1988).
65. M. F. Lyon, *Nature (London)* **320**, 313 (1986).
66. J. A. M. Graves, *Trends Genet.* **3**, 252 (1987).

The Developmental Regulation of Albumin and α-Fetoprotein Gene Expression

> SALLY A CAMPER,[1]
> ROSELINE GODBOUT*
> AND SHIRLEY M. TILGHMAN†
>
> Department of Human Genetics,
> University of Michigan, Ann
> Arbor, Michigan 48109
> * Cross Cancer Institute, Edmonton,
> Alberta, Canada T6G 1Z2, and
> † Department of Molecular Biology,
> Princeton University, Princeton,
> New Jersey 08544

Members of multigene families often have different developmental programs for gene expression. We have focused our attention on the genes that encode albumin and α-fetoprotein (AFP), the major serum proteins in the developing mammalian fetus. The genes encoding these two proteins arose from the duplication of an ancestral gene 300–500 million years ago (1). Linkage of the genes has been maintained in mammals, with the 3' terminus of the albumin gene approximately 14 kilobases (kb) upstream of the 5' end of the AFP gene (2). The activation of their transcription in fetal liver is coordinate. However, after birth these genes have independent programs for developmental regulation. Albumin transcription increases slightly from birth to adulthood, but AFP transcription undergoes a dramatic decline after birth, resulting in an approximately 10,000-fold reduction in AFP mRNA levels (3, 4).

Experiments using cell culture systems and transgenic animals have defined several regulatory elements that are required for activation and high-level expression of the AFP and albumin genes. Both genes have tissue-specific promoter proximal elements located within 200 base-pairs (bp) of the transcription start sites (5, 6). In addition, the locus contains several enhancers. One of these lies approximately 10 kb upstream of the albumin gene (7), and three others are located

[1] Speaker.

within the intergenic region, 6.5, 5.0, and 2.5 kb upstream of the AFP gene (6, 8).

In order to better understand the different developmental programs of these genes, and to explore the significance of their close linkage, the ability of each gene to function independently has been tested in transgenic mice. In addition, the developmental regulation of chimeric transgenes, consisting of the albumin enhancer paired with the AFP transcription unit and the three AFP enhancers directing transcription of the albumin gene, has been tested. Our results indicate that the AFP gene functions appropriately when separated from the albumin gene (9). Analysis of the chimeric genes suggests that the repression of the AFP gene is mediated by its promoter or by intragenic elements, but not by its enhancers.

In this article, we discuss the relative merits of the basic approaches we and others have used to assay tissue-specific gene expression, and review the current status of our knowledge of the regulatory regions of both the albumin and AFP genes. A comparison between the albumin–AFP and β-globin gene families is made in terms of the possible functions for linkage and coordinated gene activity. Finally, our current understanding of the developmental regulation of these genes is presented.

I. Models for Tissue-Specific Gene Expression

Programs for activation of gene expression must include mechanisms common to all cell types to ensure the expression of "housekeeping" genes, mechanisms that totally exclude expression except in a very specific cell type, and mechanisms that can activate expression in a limited number of distinct cell types. It is this latter case of restricted but diverse activation of gene activity that characterizes the albumin and AFP genes. Both genes are very actively transcribed in the developing liver (each constituting approximately 15% of the mRNA) and less actively transcribed in the gastrointestinal tract (0.1%) (1). In the visceral endoderm of the yolk sac, the AFP and albumin genes constitute approximately 10% and 0.1% of the mRNA, respectively (1). In addition, trace amounts of both mRNAs have been reported for a number of tissues including brain, kidney, and pancreas (5, 9).

A simplistic model to explain the regulation of genes that are moderately or strictly tissue-specific is to invoke *cis*-acting elements that are recognized by a factor or factors that are unique to the appropriate tissue. The immunoglobulin genes are the most clearly understood of the strictly tissue-specific genes (reviewed in *10*). The immunoglobulin μ-heavy-chain and κ-light-chain enhancers are each composed of multiple motifs, most of which bind proteins present in a wide variety of tissues; however, at least one motif from each enhancer is bound by proteins found primarily in the cells of the B lymphoid lineage (*10*). Interestingly, *in vivo* analyses revealed that no proteins are bound to the immunoglobulin enhancers in the nonexpressing tissues (*11, 12*), suggesting perhaps that the tissue-specific factor is required to nucleate the binding of the ubiquitous proteins.

By analogy to the immunoglobulin paradigm, one might assume that genes expressed in multiple tissues would have a *trans*-acting factor that is required for activity, but is present only in the appropriate subset of tissues. Alternatively, a unique tissue specificity could result from a specific combination of ubiquitous *trans*-acting factors (*13*). This notion is supported by the observations that combining regulatory regions of tissue-specific genes and housekeeping genes results in novel tissue specificities in transgenic mice (*14, 15*). A third possibility is that a gene must have multiple enhancers, each with a unique tissue specificity, to be expressed in multiple tissues. There are several genes of *Drosophila melanogaster* in which this model has been confirmed (*16, 17*). For example, the yolk protein genes are expressed in fat bodies and ovaries by virtue of having two enhancers, one specific for fat bodies, and one specific for ovaries (*18, 19*). The diversity in the patterns of expression of AFP and albumin provides an opportunity to test these models in a developing mammalian system.

Other strategies for diverse tissue specificity include gene duplication with divergence of the regulatory regions, and promoter duplication resulting in alternate promoters in different cell types. There are examples of these in *Drosophila*, including the alcohol dehydrogenase gene and the pattern formation genes *antennapedia* and *caudal* (*20, 21*). The amylase genes provide a good example from mammals, in which a cluster of amylase genes is expressed in the pancreas. A separate gene, present in a single copy, is activated in the liver and the parotid gland by tandem promoters of different tissue specificities (*21*). However, these models do not apply to albumin and AFP, as each gene exists as a single copy and uses the same promoter in each tissue (*1*).

II. AFP Regulatory Elements

A. Approaches for Identification of Regulatory Elements

We have relied primarily on two approaches to identify the *cis*-acting sequences that govern the tissue specificity of the AFP gene. A collection of clones bearing a systematic set of deletions in the 5′ flanking region of the gene was introduced into hepatoma cells in culture by standard transfection protocols, and the transient transcriptional activity of each was assayed (6, 8). This approach is ideal for detection and fine-mapping of *cis*-acting sequences because it lends itself to rapid analysis of multiple recombinant constructs. By comparing the activity of *cis*-acting elements in cells derived from the tissue of interest and cells that have a different tissue origin, one can often discern which elements are responsible for the tissue specificity. Indeed, this approach has been instrumental in detecting tissue-specific elements for numerous genes (10).

Transient expression assays measure the presence of excess transcription factors, as each cell is probably taking up thousands of copies of DNA. Therefore, it is not surprising that cell lines vary in their abilities to detect enhancer and promoter elements, despite the fact that each cell line expresses the endogenous gene at some level (8). Likewise, continuous culture of an individual cell line can lead to changes in the repertoire of excess factors, resulting in a loss of response to a given regulatory element. An additional problem is that any tissue-specific or developmental controls imprinted on the DNA through structuring of chromatin or DNA methylation will not be reproduced in transient expression assays in committed cells. These problems, coupled with the fact that such cell lines are usually transformed, possibly leading to an increase or change in the concentration of various factors, may result in promiscuous transcription (22).

To circumvent the problems arising from cell lines, an alternative approach, whereby DNA is introduced into the germ line of mice, has been utilized (23). The use of transgenic mice for this purpose is complicated by position effects; that is, the integration site of a transgene often markedly affects its expression. This problem can be overcome to a certain extent by producing large numbers of transgenic mice for each DNA construction, in order to separate the effects of position from the inherent behavior of the various constructs.

Another approach to identifying regulatory regions is to exploit the observation that such regions are usually associated with sites in chromatin that display hypersensitivity to the endonuclease DNase I.

This hypersensitivity is thought to arise from the binding of *trans*-acting factors that disrupt the normal spatial array of nucleosomes (24). Recently, we have shown that the DNase I hypersensitive sites that exist *in vivo* correlate well with regulatory elements we have mapped in cultured cell (25). This correlation has been noted for other genes, including albumin and globin (7, 26). Thus, new regulatory sequences could be located, especially if an appropriate cell culture model does not exist, by mapping the DNase I hypersensitive sites first and then employing transgenic mice to confirm their function.

B. Distal Regulatory Elements

The region between AFP and albumin contains three enhancers that display position and orientation independence as well as the ability to activate heterologous promoters (6). These enhancers are located at 6.5, 5 and 2.5 kb upstream of the AFP transcriptional start site (Fig. 1) and are functionally equivalent in transient expression assays (6, 8). The existence of such multiple elements is probably more common than is currently appreciated because distal regions often are not tested, and redundant elements escape detection in studies that employ simple 5' sequential deletions.

Multiple elements are emerging as a common theme in modulating gene activity. For example, responsiveness of genes to metals (27), dsRNA (28), and hormones such as estrogens (29) and glucocorticoids (30) all involve repeated elements. Perhaps the apparent redundancy of elements serves to strengthen the response, much as the repetition of enhancer segments of rRNA genes improves the ability to compete for transcription factors (31). Indeed, the function of multiple regula-

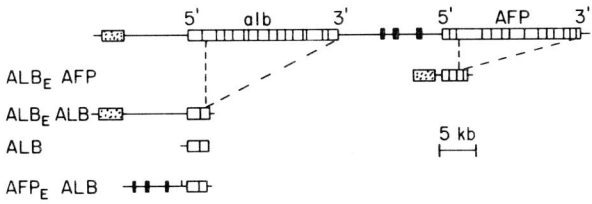

FIG. 1. Schematic representation of the albumin–AFP locus. The structural genes for albumin (alb) and AFP are depicted by the rectangles, with the direction of transcription indicated 5' to 3'. The AFP enhancers are illustrated as black boxes; enhancer I is closest to the AFP gene. The albumin enhancer is depicted as a stippled box. The chimeric constructs, ALB$_E$ AFP and AFP$_E$ ALB, each consist of the indicated enhancer fused to the alternate promoter and internally deleted structural gene (minigene). ALB$_E$ALB consists of the albumin enhancer, promoter, and minigene.

tory elements in the AFP–albumin locus may be to ensure the establishment of an active transcription complex.

We considered the possibility that the AFP enhancers are not redundant *in vivo*. A number of *Drosophila* genes with complex developmental and/or tissue-specific patterns have multiple distal regulatory elements with different functions (17–19). For example, a gene required for establishing the segmental pattern, *fushi tarazu*, has an enhancer, a separate morphogenic control element necessary for producing the "zebra" pattern of expression at the blastoderm stage, and a third element necessary for expression in the ventral nervous system (16). Perhaps some mammalian genes also achieve diverse expression through the combination of multiple nonredundant elements.

The tissue specificity of the AFP enhancers was first probed by transient expression in hepatoma cell cultures. The activity of each of the elements, when coupled to a heterologous promoter, is 5- to 10-fold greater in hepatoma cells than in nonhepatoma cells, such as HeLa cells. This result indicates that the enhancers alone are not responsible for restricting expression to the appropriate tissues (6, 8).

When the tissue specificity of each of the enhancers was tested in conjunction with the AFP promoter by germ-line transformation of mice, it was apparent that the elements are not redundant. Rather, each has a different pattern of activity in each of the three fetal tissues tested. In the yolk sac, where expression is the highest, each of the enhancers exhibited equivalent activity. In the liver, the enhancers displayed an additive behavior. All three enhancers were required to obtain the highest level of expression. Moderate levels were obtained with enhancer I or II, and enhancer III was clearly the least potent of the three. In the gastrointestinal tract, where AFP is a minor mRNA, only enhancer I exhibited marked activity (9). Thus, in contrast to the examples from *Drosophila*, the diverse tissue specificity of AFP arises not from multiple enhancers of unique specificity but rather from the combined action of enhancers with somewhat broader specificities and different "strengths."

We undertook the fine-mapping and DNA-sequence determination of each of the enhancer domains to understand the molecular basis for their diverse tissue specificities (8). Each enhancer was localized to a region of 200 to 300 bp in length. Further analysis suggested that at least enhancers I and II are made up of multiple motifs, each of which is insufficient for full activity. This kind of complexity is characteristic of the immunoglobulin and SV40 enhancers, and distinct from the simplicity of most response elements, such as heat shock (*32*), or hormone (*29, 30*) and metal (*27*) responsiveness. Enhancer I is apparently a complex enhancer composed of both positive and negative elements, similar to several other tissue-specific enhancers, including those in the β-interferon, insulin, retinol-binding protein, and immunoglobulin heavy-chain genes (reviewed in *10*). Transient expression assays in cell culture suggest that the negative element of enhancer I is not recognized in hepatoma cells, consistent with the notion that negative elements play an important role in restricting expression to the appropriate tissues (*8*).

The nucleotide sequences reveal several areas of striking homology among the AFP enhancers. Interestingly, enhancers I and II, both of which are active in the liver, share a block of 10 nucleotides (TGTTTGCWGT),[2] which has also been noted in the AFP promoter at position −78, and in the regulatory domains of several other liver-specific genes including hepatitis B virus, α_1-antitrypsin (*33*), and H19 (*34, 35*). This homology block, therefore, is an excellent candidate for the site of binding of a liver-specific *trans*-acting factor.

Another notable feature of the AFP enhancer sequences is an 18-nucleotide consensus sequence shared between all three of the enhancers (GCTTTGAGCAATRRYACA).[3] Good matches to this consensus have also been noted in the H19 gene enhancers (*35*). Experiments are in progress to determine whether tissue-specific proteins bind to these regions of homology.

In analyzing the regulatory regions of other liver-specific genes, other candidates for sequences responsible for liver-specific expression have been uncovered (*36–38*). Although still in the preliminary stages, these results suggest that multiple *trans*-acting factors are responsible for the activation of liver-specific genes. This is to be expected, since the liver is an organ in which the program for gene activity is required to undergo striking changes several times during the development and maturation of the animal (*39, 40*). Even the paradigm of the immunoglobulin heavy- and κ-light-chain enhancers indicates that the B-lymphoid lineage has at least two separate, specific *trans*-acting factors (*10*).

A complete understanding of the molecular basis of tissue specificity requires that the *trans*-acting factors utilized in each tissue be identified. As this work progresses, we may eventually be able to

---

[2] W = A or T [Recommendations of the Nomenclature Committee of the International Union of Biochemistry, *Eur. J. Biochem.* **150**, 1 (1985)]. [Eds.]

[3] R = puRine, Y = pYrimidine [Recommendations of the Nomenclature Committee of the International Union of Biochemistry, *Eur. J. Biochem.* **150**, 1 (1985)]. [Eds.]

construct artificial enhancers expected to exhibit unique specificities, rather than diverse specificities, and confirm this tissue specificity with transgenic mice.

C. AFP Promoter Proximal Elements

A 1-kb region of 5'-flanking DNA containing the AFP promoter is transcriptionally active in hepatoma cells, but not in heterologous cells such as HeLa or L cells (6). The lack of activity in HeLa cells can be overcome if a strong enhancer, such as the SV40 enhancer, is added (41). However, in this case, the only sequence in the promoter that is required for activity is the TATAA motif at −33 bp. In hepatoma cells, however, a region spanning −250 to −33 bp is absolutely required for transcriptional activity (M. Feuerman, R. Godbout, and S. M. Tilghman, unpublished results). The strict requirement for this proximal regulatory region in hepatoma cells, and its repressive effects in heterologous cells in the absence of the SV40 enhancer, suggest that it may play a crucial role in restricting the activation of the AFP gene to the appropriate tissues.

III. Albumin Regulatory Elements

A. Promoter-Proximal Elements

The albumin promoter has been the focus of considerable attention. A 400-bp region of the rat albumin promoter is sufficient for tissue specificity in that it directed high levels of expression of a reporter gene in a variety of rat hepatoma lines expressing the endogenous albumin gene, but the same construction exhibited undetectable or weak activity in mouse L cells and hepatoma cell lines that had lost the capacity for albumin production (42). The tissue specificity of the albumin promoter can be overcome with the strong SV40 enhancer (43), just as we observed with the AFP promoter (41).

A deletion analysis of the rat albumin promoter revealed the presence of at least two proximal domains important for high level transcription (43). Deletion of sequences between −151 and −118 bp (DE-II) resulted in approximately a 17-fold drop in transient expression of a reporter gene in albumin-positive cells, but had no effect on basal transcription of the reporter gene in the albumin-negative cells. A less tissue-specific element lies between −118 and −93 bp (DE-I), because deletion of that region lowers transcription 15-fold in both expressing and nonexpressing cells. Finally, a region (−81 to −34 bp) just 5' of the TATA motif may also have significance for tissue-specific expression. All of these regions are highly conserved in both sequence and position between human, rodent, and chicken albumin genes. Consistent with the deletion analysis, the region between −170 and −55 of the mouse promoter is required for efficient *in vitro* transcription, and that region is sufficient to confer liver-specific transcription on a heterologous promoter *in vitro*. (5).

The issue of which sequences are bound by factors has also been addressed by several groups. Combining the results of exonuclease III and DNase I "footprinting," methylation protection, and gel mobility shift experiments, two regions of protein binding have been detected spanning −134 to −110 and −105 to −85, respectively (44). These regions correspond well with the previously defined elements DE-II and DE-I. However, while factors interacting with these regions are enriched in liver extracts, binding activity is also present in extracts of nonliver cells. Further analysis has demonstrated additional binding sites in the −164 to −32 bp region (38). Liver extracts are much more enriched than brain or spleen extracts for a factor that apparently occupies three separate sites in the region. In addition, a site that has the potential for binding the ubiquitous transcription factor NF1 is bound by a distinct factor from liver extracts (ϕNF1). The CCAAT motif (−80) is bound by a protein equally abundant in all of the extracts tested. The organization of binding sites for ubiquitous proteins and proteins enriched in liver corresponds with the pattern expected from the deletion analysis. The tight clustering of factors abundant in the liver around the albumin promoter suggests that liver-specific expression could result from cooperative protein–protein interactions.

B. Distal Elements

Transgenic mice have been used to assay the transcriptional activity of the albumin 5′-flanking region (7). In spite of the high activity of the albumin promoter in cell culture (42), the same promoter fused to a human growth hormone reporter gene was essentially inactive in transgenic mice, although an occasional founder animal grew larger than normal. The inclusion of a region spanning −8.5 to −12 kb upstream of the albumin transcription start site resulted in a significant increase in the percentage of transgenic animals that grew larger than normal; however, the absolute level of expression was variable. This element has some characteristics of an enhancer in that it functions in an orientation- and position-independent manner, although it failed to activate a heterologous promoter. There are two DNase-I-hypersensitive sites in the 5′ flanking DNA between the promoter and the distal enhancer-containing domain (at approximately −3.5 and −8 kb). While no function could be ascribed to these regions, they may indeed have some role, because their inclusion resulted in a progressive increase in the level of expression of the reporter gene. The developmental regulation of the albumin gene remains to be examined.

IV. Function of Albumin and α-Fetoprotein Linkage

Many evolutionarily related genes are clustered, including the α-globins, β-globins, growth hormones, keratins, actin, myosin, and ovalbumin genes. Is there a purpose to this clustering, or is it simply a remnant of their generation by duplication? It has been hypothesized that genes that have remained clustered share regulatory elements (2).

The β-globin gene cluster provides a good example of shared regulatory regions. The human β-globin cluster contains the ε, $γ^G$, $γ^A$, δ, and β genes oriented 5′ to 3′, corresponding to their order of expression in development. The genes span approximately 45 kb of DNA, and the cluster is flanked by DNase-I-hypersensitive sites located 17 and 18 kb from the 5′ and 3′ ends, respectively, of the ε and β genes (26). These outlying hypersensitive sites are probably important for activating transcription because a class of β-thalassemias apparently results from a deletion encompassing the 5′ distal group of DNase-I-hypersensitive sites (45). The dominance of these regulatory regions is evident from the demonstration that they can confer position-independent transcription on the human β-globin gene in transgenic mice (46). This position-independent, high-level expression is not observed with the region surrounding the β-globin gene that contains at least one enhancer element (47). Thus, the 5′ distal element, 50 kb away from the β-globin gene, is presumed to be important for adult β-globin gene expression.

In the chicken, the β-globin gene expressed in the adult is situated 5′ of the embryonic ε-globin gene. An enhancer that lies in the intergenic region is apparently responsible for activation of both genes (48). Thus, the clustering of both human and chicken β-globin genes has functional significance for the activation of transcription.

The AFP and albumin genes are estimated to have diverged 300–500 million years ago by a gene duplication (1). If the duplication of the genes did not also duplicate the regulatory regions, evolutionary pressure would be exerted to maintain linkage. Our experiments with transgenic mice have shown that AFP can function independently of the albumin gene in terms of activation of transcription, developmental regulation, and reinduction during liver regeneration (9). The expression of the albumin gene in transgenic mice, on the other hand, appears to be very sensitive to the site of integration (7), but this can be overcome by inclusion of the three AFP enhancers (S. A. Camper and S. M. Tilghman, unpublished results), implying that the AFP-enhancer domain is required for expression of both genes.

V. Developmental Regulation of α-Fetoprotein

The developmental programs for AFP and albumin differ after birth. Albumin expression increases slightly, whereas AFP transcription declines dramatically until a low basal level of transcription is reached in adult animals (3). Previous experiments from both transgenic mice and F9 teratocarcinoma stem cells had demonstrated an obligate requirement for the enhancer domain for activation of AFP transcription (9, 49, 50).

We have used transgenic mice to map the regions required for the transcriptional inactivation of AFP (9, 51). (S. A. Camper and S. M. Tilghman, unpublished results). To determine whether developmental regulation results from stage-specific enhancers, we tested the capacity of the AFP enhancers to direct the expression of albumin (AFP_E ALB, Fig. 1). In three separate lines of transgenic mice, expression in fetal livers was high, and this level persisted into adulthood (data not shown). In contrast, an AFP gene activated by the albumin enhancer (ALB_E AFP, Fig. 1) was repressed after birth. These results prove that the enhancer domain is not sufficient to direct the developmental program of AFP; that is, the enhancers themselves are not stage-specific. The transcriptional decline after birth is mediated by the proximal 1 kb of 5'-flanking DNA, or intragenic sequences. Further experiments are necessary to localize this interesting domain.

In conclusion, the albumin and AFP genes contain a number of regulatory sequences important for their expression. We are now in the process of determining how these regulatory sequences could simultaneously or alternately affect the expression of both genes.

Acknowledgments

This work was supported by National Institutes of Health Postdoctoral Fellowship GM10237 (S.A.C.), by a postdoctoral fellowship from the Medical Research Council of Canada (R.G.), and by U.S. Public Health Service Grant CA 44976 (S.M.T.).

References

1. S. M. Tilghman, *Oxford Surv. Eukaryotic Genes* **2**, 160 (1985).
2. R. S. Ingram, R. W. Scott, and S. M. Tilghman, *PNAS* **78**, 4694 (1981).
3. A. Belayew and S. M. Tilghman, *MCBiol* **2**, 1427 (1982).
4. S. M. Tilghman and A. Belayew, *PNAS* **79**, 5254 (1982).
5. K. Gorski, M. Carneiro, and U. Schibler, *Cell* **47**, 767 (1986).
6. R. Godbout, R. S. Ingram, and S. M. Tilghman, *MCBiol* **6**, 477 (1986).
7. C. A. Pinkert, D. M. Ornitz, R. L. Brinster, and R. D. Palmiter, *Genes Dev.* **1**, 268 (1987).
8. R. Godbout, R. S. Ingram, and S. M. Tilghman, *MCBiol* **8**, 1169 (1988).
9. R. E. Hammer, R. Krumlauf, S. A. Camper, R. L. Brinster, and S. M. Tilghman, *Science* **235**, 53 (1987).
10. M. L. Atchison, *ARCell Biol* **4**, 127 (1988).
11. G. M. Church, A. Ephrussi, W. Gilbert, and S. Tonegawa, *Nature (London)* **313**, 798 (1985).
12. R. Hromas, U. Pauli, A. Marcuzzi, D. Lafrenz, H. Nick, J. Stein, G. Stein, and B. Van Ness, *NARes* **16**, 953 (1988).
13. S. D. Voss, U. Schlokat, and P. Gruss, *TIBS* **11**, 287 (1986).
14. M. J. Low, R. M. Lechan, R. E. Hammer, R. L. Brinster, J. F. Habener, G. Mandel, and R. H. Goodman, *Science* **231**, 1002 (1986).
15. L. W. Swanson, S. M. Simmon, J. Arriza, R. Hammer, R. Brinster, M. G. Rosenfeld, and R. M. Evans, *Nature (London)* **317**, 363 (1985).
16. Y. Hiromi, A. Kuroiwa, and W. J. Gehring, *Cell* **43**, 603 (1985).
17. P. K. Geyer and V. G. Corces, *Genes Dev.* **1**, 996 (1987).
18. M. J. Garabedian, M.-C. Hung, and P. C. Wensink, *PNAS* **82**, 1396 (1985).
19. M. J. Garabedian, B. M. Shepard, and P. C. Wensink, *Cell* **45**, 859 (1986).
20. J. A. Fischer and T. Maniatis, *EMBO J.* **5**, 1275 (1986).
21. U. Schibler and F. Sierra, *ARGen* **21**, 237 (1987).
22. L. Dente, U. Rüther, M. Tripodi, E. F. Wagner, and R. Cortese, *Genes Dev.* **2**, 259 (1988).
23. R. D. Palmiter and R. L. Brinster, *ARGen* **20**, 465 (1986).
24. S. Weisbrod, *Nature* **297**, 289 (1982).
25. R. Godbout and S. M. Tilghman, *Genes Dev.* **2**, 949 (1988).
26. D. Tuan, W. Solomon, Q. Li and I. M. London, *PNAS* **82**, 6384 (1985).
27. A. D. Carter, B. K. Felber, M. J. Walling, M.-F. Jubier, C. J. Schmidt, and D. H. Hamer, *PNAS* **81**, 7392 (1984).
28. S. Goodbourn, K. Zinn, and T. Maniatis, *Cell* **41**, 509 (1985).
29. J. B. E. Burch, M. I. Evans, T. M. Friedman, and P. J. O'Malley, *MCBiol* **8**, 1123 (1988).
30. F. Payvar, D. DeFranco, G. L. Firestone, B. Edgar, O. Wrange, S. Okret, J.-A. Gustafsson, and K. R. Yamamoto, *Cell* **35**, 381 (1983).
31. P. Labhart and R. Reeder, *Cell* **37**, 285 (1984).
32. H. R. B. Pelham, *Cell* **30**, 517 (1982).
33. Y. Schaul and R. Ben-Levy, *EMBO J.* **6**, 1913 (1987).
34. V. Pachnis, C. I. Brannan, and S. M. Tilghman, *EMBO J.* **7**, 673 (1988).
35. H. Yoo-Warren, V. Pachnis, R. Ingram, and S. M. Tilghman, *McBiol.* **8**, 4707 (1988).
36. G. Courtois, J. G. Morgan, L. A. Campbell, G. Fourel, and G. R. Crabtree, *Science* **238**, 688 (1987).

37. D. R. Grayson, R. H. Costa, K. G. Xanthopoulos, and J. E. Darnell, *Science* **239**, 786 (1988).
38. S. Lichtsteiner, J. Wuarin, and U. Schibler, *Cell* **51**, 963 (1987).
39. O. Greengard, *Essays Biochem.* **7**, 159 (1971).
40. A. Panduro, R. Shalaby, and D. A. Shafritz, *Genes Dev.* **1**, 1172 (1987).
41. R. W. Scott and S. M. Tilghman, *MCBiol* **3**, 1295 (1983).
42. M.-O. Ott, L. Sperling, P. Herbomel, M. Yaniv, and M. C. Weiss, *EMBO J.* **3**, 2505 (1984).
43. J.-M. Heard, P. Herbomel, M.-O. Ott, A. Mottura-Rollier, M. Weiss, and M. Yaniv, *MCBiol* **7**, 2425 (1987).
44. L. E. Babiss, R. S. Herbst, A. L. Bennett, and J. E. Darnell, *Genes Dev.* **1**, 256 (1987).
45. D. Kioussis, E. Vanin, T. de Lange, R. A. Flavell, and F. Grosveld, *Nature (London)* **306**, 662 (1983).
46. F. Grosveld, G. Blom van Assendelft, D. R. Greaves, and G. Kollias, *Cell* **51**, 975 (1987).
47. M. Trudel and F. Costantini, *Genes Dev.* **1**, 954 (1987).
48. J. M. Nickol and G. Felsenfeld, *PNAS* **85**, 2548 (1988).
49. R. W. Scott, T. F. Vogt, M. E. Croke, and S. M. Tilghman, *Nature (London)* **310**, 562 (1984).
50. T. F. Vogt, R. S. Compton, R. W. Scott, and S. M. Tilghman, *NARes* **16**, 487 (1988).
51. R. Krumlauf, R. E. Hammer, S. M. Tilghman, and R. L. Brinster, *MCBiol* **5**, 1639 (1985).

A Methylation Mosaic Model for Mammalian Genome Imprinting

Carmen Sapienza[1]
Thu-Hang Tran
Jean Paquette
Ross McGowan and
Alan Peterson

Ludwig Institute for Cancer Research–Montreal Branch, Montreal, Quebec, Canada H3A 1A1

All but one of the major eukaryotic taxonomic groups contain species that reproduce sexually, as well as species capable of reproducing asexually (*1*). Among vertebrates, there are relatively few species that reproduce asexually, and no instances of such are ascribed to mammals. The development of the pronuclear transplantation technique (*2*) and its subsequent use to produce diploid gynogenetic and androgenetic mouse embryos (*3, 4*) have directly addressed the question of whether asexual reproduction is possible in mammals. The developmental failure of all gynogenotes (embryos with two maternally derived pronuclei) and androgenotes (embryos with two paternally derived pronuclei) indicates that not only is asexual reproduction not observed, but that it is not possible (*5, 6*). The generally accepted reason for the developmental failure of these reconstituted zygotes [and parthenotes (*6–9*)] is that male- and female-derived genetic information is differentially "imprinted" prior to formation of male and female pronuclei in the zygote (*3, 4*). This definition implies that identical genetic information is rendered functionally different as a result of passage through the male or female germ line (Fig. 1). This model predicts that the paternally transmitted allele at some loci will be unavailable for appropriate expression, while at other loci the expression of the maternally transmitted allele will be affected.

[1] Speaker.

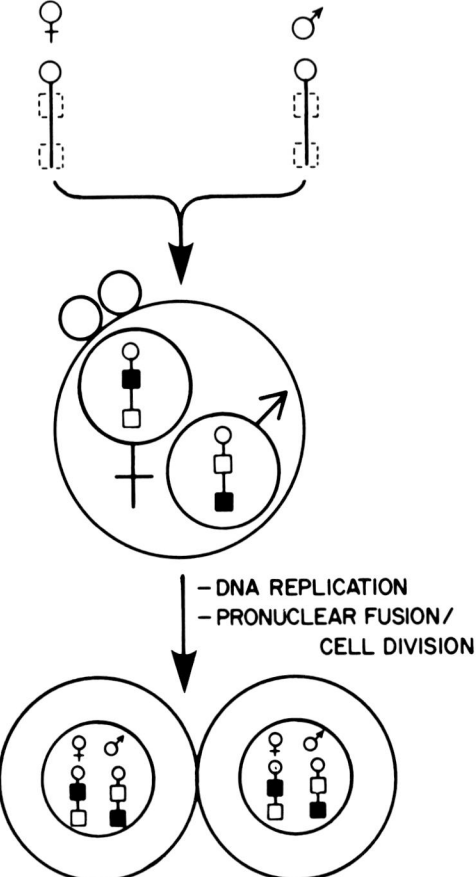

FIG. 1. Differential maternal and paternal imprinting of alleles at two loci on a single chromosome. Open squares represent alleles that are expressed. Solid squares represent alleles that are not expressed.

The mechanism by which these differences are established is unknown, but the available data require (*10*): (*a*) the imprint to be physically linked to, or contained within, the pronucleus (*3, 4*); (*b*) the imprint to survive several rounds of DNA replication and cell division (*5*); (*c*) the imprint to affect gene expression (*5, 6, 11*); and (*d*) the imprint to be capable of having its gamete-of-origin-dependent identity switched in successive generations (*3, 4*).

Several groups have recently provided evidence that differential DNA methylation is capable of fulfilling all of the above requirements (*10, 12, 13*).

I. Gamete-of-Origin-Dependent DNA Methylation

The introduction of new genetic material into the mouse genome by microinjection of DNA has provided a powerful tool with which to investigate gamete-of-origin-dependent DNA methylation. Matings between transgenic and nontransgenic animals yield offspring in which the transgene is inherited from only one parent. Therefore, potential gamete-of-origin-dependent differences in DNA methylation at these loci can be directly assayed in such offspring. Such differences have, in fact, been found (10, 12, 13). If the gamete-of-origin-dependent methylation changes demonstrated for a number of hemizygous transgene loci are directly related to the developmental failure of gynogenotes and androgenotes, several additional observations made in these experiments are unexpected. In the simplest model, hypomethylation of an allele would result in the expression of that allele, while hypermethylation of that allele would result in no expression. However, there is little correlation between the gamete-of-origin-dependent methylation of transgenes and the expression of these sequences. The methylation imprint is detectable independent of whether the transgene is not expressed in any tissue (12), expressed tissue-specifically and independent of gamete-of-origin (our unpublished data), or expressed in a gamete-of-origin-dependent fashion (13). The latter observation has been made only once, and, while expression was limited to only one tissue, the methylation pattern of the transgene was identical in all tissues (13).

The second unexpected observation concerns the relatively high frequency with which gamete-of-origin-dependent methylation differences are observed in transgenic lines, compared to the small number of loci deduced to be imprinted by genetic criteria. The use of Robertsonian translocation chromosomes in classical genetic experiments has demonstrated that, at the level of gross morphological or behavioral phenotype, imprinted loci are present on parts of six mouse autosomes (14 and 15 compiled in 16). More importantly, no phenotypic effect of deriving both copies of seven other mouse chromosomes through one gamete or the other could be detected (16). These data imply that only a small number of loci, perhaps as few as six, are affected by imprinting. This conclusion must also be reached by reflecting that, almost without exception, mutant alleles at a large number of loci display Mendelian behavior in genetic crosses. The best described exception to this rule is the maternal-lethal, hairpin-tail allele of the mouse T complex (T^{hp}) (11). In humans, the preferential (but not exclusive) paternal inheritance of the juvenile-onset form of Huntington's chorea (17) and the preferential (but not exclusive) maternal inheritance of the neonatal form of myotonic dystrophy (18) may provide further examples of such exceptional loci.

In the strictest sense, gamete-of-origin-dependent transgene methylation may be defined as a form of imprinting, independent of any other criteria. But given the paucity of genetic evidence for large numbers of imprinted loci, one is forced to ask whether the discovery of gamete-of-origin-dependent methylation changes by five different groups using five different transgene constructs, involving at least eight loci (10, 12, 13, 19) reflects genome imprinting as defined by either pronuclear transplantation or genetic criteria. Imprint-based control of a small number of loci on only a few chromosomes seems to demand a mechanism with a high degree of specificity, yet differential transgene methylation (which fulfills all of the known criteria for an imprinting mechanism) is relatively easily identified and seemingly global.

II. Models That Accommodate the Data

There are two easily envisioned explanations for this paradox. (a) Differential DNA methylation is, at best, indirectly related to genome imprinting as defined by phenotypic criteria. (b) Gamete-of-origin-specific methylation is directly related to genome imprinting. But these methylation differences do not generally persist beyond early embryogenesis. The first explanation presupposes unspecified elements of DNA sequence or protein interaction as being responsible for imprinting. Further, it does not explain the persistence of a transgene methylation imprint at a multitude of assayable sites (10, 12, 13).

There is evidence that the second explanation is at least partly true. The apparent erasure of gamete-specific methylation has been demonstrated previously at endogenous loci (20); numerous loci exhibit differences between the methylation pattern observed for those sequences in sperm versus the pattern found in oocytes. Such differences remain detectable by blot-hybridization during preimplantation stages of embryogenesis, but are not observable in somatic tissues derived from later-stage embryos or adults (20).

Two alternative interpretations of these data are illustrated in Figs. 2 and 3. In Fig. 2, ova-specific and sperm-specific methylation differences are established prior to the completion of gametogenesis. These gamete-of-origin-specific differences are faithfully propagated

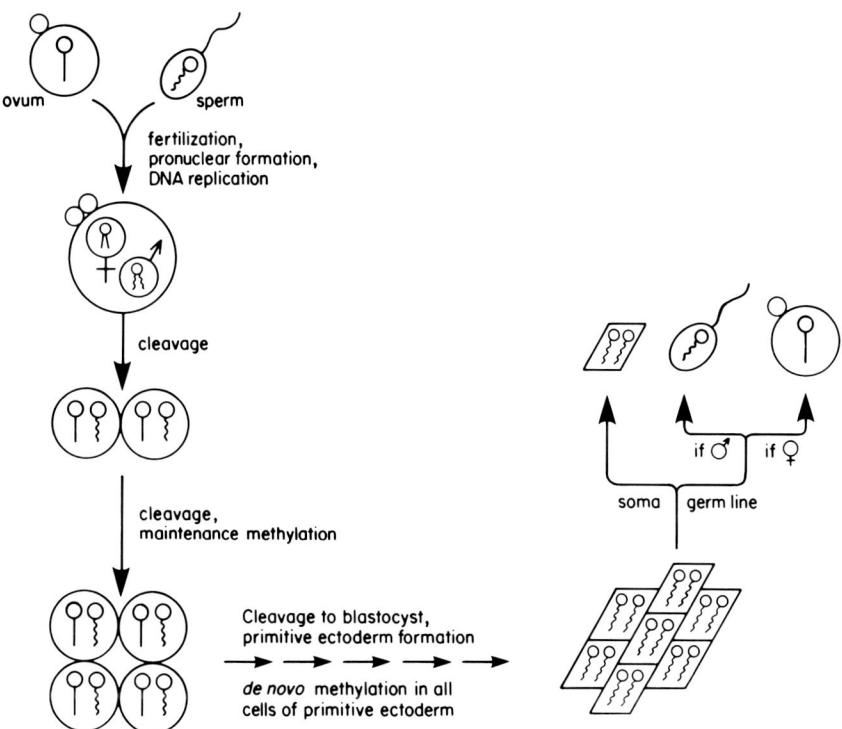

FIG. 2. Behavior of differentially methylated sequences in a transient methylation maintenance model (see 20 and text). Straight lines represent chromosomes or chromatids containing unmethylated DNA. Wavy lines represent methylated DNA.

in each cell throughout preimplantation development, then erased through the action of a *de novo* methylase which operates in all cells that give rise to primitive ectoderm. If the embryo is male, germ cells may be derived directly from this population without substantial methylation changes occurring. If the embryo is female, the formerly methylated sequences in these cells must become demethylated at both alleles in cells that become the germ line.

In the second interpretation (Fig. 3), ova-specific and sperm-specific methylation patterns are also established prior to the completion of gametogenesis, but these differences are not faithfully propagated in every cell of the early embryo. Each of the four-cell embryos shown in Figs. 2 and 3 contain exactly the same number of chromosomes bearing a methylated or unmethylated DNA sequence [hence, a blot-hybridization assay on each embryo will give identical

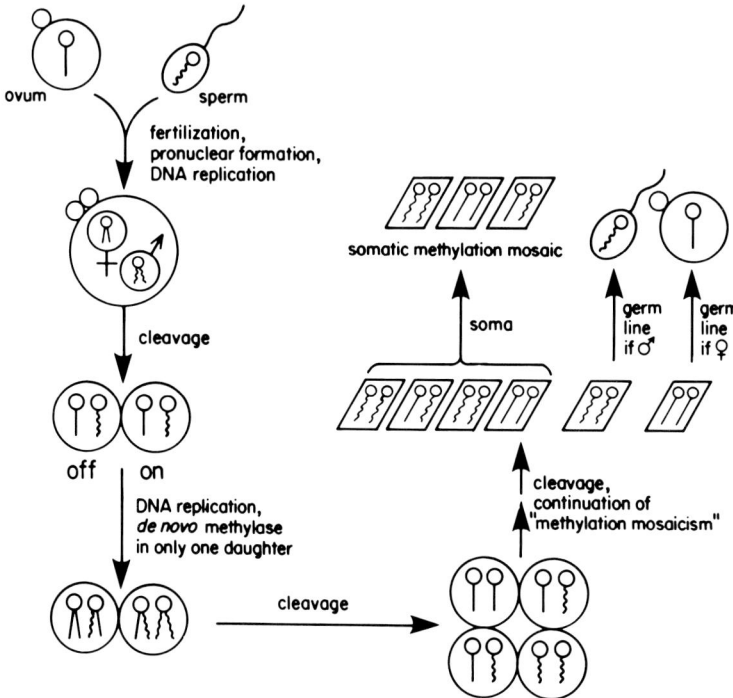

Fig. 3. Behavior of differentially methylated sequences in a methylation mosaic model (see text). Wavy lines denote methylated DNA sequences and straight lines denote unmethylated DNA sequences.

results (20)], but their distribution among the cells of each embryo is different. A simple mechanism by which to create this difference is to invoke the activity of a *de novo* methylase in only one cell of a two-cell embryo. If the DNA sequence carried by the sperm is methylated on both strands, semiconservative replication in the male pronuclei will yield hemimethylated homologs. Segregation of these homologs into two daughter cells, only one of which expresses a methylating activity, will result in the creation of newly methylated homologs in one cell, and unmethylated homologs in the other after the second DNA replication cycle. Normal random segregation will give rise to a four-cell embryo that is mosaic with respect of methylation of alleles at imprinted loci. In one cell, both alleles will be methylated; in another cell, neither allele will be methylated; and the remaining two cells will exhibit allele-specific methylation. In the model shown in Fig. 3, cells bearing both the male and female gamete-specific methylation pattern are established early. It is therefore possible to allocate cells to either germ line without later sex-dependent methylation modification.

For the purpose of illustration, the generation of methylation mosaicism is depicted as beginning at the second DNA replication cycle, but a mosaic effect will be achieved by imposing this distinction on any two cells.

If this model is applied to a locus that is transmitted through only one gamete (i.e., a hemizygous transgene), the net effect will be to skew the number of cells that contain a methylated or unmethylated locus in the direction of its methylation state in the gamete of origin (Fig. 4). The two embryos depicted in Fig. 4 differ only in the relative proportion of cells that contain a methylated or an unmethylated transgene. Each of these populations may be uniformly expanded. If one then selects a small number of cells from each of these embryos, and these cells become all of the somatic tissues of the mouse, the somatic tissues from each of these embryos will almost always be different with respect to transgene methylation. However, these differences will only reflect differences in the relative proportion of cells in the original population that contained a methylated or unmethylated transgene.

In the model in Fig. 2, it is difficult to explain why the methylation activity that results in the apparent "erasure" of gamete-specific methylation differences at endogenous loci does not also "erase" the methylation imprint of transgenes. In the model in Figs. 3 and 4, "maintenance" of a transgene methylation imprint requires only hemizygosity and a methylation difference between sperm and ova.

III. Predicted Consequences of the Mosaic Methylation Model

A. Allele-Specific Methylation

An important prediction of the mosaic methylation model is that, for endogenous loci with two alleles, allele-specific methylation differences will not be observed in tissue samples, but only in populations of cells that are clonal derivatives of allele-specifically methylated progenitors. Furthermore, among multiple clones derived from a mosaic tissue, more than one methylation configuration should be found.

Evidence that this is the case comes from two different sets of experiments. Allele-specific methylation of c-Ha-*ras* alleles in clonal

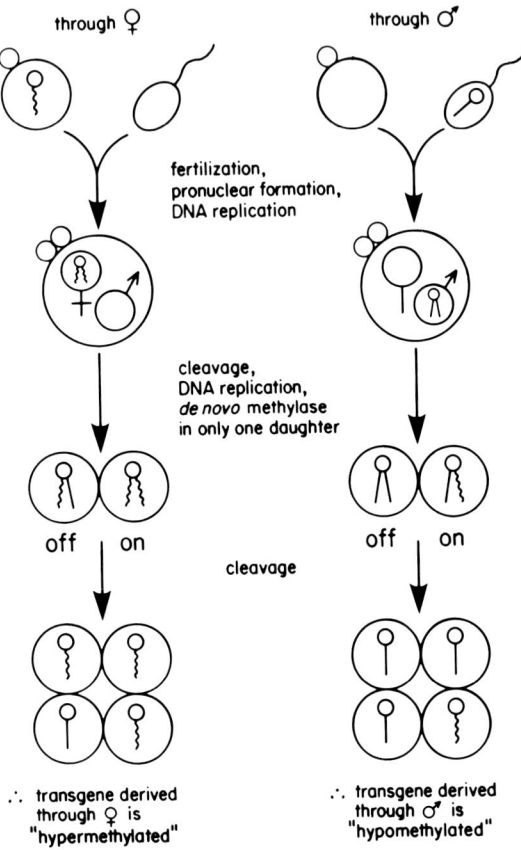

FIG. 4. Application of the methylation mosaic model to a differentially methylated DNA sequence inherited from only one parent. Wavy lines denote methylated DNA sequences and straight lines denote unmethylated DNA sequences; the gamete from which they are inherited is reversed to reflect actual transgene data (*10, 12, 13, 19*).

derivatives of human fetal fibroblasts and in different osteogenic sarcoma cell lines has been observed (*21*). Four of seven fetal fibroblast lines (all derived from the same fetus) that were heterozygous for an *Msp*I restriction-fragment-length polymorphism (RFLP) (*22*) exhibited allele-specific methylation, while the remaining three did not. In contrast, no allele-specific methylation could be demonstrated in DNA samples of peripheral blood leukocytes taken from different individuals (*21*).

We have also examined the methylation of C-Ha-*ras* alleles in a series of human osteosarcomas. This chromosome-11 locus is not thought to be involved in the formation of these osteosarcomas. Such tumors are thought to be clonal derivatives of a single progenitor cell in which a mutation event has occurred on chromosome 13 (23). Among seven tumors heterozygous for an *Msp*I RFLP (22), one exhibited allele-specific methylation, two were unmethylated at both alleles, and four were methylated at both alleles (unpublished results). DNAs from normal tissue taken from each of these individuals displayed different methylation patterns, but both alleles were predominantly methylated. Differences between these individuals are consistent with the interpretation that the somatic tissues of each individual are derived from progenitor populations that contained variable numbers of cells of each of the three possible methylation phenotypes (see Fig. 3).

B. Effects on the Penetrance of Genetic Traits

Cells that bear an allele-specific methylation imprint inactivating that allele will be subject to somatic mutation events that result in a null phenotype at "one-hit" mutational frequency. For example, if one or more such loci are involved in the initiation or progression of tumorigenesis, then a mitotic recombination event proximal to this locus will result in a daughter cell containing two genetically wild-type, but epigenetically inactive, alleles at that locus. A prediction of this hypothesis is that if gamete-specific imprinting always occurs in the same way at this locus (i.e., if alleles are inactivated through sperm and not through ova), then, in those cases of sporadic tumors in which reduction to homozygosity occurs over a particular chromosome (23), the chromosome that is left in the tumor should always be from the same parent. In the small number of cases for which this has been determined [five cases of Wilms' tumor (24)], the chromosome that is reduced to homozygosity in the tumor is always derived from the father.

Two variables will operate in the genesis of such a tumor: the number of cells that are allele-specifically inactivated, and the frequency with which the mitotic recombination event occurs. If the number of allele-specifically inactivated susceptible cells significantly exceeds the frequency of the recombination event, the individual in question is at high risk for the development of that tumor. If the reverse is true, then the risk is correspondingly reduced.

An extension of this argument is that if there are genetic factors that affect the degree to which a particular individual is mosaic for each methylation-imprint phenotype depicted in Fig. 3, this will also affect the penetrance of genetic traits. In the tumorigenesis example given above, genotype-dependent differences between individuals in early embryo mosaicism would give rise to "host resistance" (25) or "premutation" (26) phenomena.

IV. Effects of Genotype on Methylation Imprinting

For the well-described mosaicism represented by X-chromosome inactivation in the somatic tissues of female humans, there are individuals in which one X chromosome or the other is more often inactivated, and the ratio between active and inactive chromosomes may be as high as 4:1 in favor of one X chromosome (27). It is not known whether this particular form of variation is due to stochastic or genetic factors.

Genetic factors apparently do influence the level of methylation of hemizygous transgenes. As described in Section I, several characteristics of transgene loci make them amenable to the study of gamete-of-origin effects. In pedigree 379 (10), in which a quail troponin I transgene was segregating, we observed the transgene locus become less methylated in the somatic tissue of offspring that inherited the locus from a male, and more methylated in offspring that inherited the locus from a female. One puzzling aspect of this study was that not all offspring of transgenic males showed a decrease in transgene methylation (10). After a careful examination of breeding records, it became apparent that the transgene methylation pattern observed in the somatic tissue of offspring was, to a large degree, dependent on the genotype of the nontransgenic mother. When the same transgenic male was mated to C57BL/6J and DBA/2J females, the offspring from C57BL/6J females had a transgene locus more methylated than the offspring from DBA/2J females. (C57BL/6J × DBA/2J)F_1 females, however, gave rise to progeny which had either one or the other of the two distinct methylation patterns (unpublished results).

The biological mechanism by which these differences are achieved is not known, but significant differences do exist between inbred strains in at least the number of cells making up a preimplantation embryo. These genetic differences are primarily [but not exclusively (28)] dependent on the genotype of the ova (29, 30). Taken together, these observations indicate that (1) not only does genetic background have an effect on the degree of methylation observed in an individual, but (2) because these genetic differences segregate in interstrain F1 ova, they are not expressed until after the second meiotic division, i.e., postfertilization. The latter time limit by which these patterns must be established is at, or prior to, primitive ectoderm formation, because all somatic tissues of any individual show the same transgene methylation pattern (10). Establishment of the pattern at later than this stage would require multiple, independent methylation and demethylation events to occur to the same extent in derivatives of all primary germ layers. The timing of these events is also consistent with the model presented in Fig. 3.

V. Why Imprint?

At this point, we are left with the paradox that mammals establish widespread genomic DNA methylation differences between gametes in a sex-specific way. These differences are then eradicated in a sufficiently large proportion of embryonic cells so that they are detectable only in clonally derived cell populations. Thus, a mutant allele present at an imprinted locus would manifest a mutant phenotype only in that fraction of cells in which the wild-type allele remained epigenetically inactivated.

If these cells are a sufficiently small fraction of the total, or intercellular communication permits phenotypic rescue of these cells, the overall result will be a normal phenotype. The mutant allele will behave as a standard recessive mutation in genetic crosses, even though the locus at which it resides is affected by imprinting. Therefore, imprinting will have phenotypic consequences only under the exceptional circumstances detailed in Sections I and III. What, then, might be the function of establishing a gamete-of-origin-specific imprint?

It is of interest that genomic imprinting of one form or another has been demonstrated in sexually reproducing organisms as phylogenetically divergent as mammals, protists, and insects (3, 4, 31, 32). On this basis, it seems likely that imprinting, as a gamete-specific marking process, is as old as sex itself. The fixation of sexual reproduction as the only reproductive mode possible may reflect processes peculiar to the ontogeny of placental mammals. The phenotypes exhibited by gynogenotes and androgenotes certainly support this idea. Under the mosaic methylation model (Fig. 3), the early creation of differentially methylated cell types could serve as a cell lineage allocation system based on phenotypic differences (33) that may arise as a result of methylation mosaicism. The overall undermethylation of DNA isolated from trophectoderm tissues (placenta), as compared to DNA isolated from somatic tissues of the embryo proper (34), is consistent with this interpretation. Diploid genomes derived only through oogenesis (gynogenotes) or only through spermatogenesis (androgenotes) should behave overall much like the transgene locus depicted in Fig. 4. The embryonic cell population would then be skewed in one direction, such that gynogenotes would be unable to allocate sufficient cells to extraembryonic lineages, and androgenotes would be unable to allocate sufficient cells to embryonic lineages.

Although this scenario can explain what happens as a result of genome imprinting, it begs the question of what the function of imprinting might be. In the simplest view, the "function" of imprinting in the methylation mosaic model is the early establishment of cells bearing the germ line methylation configuration. *A priori*, the establishment of the germ line must be an important consideration in the development of sexually reproducing organisms. In fact, most organisms do allocate the germ line at very early stages of embryogenesis and, even though mammals are generally thought not to allocate the germ line early, more recent data indicate that they may (35). Future experiments that unequivocally demonstrate early or late germ-cell allocation will, obviously, be of great importance in testing this notion.

Acknowledgments

We are grateful to the many colleagues who took the time to discuss either the ideas presented in this manuscript or the manuscript itself: Alain Nepveu, Morag Park, Web Cavenee, Marc Hansen, Alex Koufos, Paul Grundy, Joe Nadeau, Alan Berstein, Mike McBurney, Janet Rossant, Sue Varmuza, Terry Magnuson, and Vern Chapman. We are especially grateful to Susan Caluori, who amended the typescript throughout all of these discussions.

References

1. G. Bell, "The Masterpiece of Nature: The Evolution and Genetics of Sexuality," Croom Helm London, 1982.
2. J. McGrath and D. Solter, *Science* **220**, 1034 (1983).
3. M. A. H. Surani, S. C. Barton, and M. L. Norris, *Nature (London)* **308**, 548 (1984).
4. J. McGrath and D. Solter, *Cell* **37**, 179 (1984).
5. M. A. H. Surani, S. C. Barton, and M. L. Norris, *Cell* **45**, 127 (1986).
6. M. A. H. Surani, S. C. Barton, and M. L. Norris, *Nature (London)* **326**, 395 (1987).
7. A. Witowska, *J. Embryol. Exp. Morphol.* **30**, 547 (1973).
8. M. A. Surani and S. C. Barton, *Science* **222**, 1034 (1983).
9. L. C. Stevens and D. S. Varnum, *Dev. Biol.* **37**, 369 (1984).
10. C. Sapienza, A. C. Peterson, J. Rossant, and R. Balling, *Nature (London)* **328**, 251 (1987).
11. J. McGrath and D. Solter, *Nature (London)* **308**, 550 (1984).

12. W. Reik, A. Collick, M. L. Norris, S. C. Barton, and M. A. H. Surani, *Nature (London)* **328**, 248 (1987).
13. J. L. Swain, T. A. Stewart, and P. Leder, *Cell* **50**, 719 (1987).
14. B. M. Cattanach and M. Kirk, *Nature (London)* **315**, 496 (1985).
15. C. V. Beechey and A. G. Searle, *Mouse News Lett.* **77**, 126 (1987).
16. M. A. H. Surani, *in* "Experimental Approaches to Mammalian Embryonic Development" (J. Rossant and R. A. Pedersen, eds.), P. 401. Cambridge Univ. Press, Cambridge, England, 1986.
17. L. A. Farrer and P. M. Connealy, *Am. J. Hum. Genet.* **37**, 350 (1985).
18. A. Glanz and F. C. Fraser, *J. Med. Genet.* **21**, 186 (1984).
19. M. Hadchouel, H. Farza, D. Simon, P. Tiollais, and C. Pourcel, *Nature (London)* **329**, 454 (1987).
20. J. P. Sanford, H. J. Clark, V. M. Chapman, and J. Rossant, *Genes Dev.* **1**, 1039 (1987).
21. L. A. Chandler, H. Ghazi, P. A. Jones, P. Boukamp, and N. E. Fusenig, *Cell* **50**, 711 (1987).
22. T. G. Krontiris, N. A. DiMartino, M. Colb, and D. R. Parkinson, *Nature (London)* **313**, 369 (1985).
23. M. F. Hansen and W. K. Cavenee, *Trends Genet.* **4**, 125 (1988).
24. W. T. Schroeder, L. Y. Chao, D. D. Dao, L. C. Strong, L. Pathak, V. Riccardi, W. H. Lewis, and G. F. Saunders, *Am. J. Hum. Genet.* **40**, 413 (1987).
25. E. Matsunaga, *Am. J. Hum. Genet.* **30**, 406 (1978).
26. J. Herrmann, *in* "Genetics of Human Cancer" (J. J. Mulvihill, R. W. Miller, and J. F. Fraumeni, Jr., eds.), P. 417. Raven, New York, 1977.
27. B. Vogelstein, E. R. Fearon, S. R. Hamilton, A. C. Preisinger, H. G. Willard, A. M. Michaelson, A. D. Riggs, and S. H. Orkin, *Cancer Res.* **47**, 4806 (1987).
28. W. K. Whitten and C. P. Dagg, *J. Exp. Zool.* **148**, 173 (1962).
29. A. McLaren and P. Bowman, *J. Embryol. Exp. Morphol.* **30**, 391 (1973).
30. N. Wakasugi and M. Morita, *J. Embryol. Exp. Morphol.* **38**, 211 (1973).
31. A. J. S. Klar, *Nature (London)* **326**, 466 (1987).
32. J. B. Spofford, *Genetics* **46**, 1151 (1961).
33. M. H. Johnson and C. A. Ziomek, *Cell* **24**, 71 (1981).
34. V. Chapman, L. Forrester, J. Sanford, N. Hastie, and J. Rossant, *Nature (London)* **307**, 284 (1984).
35. P. Soriano and R. Jaenisch, *Cell* **46**, 19 (1986).

Insertional Mutations in Transgenic Mice

FRANK COSTANTINI[1]
GLENN RADICE
JAMES J. LEE
KIRAN K. CHADA*
WILLIAM PERRY AND
HYEUNG JIN SON

Department of Genetics and Development, College of Physicians and Surgeons, Columbia University, New York, New York 10032, and
* Department of Biochemistry, Robert Wood Johnson Medical College, University of Medicine and Dentistry of New Jersey, Piscataway, New Jersey 08854

In contrast to the rapid rate of progress in the developmental genetics of invertebrate species such as *Drosophila melanogaster* or the nematode *C. elegans*, the genetic analysis of mammalian development remains a more difficult undertaking. Nevertheless, several new approaches are now making it possible to identify, and to analyze at the molecular level, genes that are potentially important in mammalian development. One approach is to isolate and characterize mammalian genes that are homologous to developmentally important genes in other species, such as the homeobox[2]-containing genes (1, 2). A different strategy is to select an interesting mutation from the large collection of available mouse mutants (3), and attempt to clone the gene involved through a combination of genetic and molecular methods. A third approach, which is the subject of this paper, involves the analysis of mutations caused by the insertion of foreign DNA or viruses into the mouse germ line.

Approximately 5–10% of transgenic mouse lines carry recessive mutations that cosegregate with the integrated foreign DNA or virus,

---

[1] Speaker.
[2] A conserved sequence found in homoetic and related genes. [Eds.]

and thus appear to represent insertional mutations (reviewed in 4 and 5). The phenotypes associated with these mutations include embryonic lethality at various stages of gestation (6–10), limb deformities (11, 12; S. Potter, personal communication), transmission distortion (13), neurological defects (J. Rossant, personal communication), fertility defects, and growth deficiencies (K. K. Chada, unpublished data). The variety of insertional mutants available for analysis remains limited, in comparison with the hundreds of different mouse mutants obtained by conventional means over the past several decades (3). However, insertional mutants offer the considerable advantage that the affected gene can be cloned relatively easily, by virtue of its linkage to the introduced DNA. Given the rate at which new transgenic lines are being generated, the range of mutant phenotypes available should increase rapidly in the future.

Insertional mutations caused by retroviruses are somewhat easier to analyze than those caused by the integration of microinjected DNA fragments, because the retrovirus integrates as a single copy and does not cause deletions or rearrangements of the host DNA. Two such insertional mutations have been characterized by Jaenisch and colleagues. (10, 14). In the earliest of these to be described, Mov13, the retrovirus was inserted into the $\alpha 1(I)$ collagen gene, resulting in the death of homozygous embryos at around day 13 of gestation (14). The second of these mutations, Mov34, resulted in the death of homozygous embryos shortly after implantation (10); while the Mov34 locus has also been cloned and contains a gene that is expressed in almost all tissues, the sequence or identity of this gene has not yet been reported. Additional recessive mutations caused by retroviral insertions have been observed, and remain to be characterized (personal communications from R. Jaenisch; N. Jenkins and N. Copeland; E. Robertson). It has been suggested that retroviruses may integrate preferentially (or exclusively) in active genes (5); if this is the case, germ-line mutagenesis using retroviruses may be limited to those genes active in the early embryo, or in embryonic stem-cell lines. In contrast, microinjected DNA appears to integrate by an entirely different mechanism, and may not be limited to the same set of potential integration sites. Thus, microinjection of DNA into the mouse zygote may induce mutations in a different set of genes than that accessible to retroviral insertion.

In transgenic mice generated by the microinjection of DNA fragments, the insertion of the foreign DNA has often been accompanied by alteration of the host DNA. Deletions ranging from 1 to 20 kb (8, 11, 15), duplication of host DNA sequences (16), co-integration of unlinked mouse DNA sequences (15, 16), and translocations (12) have been reported. Nevertheless, in several cases it has been possible to clone the locus of insertion, to determine its structure relative to that of the wild-type locus, and to identify a gene whose disruption may be associated with the mutant phenotype. Woychik *et al.* have cloned the locus of a DNA insertion that causes a recessive limb deformity, and found that the insertion was associated with a 1-kb deletion of host DNA sequences, but no other obvious rearrangements (11). Using interspecies conservation as an indication of potential exon sequences, they detected several transcription units at the insertion locus (P. Leder and R. Woychik, personal communication). Mark *et al.* (8) have cloned an insertion associated with a peri-implantation recessive lethal mutation, and have also obtained evidence for transcripts encoded at the locus (E. Lacy, personal communication). Wilkie and Palmiter (16) have analyzed an insertion that failed to be transmitted through the male germ line. The locus had a complicated structure, in which several fragments of the injected DNA molecule were inserted together with an unidentified 532-bp mouse repetitive element, and a 5-kb host DNA sequence was duplicated on either side of the insertion. While it has been shown that the insertion is unstable, because of high-frequency recombination between the duplicated flanking sequence, a complete explanation of the observed transmission distortion remains to be obtained.

In conclusion, studies to date have established that many insertional mutations caused by DNA insertion as well as by retroviral infection are amenable to analysis at the molecular level. In this paper, we describe the methods we have used to identify several new insertional mutations causing recessive prenatal lethality, and summarize what we have learned so far about these mutations and their phenotypic effects.

I. Identification of Prenatal Recessive Lethal Mutations in Transgenic Lines

The presence of a prenatal recessive lethal mutation in a transgenic mouse line can be discerned by the failure to obtain homozygous animals among the progeny of an intercross (i.e., a cross between two heterozygotes). In cases in which host DNA sequences flanking a transgene insertion have already been cloned, a flanking probe may be used to identify homozygotes, based on Southern blot analysis (14). In other cases, the identification of homozygous transgenic mice has usually relied on quantitation of the transgene copy number: homozygous mice carry twice the number of copies as do heterozygous mice. The relative transgene copy number can be estimated either by Southern blot (7) or dot blot hybridization (8), in each case comparing the signal obtained with a probe for the transgene to the signal obtained with a probe for an endogenous gene. In our experience, while both of these methods are fairly reliable, they are subject to experimental error, and it is frequently necessary to confirm the diagnosis of homozygosity or heterozygosity by progeny testing, a time-consuming procedure.

A second method that we have also employed involves the quantitation of a gene product produced by the transgene. In the course of studies of β-globin gene expression in transgenic mice, we produced several transgenic lines that carried and expressed the human β-globin gene (17). These mice produced a novel form of hemoglobin containing human β-globin and mouse α-globin chains, which could be easily distinguished from mouse hemoglobins by electrophoresis on cellulose acetate plates (18, 19). The presence of this novel hemoglobin was used as a rapid screen for the inheritance of the transgene during the propagation of these lines. It was also used to distinguish homozygotes and heterozygotes, as homozygotes produced approximately twice as much of the novel hemoglobin, relative to mouse hemoglobin, as did heterozygotes (Fig. 1). Like the quantitation of transgene copy number, this method frequently requires confirmation by progeny testing. In addition, its application is limited to transgenic lines that produce substantial amounts of a foreign globin chain (or another easily measured gene product).

A different type of diagnostic method we have found extremely valuable employs *in situ* hybridization to detect the transgene in cytological preparations from transgenic animals (20, 21). If a sufficient number of copies of the transgene are present in tandem (approximately 20–40 kb total length), the transgene locus can be easily detected by *in situ* hybridization with a biotin-labeled probe (21, 22) followed by incubation with streptavidin-conjugated horseradish peroxidase (HRP) and histochemical detection of the HRP enzyme (23). Most cells from mice homozygous for a transgene array show two spots of hybridization, while those from heterozygous mice show only one spot (21). This technique can easily be applied to small samples of nucleated white blood cells obtained from juvenile or adult mice (Fig. 2). It can also be used to distinguish homozygous from heterozygous transgenic embryos or fetuses, by hybridization to trophoblast giant cells (21) or the amnion (J. Rossant, personal communication). While this technique is quite simple and much faster

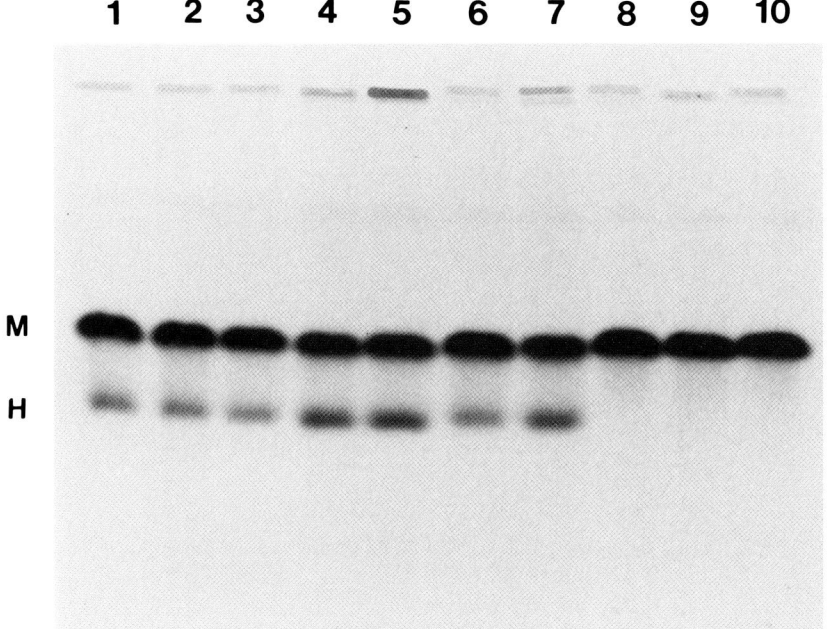

FIG. 1. Shown is cellulose acetate electrophoresis of hemoglobins to distinguish heterozygous and homozygous transgenic mice carrying the human β-globin gene. Hemolysates were prepared from 10 progeny of an intercross between heterozygous transgenic mice from line HB56 (a line in which homozygotes are viable), analyzed by cellulose acetate electrophoresis, as described (18, 19). In this line, the human β-globin transgene is carried on the C57BL/6J background. The upper band is mouse (M) single hemoglobin, while the lower band (seen in lanes 1–7) is a hybrid (H) hemoglobin containing human β- and mouse α-globin chains. The animals examined in lanes 4, 5, and 7 contain approximately twice the amount of the hybrid hemoglobin, relative to endogenous single hemoglobin, indicating that they are homozygous for the transgene. The mice examined in lanes 1, 2, 3, and 6 are heterozygous, while those in lanes 8–10 failed to inherit the transgene.

than DNA analysis, its greatest advantage is that it provides an unambiguous distinction between homozygotes and heterozygotes, and one that does not rely on accurate quantitation.

Using these techniques, we have tested 20 different transgenic lines for recessive lethal mutations. Ten lines carrying a hybrid mouse/human β-globin gene (24) and two lines carrying a mouse immunoglobulin heavy-chain gene (25) were tested using the dot blot method of transgene quantitation, and apparently normal homozygous offspring were obtained in all of these lines. Six transgenic lines

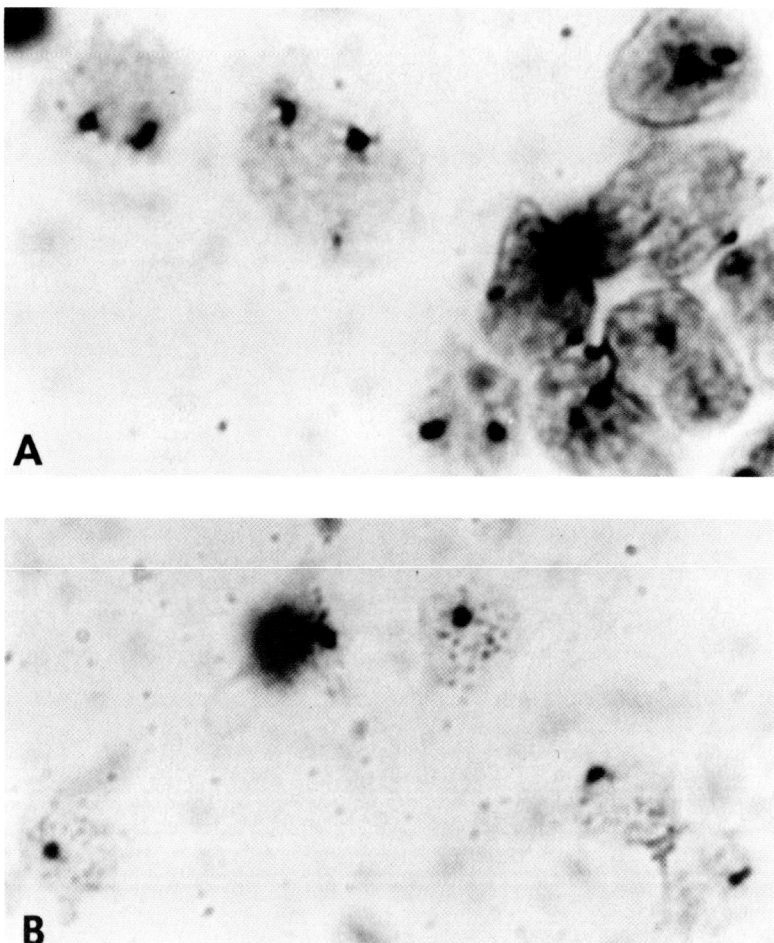

FIG. 2. Identification of homozygous versus heterozygous transgenic mice by *in situ* hybridization to nucleated peripheral blood cells. Blood was collected from a homozygous (A) and a heterozygous (B) transgenic mouse in line HB56, centrifuged in a microcrit tube, and the buffy coat was resuspended in serum and smeared on a slide. A probe specific for the human β-globin transgene was labeled by nick translation with biotinylated nucleoside triphosphates, and hybridized to the slides. The biotinylated probe was then visualized using streptavidin–horseradish peroxidase (Enzo Biochemicals), histochemical staining with diaminobenzidine, and silver amplification (23), resulting in the deposition of silver grains over areas where the probe has hybridized.

carrying and expressing exogenous β-globin genes were tested by measuring the amount of the novel hemoglobin band produced. In two of these lines, we observed no homozygotes among at least 60 intercross progeny, suggesting that the transgene insertion had caused a prenatal recessive lethal mutation. One line (HB58) carried the normal human β-globin gene and the other (line BS12) the human variant $β^S$-globin gene. Finally, two transgenic lines carrying the human ε-globin gene were tested, using the *in situ* hybridization procedure, and one of these two lines (HE46) failed to produce homozygous offspring. Thus, we have identified three prenatal recessive lethal mutations among 20 transgenic lines tested, a frequency slightly higher than, but not inconsistent with, previous estimates of the frequency of such mutations in transgenic mice (4).

II. Analysis of Prenatal Recessive Lethal Mutations

We have analyzed one of the insertional mutations carried by the HB58 line in some detail, and the following is a summary of our findings to date (unpublished results). The HB58 mutation is associated with an insertion of approximately 10–20 copies of the human β-globin gene, in the form of a 7.7-kb *Hin*dIII fragment, into the mouse genome. Although the transgene is expressed at significant levels (it encodes about 15% of total β-globin chains in heterozygous transgenic mice), the death of homozygotes seemed unlikely to be related to an increased level of expression of human β-globin, because other transgenic lines that expressed even higher levels of human β-globin were completely viable and fertile in the homozygous state. We therefore concluded that the apparent recessive lethality of the HB58 transgene was caused by a recessive lethal insertional mutation.

We subsequently analyzed several litters of embryos, generated by an intercross between two heterozygous animals, at 9.5 and 10.5 days of gestation. Approximately 25% of the embryos appeared grossly abnormal, and these were found to be homozygous for the transgene, while their normal "sibling" embryos were either heterozygous or negative for the transgene. This confirmed that the homozygous embryos died *in utero*, and explained the absence of homozygotes among the viable progeny of such a cross. To examine the mutant phenotype in more detail, and to determine the onset of developmental abnormalities, we performed a histological study of embryos derived from intercrosses, and recovered at 6.5 through 10.5 days of gestation. At 6.5 days of gestation (egg cylinder stage), we saw no clearly abnormal embryos in several litters examined. However, by 7.5 days (late primitive streak stage) and thereafter, we did see an abnormal class (~25%) of embryos, which presumably represents the homozygotes. At 7.5 days, the abnormal embryos had a small, conical embryonic ectoderm region and a chorion that appeared to have proliferated excessively and that protruded into the exocoelom. At 8.5 days, the presumptive mutants had formed a rudimentary neural fold, as well as a normal amnion and allantois. However, these embryos were already considerably smaller than their normal littermates, and failed to develop visible somites, although regions of mesenchymal cells flanked the neural ectoderm. By 9.5 or 10.5 days, the mutant embryos (which, by this time, could be identified by DNA analysis) had developed a normal yolk sac, complete with blood islands, but the embryo proper had grown little since day 8.5. Somites were still absent, but the embryos had developed a foregut and the beginnings of a heart, which beat in about 50% of the embryos. In many cases, a large, spherical structure resulting from proliferation of the allantois remained connected to the posterior end of the embryo. Some of the embryos were already disintegrating by day 10.5, and we assume that they do not survive long beyond this time.

In summary, it appears that embryos homozygous for the HB58 insertion develop normally through the egg cylinder stage, undergo gastrulation, but then develop severe abnormalities affecting several embryonic as well as extraembryonic lineages. The nature of the primary defect is difficult to ascertain, based on this analysis. We expect that the use of the *in situ* hybridization scheme described above will allow us to identify homozygous embryos definitively at an early stage, facilitating analysis of the primary phenotypic defect. Based on our current knowledge of the phenotype, the extent of development and the apparently normal differentiation of some cell types (for example, blood and cardiac muscle) indicate that the mutation is not in a gene essential for cell division, metabolism, or other "housekeeping" functions. Rather, it appears to be in a gene required for more specialized morphogenetic processes.

The HB58 insertion site was mapped to chromosome 10 (bands B3–B5) by *in situ* hybridization to metaphase chromosome spreads (26), using a biotin-labeled probe and the detection system described above (21, 23). This analysis also indicated a normal karyotype, with no indications of gross deletions or rearrangements. As no other previously described embryonic lethal mutations with similar phenotypes map to this region, we believe that this mutation defines a new genetic locus. Experiments to localize this mutation on the genetic map are in progress.

To begin to analyze this mutation at the molecular level, we isolated λ clones containing one of the junctions between the foreign and host DNAs. Two overlapping clones representing the same junction were found, and this was arbitrarily designated the left junction. Unique sequence probes were identified, and used to confirm, by Southern blot analysis of wild-type versus HB58 genomic DNA, that the λ clone did indeed contain one of the junctions. The probes were then used to isolate λ-clones containing the wild-type (i.e., preinsertion) locus from a library of normal C57BL/6J mouse DNA. Analysis of these λ clones indicated that the foreign DNA had inserted without causing any gross rearrangements at the left junction. Attempts to clone and characterize the right junction directly have so far been unsuccessful. Data currently available from a variety of other experiments suggest that additional mouse DNA sequences from elsewhere in the genome may have been inserted between the transgene and the host genome at the right junction. In addition, a small segment (2–3 kb) of the host DNA at this locus has been deleted.

The cloning of wild-type DNA from the locus of insertion allowed us to begin to look for a gene whose disruption might be responsible for the mutant phenotype. For this purpose, we employed several unique sequence probes which spanned the region surrounding the insertion site. These probes were hybridized to Northern blots of RNA from mouse embryos, several embryonal carcinoma cell lines, and a variety of adult mouse tissues, and one of the probes was also used to screen several cDNA libraries. Together, these analyses have identified a transcription unit at the site of the insertion with the following properties. The major poly(A)$^+$ RNA encoded at the locus is approximately 3.6 kb in length, and is found in many adult tissues as well as in embryos after 8.5 days of gestation (earlier stages have not yet been examined). The transcription unit, as defined by cDNA clones, spans the site of the transgene insertion, and is thus physically disrupted by the insertion. The level of this transcript in the liver of heterozygous HB58 mice is approximately one half that of the wild-type level, indicating that expression of the gene is affected by the transgene insertion. Thus, these studies have identified a candidate gene at the HB58 locus whose physical disruption and consequent inactivation is likely to be responsible for the mutant phenotype.

Definitive proof that the disruption of this gene is responsible for the mutant phenotype will require additional experiments, such as rescue of the mutation by reintroduction of the wild-type gene.

The other two recessive lethal mutations, carried in the BS12 and HE46 transgenic lines, are only beginning to be characterized. The BS12 line carries approximately 30–40 copies of the human β^S-globin gene on mouse chromosome 1 (as determined by *in situ* hybridization). Heterozygotes are normal and fertile, yet no homozygotes were observed among over 60 progeny of intercrosses. At 9.5 days of gestation, all embryos appeared normal, and no homozygous embryos nor empty decidua were found. Therefore, the BS12 mutation appears to be lethal prior to implantation. The HE46 transgenic line carries 70–80 copies of the human ε-globin gene, introduced as an 8-kb *Hin*dIII fragment. The copies are integrated at a single locus, based on Mendelian transmission, although the locus has not yet been assigned to a chromosome. Fifty-six adult progeny of intercrosses have been analyzed by the *in situ* hybridization method, but no homozygotes were observed. Subsequently, five litters of fetuses were recovered at 12.5 or 13.5 days of gestation, and their genotypes were determined by *in situ* hybridization to the amnion, using the human ε-globin gene as a probe. Forty-one normal fetuses and two small fetuses were found to be either heterozygous or negative for the transgene insertion, while the remaining nine decidua contained only resorbed embryos. This suggests that the homozygous embryos survived at least long enough to cause a decidual response, and died sometime between the blastocyst stage (day 4.5) and day 9 or 10 of gestation.

III. Summary

Insertional mutagenesis represents a promising approach to the identification of new genes involved in mammalian development. In this paper, we have presented a brief review of the literature on the analysis of mutations caused by DNA and retroviral insertion into the mouse genome. We have discussed several methods that we and others have used to identify recessive insertional mutations among transgenic mouse lines. Finally, we have summarized the results of our studies to date on three recessive prenatal lethal mutations that we have identified.

Acknowledgments

This work was supported by grants to F.C. from the National Institutes of Health (HD17704) and the March of Dimes Birth Defects Foundation (1-1016). F.C. is a Pew Scholar in the Biomedical Sciences and J.J.L. is an American Cancer Society Postdoctoral Fellow. We thank Chu Hui Peng for excellent technical assistance, and Janet Rossant and Valerie Prideaux for advice on *in situ* hybridization.

References

1. W. McGinnis, R. L. Garber, J. Wirz, A. Kuroiwa, and W. Gehring, *Cell* **37**, 403 (1984).
2. W. McGinnis, C. P. Hart, W. J. Gehring, and F. H. Ruddle, *Cell* **38**, 675 (1984).
3. M.C. Green (ed.), "Genetic Variants and Strains of the Laboratory Mouse." Gustav Fischer Verlag, Stuttgart, Federal Republic of Germany, 1981.
4. R. D. Palmiter and R. L. Brinster, *ARGen* **20**, 465 (1986).
5. T. Grindley, P. Soriano, and R. Jaenisch, *Trends Genet.* **3**, 162 (1987).
6. R. Jaenisch, K. Harbers, A. Schnieke, J. Lohler, I. Chumakov, D. Jahner, D. Grotkopp, and E. Hoffmann, *Cell* **32**, 209 (1983).
7. E. F. Wagner, L. Covarrubias, T. A. Stewart, and B. Mintz, *Cell* **35**, 647 (1983).
8. W. H. Mark, K. Signorelli, and E. Lacy, *CSHSQB* **50**, 453 (1985).
9. M. Shani, *MCBiol* **6**, 2624 (1986).
10. P. Soriano, T. Grindley, and R. Jaenisch, *Genes Dev.* **1**, 366 (1987).
11. R. P. Woychik, T. A. Stewart, L. G. Davis, P. D'Eustachio, and P. Leder, *Nature (London)* **318**, 36 (1985).
12. P. A. Overbeek, S.-P. Lai, K. R. Van Quill, and H. Westphal, *Science* **231**, 1574 (1986).
13. R. D. Palmiter, T. M. Wilkie, H. Y. Chen, and R. L. Brinster, *Cell* **36**, 869 (1984).
14. A. Schnieke, K. Harbers, and R. Jaenisch, *Nature (London)* **304**, 315 (1983).
15. L. Covarrubias, Y. Nishida, and B. Mintz, *PNAS* **83**, 6020 (1986).
16. T. M. Wilkie and R. D. Palmiter, *MCBiol* **7**, 1646 (1987).
17. F. Costantini, G. Radice, J. Magram, G. Stamatoyannopoulos, T. Papayannopoulou, and K. Chada, *CSHSQB* **50**, 361 (1985).
18. J. B. Whitney III, *Biochem. Genet.* **16**, 667 (1978).
19. F. Costantini, K. Chada, and J. Magram, *Science* **233**, 1192 (1986).
20. C. Lo, *J. Cell Sci.* **81**, 143 (1986).
21. S. Varmuza, V. Prideaux, R. Kothary, and J. Rossant, *Development* **102**, 127 (1988).
22. L. Manuelidis, P. Langer-Safer, and D. Ward, *J. Cell Biol.* **95**, 619 (1982).
23. J. Burns, V. T. W. Chan, J. A. Jonasson, K. A. Fleming, S. Taylor, and J. O. McGee, *J. Clin. Pathol.* **38**, 1085 (1985).
24. K. Chada, J. Magram, K. Raphael, G. Radice, E. Lacy, and F. Costantini, *Nature (London)* **314**, 377 (1985).
25. R. Grosschedl, D. Weaver, D. Baltimore, and F. Costantini, *Cell* **38**, 647 (1984).
26. E. Lacy, S. Roberts, E. P. Evans, M. D. Burtenshaw, and F. D. Costantini, *Cell* **34**, 343 (1983).

IV. Structure and Function of Repetitive and Unusual Sequences

The L1 Family of Repetitive Sequences in Mammals (Abstract only, p. 327)
M. H. Edgell, D. D. Loeb, R. Shehee, M. B. Comer, N. C. Casavant, and C. A. Hutchinson III

Repetitive Sequences in the Human Genome (Abstract only, p. 329)
R. K. Moyzis

Transposition of Intracisternal A-Particle Genes   173
Kira Lueders and Edward Kuff

Use of Variable Number of Tandem Repeat (VNTR) Sequences for Monitoring Chromosomal Instability   187
Paul M. Kraemer, Robert L. Ratliff, Marty F. Bartholdi, Nancy C. Brown, and Jonathan L. Longmire

Transposition of Intracisternal A-Particle Genes

KIRA LUEDERS[1] AND
EDWARD KUFF

Laboratory of Biochemistry,
National Cancer Institute,
National Institutes of Health,
Bethesda, Maryland 20892

Murine intracisternal A-particles (IAPs) are defective retrovirions abundantly expressed in neoplastic cells (reviewed in *1*). The particles assemble on the endoplasmic reticulum membranes, bud into the cisternae, and seem to undergo an entirely intracellular life cycle (*2*). IAPs are encoded by members of a large family of endogenous proviral elements, numbering about 1000 per haploid genome in the nuclear DNAs of *Mus musculus* and closely related species (*1, 3, 4*). The elements have properties of integrated retroviruses, such as colinearity with the particle genomic RNA, the presence of long-terminal-repeat sequences (LTRs), and a tRNA primer-binding site. One randomly selected full-sized IAP element, MIA14, has been completely sequenced (*5*). On the basis of this sequence and homologies with an IAP-related element in the Syrian hamster genome and with the type-D simian retroviruses, the organization of the IAP genome has been defined. It is apparent that some characteristic properties of mouse IAPs are determined by the peptide encoded in the 5′ *gag* domain unique to the mouse elements.

In addition to full-sized, 7.2-kilobase (kb) IAP genetic elements, the mouse genome also contains a large number of deleted variants that may represent as many as 40% of the family members (*1*). Full-sized IAP elements are designated type I (Fig. 1), and deleted forms derived from such elements as IΔ elements based on the sizes of the deletions. Type-II elements (*6*) are distinguished from the type-I elements by a 270-base-pair (bp) insertion, AIIins (*7*); this class of elements is further divided into three subclasses—A, B, and C—on the basis of the sizes of the deletions they contain relative to the type-I elements. Table I summarizes examples (*3, 6–21*) of genomic clones

[1] Speaker.

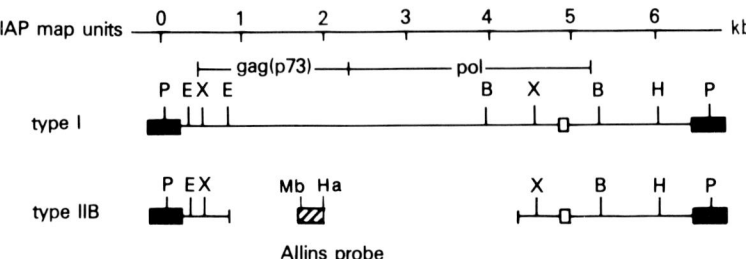

FIG. 1. Physical maps of the type-I and type-IIB IAP elements. The ends of the elements are delineated by long terminal repeats, indicated by black boxes. The type-II IAP-specific insertion, AIIins, is shown by a hatched rectangle, which is placed at an arbitrary position within the deletion in the type-IIB element, indicated by a blank region within the map; the open box shows the location of the IAP endonuclease core sequence (1). Restriction sites are as follows: P, *Pst*I; E, *Eco*RI; X, *Xba*I; B, *Bam*HI; H, *Hin*dIII; Mb, *Mbo*I; Ha, *Hae*III.

as well as of cDNA clones that have been characterized sufficiently to permit classification by structural type. The elements involved in transpositions are indicated.

IAP genetic elements transpose in the genome of mouse tumor cells and also in the germ line of several mouse strains (reviewed in *1*). IAP element transpositions have resulted in both activation and inactivation of target genes (reviewed in detail in *1*). The majority of transpositions have occurred in tumor cell lines with high levels of IAP RNA expression, and have frequently involved deleted forms of the IAP elements (see Table I). The first case of IAP element transposition (*11*) to be reported occurred in a mouse hybridoma derived from MOPC-21 myeloma, and resulted in defective κ-light-chain gene expression. Two distinct transpositions involving different types of IAP elements (igk-1 and igk-20, see Table I) reflected the major IAP transcripts in this myeloma (*22*). Other effects of IAP proviral insertions have included the constitutive activation of c-*mos* (*16*) and interleukin-3 (*12*) genes. Insertions into the c-*mos* gene also occurred in myeloma cell lines abundantly expressing IAP RNAs (E. L. Kuff, unpublished results).

An increase in the number of genomic type-II IAP elements was found in MOPC-315 and MOPC-104E myeloma cells where type-IIB elements are the primary source of RNA (*21*). From restriction analysis, the newly acquired elements appeared to be very similar to one another, suggesting that they resulted from reinsertion of copies derived from a limited number of transcripts. However, previous

TABLE I
Structural Variants of Mouse IAP Elements

Element type	Size (kb)	Examples[a]
I	7.1	MIA14 (8), MIA2 (8), MIA21 (8), IAP81A (9), IAP17 (10), (T) igk-20 (11)
IΔ1	5.2	DNAX 10.2,[b] IAP19A (9), IAP71 (10), (T) IL-3 insertion (12), pMIA34 (13), (T) igk-1 (11), cDNA D20 (14), clone G2 (6)
IΔ2	4.9	DNAX 9.5,[c] MIA48 (3), VL30 clone[d]
IΔ3	4.1	Pseudo-α-globin (15), (T) rc-mos X24 (16), (T) rc-mos NSI (17), L31 (18)
IΔ4	2.8–3.0	IAP81B (9), (T) MIARN (19), genomic clones BALB/c embryo (20)
?	3.0	DNAX 8.3[b]
IIA	4.6	IAP19B (9), IAP14 (10), genomic clones BALB/c embryo (7)
IIB	3.9	Clone 122 (6), genomic clones (T) MOPC-315 (21), MOPC104E (21), BALB/c embryo[d]
IIC	3.1	IAP62 (10), genomic clones BALB/c embryo (7)

[a] Reference numbers are in italics and in parentheses. (T) indicates the element was involved in a transposition.
[b] M. Trounstine, DNAX, personal communication.
[c] J. Mietz, personal communication.
[d] K. Lueders, unpublished sequence.

efforts to demonstrate unintegrated provirus in MOPC-315 cells have been unsuccessful (21).

To define some of the factors involved in IAP element transposition, we have studied the MOPC-315 and MOPC-104E myeloma cell lines in which the type-IIB subclass of IAP elements has undergone a marked amplification compared with embryonic cells and other myelomas. These elements contain an extensive deletion encompassing the *gag* and polymerase genes, but have an intact endonuclease-coding region. We have found endonuclease activity in

IAP fractions from both myelomas as demonstrated by conversion of supercoiled plasmid DNA to relaxed and linear forms. DNA molecules consistent in structure with a linear type-IIB IAP provirus were present in the cytoplasm of MOPC-315 cells.

Rabbits were immunized with an oligopeptide predicted from the nucleotide sequence of the endonuclease gene in a full-length genomic IAP element (5). The antisera reacted on Western blots with a protein present in gradient-purified IAPs from MOPC-104E cells. The weakness of the reaction suggested that the concentration of this protein was extremely low. The antisera also reacted with a fusion protein produced in bacteria transfected with a randomly selected genomic type-IIB IAP element, demonstrating that the endonuclease reading frame was open in this otherwise defective endogenous provirus.

I. Results

A. Amplification of Type-II IAP Elements Correlates with Expression of Type-II RNA

Sequencing of type-II IAP clones isolated from a mouse genomic DNA library (7) revealed that there are *Mbo*I and *Hae*III sites bracketing the type-II IAP-element-specific insertion AIIins (Fig. 1). Mouse genomic DNA from three myeloma cell lines and from embryo was cut with a combination of these restriction enzymes and run on an agarose gel, and the amount of DNA representing type-II sequences was assessed by hybridization of the resulting blotted DNA with AIIins probe. A prominent 250-bp fragment was seen in all of the DNAs. The small 340-bp band derives from elements cut with *Hae*III that do not have the *Mbo*I site. Scans of the autoradiograph (Fig. 2) show that both the MOPC-315 and MOPC-104E myeloma cell lines have an increased number of copies ($5\times$ and $8\times$, respectively) of type-II sequences compared with those in embryo and MOPC-21 myeloma cells. The original estimates of the level of type-IIB element amplification were fourfold for MOPC-315 and twofold for MOPC-104E (21). The present analysis does not distinguish between the subclasses of type-II elements; the higher copy numbers obtained may be due to the more specific type-II probe used here, or the result of tumor growth with continued reinsertion of new copies.

IAP RNA transcripts of various sizes are found in cells that express the elements, and the amounts and proportion of different-sized IAP RNAs appear to be characteristic for each cell type (22). Figure 3

INTRACISTERNAL A-PARTICLE GENES 177

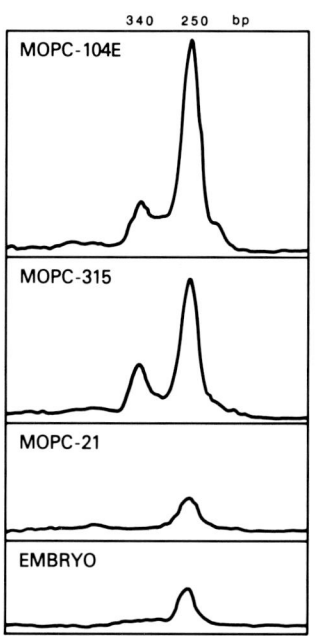

FIG. 2. Scans of autoradiographs showing amplification of type-II IAP sequences. Aliquots (10 μg) of DNA from the indicated cells were digested with MboI/HaeIII, fractionated by electrophoresis on a 1.8% agarose gel, blotted to a nitrocellulose membrane, and hybridized to radioactively labeled AIIins probe. The X-ray film exposed to the resulting blot was scanned to quantitate hybridization to the various DNAs.

shows Northern blots of poly(A)$^+$ RNAs from MOPC-21 and MOPC-315 myeloma cells hybridized to type-II and type-I IAP probes. Hybridization of MOPC-315 RNA with the two probes reveals that the type-IIB transcript (4 kb) is the primary RNA expressed in these cells, although some type-IIA transcript (4.7 kb) is also detected. The type-I probe detects the same type-II RNAs and low levels of full-sized (7.2-kb) and type-IΔ (5.4-kb) RNAs. The type-II transcripts are present at low levels in the MOPC-21 cell line in which the primary IAP RNAs are the 7.2- and 5.4-kb transcripts detected with the type-I probe (the high level of 7.2-kb transcript is also detected with the type-II probe because of the presence of 19 bp of type-I sequence in the 303-bp AIIins probe). IAP RNA expression in the MOPC-104E myeloma cell line is very similar to that in MOPC-315, except that the level of 7.2-kb RNA is lower and the 4.7-kb IIA RNA is

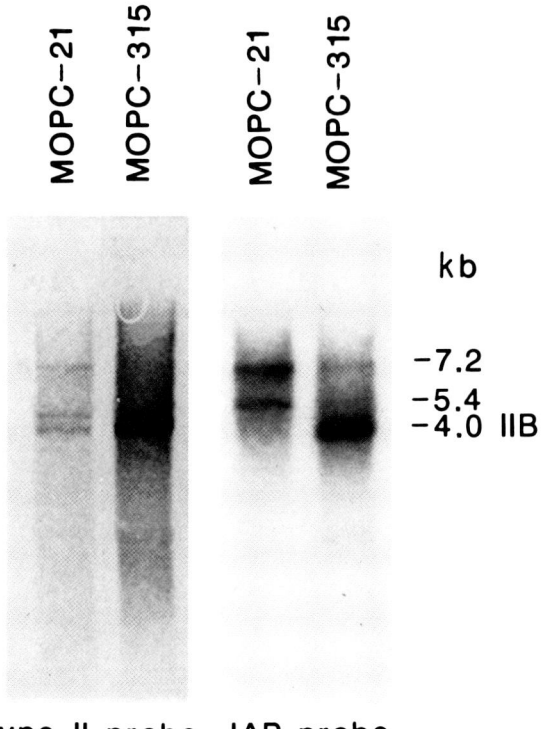

type-II probe   IAP probe

FIG. 3. Electrophoretic analysis of IAP-specific transcripts in RNA fractions from myeloma cell lines. Poly(A)$^+$ RNAs (1 μg) were fractionated and transferred to nitrocellulose as previously described (22) and hybridized with AIIins (type-II probe) or pMIA1 (IAP probe).

higher. Thus, there is a good correlation between the major IAP RNA transcript and the type of IAP element that is amplified in the MOPC-104E and MOPC-315 cell lines.

B. Unintegrated Type-IIB Proviral Elements Present in MOPC-315 Cells

Until recently, it had been generally accepted that proviral integration takes place in the nucleus with a two-LTR circular form of unintegrated viral DNA as the precursor (23). Efforts to detect circular forms of IAP elements had been unsuccessful (21; A. Feenstra, unpublished results) or had led to isolation of rare forms whose structure was not entirely consistent with that expected for an IAP provirus (24). The development of an *in vitro* integration system for murine leukemia virus (25) showed that integration activity was proportional to the level of linear viral DNA in cytoplasmic extracts rather than to the levels of circular forms in the nucleus. These experiments led us to examine MOPC-315 cells for the presence of linear type-IIB proviral elements in the cytoplasm as precursors for integration.

Cells were lysed in the absence of detergent and the nuclear supernatant was centrifuged at 10,000 rpm to prepare membrane (10K pellet) and supernatant (10K super) fractions. The 10K super was further fractionated by centrifugation at 40,000 rpm into pellet (40K pellet) and supernatant (40K super). DNA was extracted from the fractions and analyzed on a Southern blot without restriction endonuclease digestion. A discrete 4-kb DNA band corresponding in size to that expected for a linear type-IIB element was detected by hybridization of the subsequent blot with an IAP element probe in the 10K pellet (Fig. 4). The 10K supernatant contained none of this DNA even when the particulate fraction was concentrated (40K pellet). When the cells were lysed in the presence of detergent, the 4-kb DNA did not sediment at 10K but was sedimentable in the 40K pellet, consistent with the DNA in a particulate form. The 4-kb DNA fraction could be isolated from an agarose gel and cut with restriction enzymes to give fragments predicted from the known structure of type-IIB elements (not shown).

C. A Genomic Type-IIB Element Having an Open Reading Frame (ORF) for Endonuclease

Since the type-IIB elements are apparently the source of both the primary type-II RNA and linear proviral DNA, we asked whether they are capable of encoding an endonuclease activity. In order to test for the coding capacity of such elements, a type-IIB element was cloned as a 2.8-kb *Eco*RI/*Hin*dIII fragment from BALB/c mouse embryo DNA. A fragment corresponding to the endonuclease coding region predicted from the sequence of IAP element MIA14 (the *Xba*I/*Hin*dIII fragment, see Fig. 1) was cloned into a bacterial expression vector in which the protein would be synthesized as a fusion protein with *trpE* (26). As shown in Fig. 5, this construct produced high levels of a 68-kDa protein when the *trpE* promoter was induced (*trpE* codes for a 37-kDa protein in this construct). This result indicated that a randomly selected type-IIB element had an ORF corresponding to a 31-kDa protein that included the retroviral endonuclease gene. The

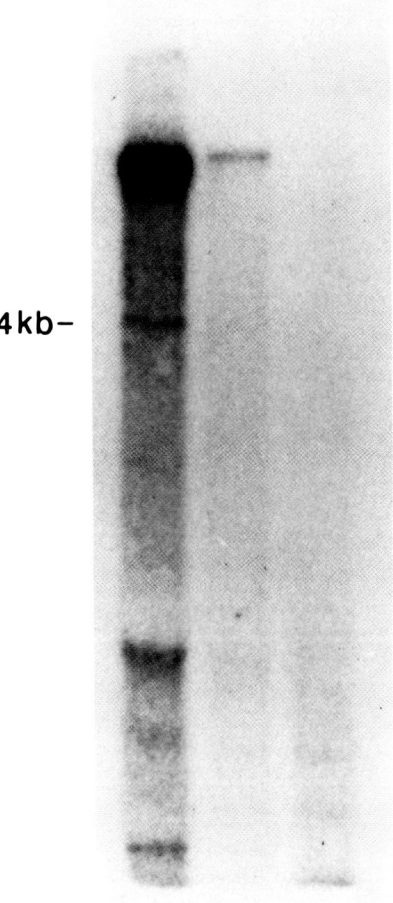

FIG. 4. Detection of linear type-IIB proviral DNA in MOPC-315 myeloma cytoplasm. The nuclear supernatant from cells lysed in the absence of detergent was fractionated into a 10K pellet; the 10K supernatant was further fractionated by centrifugation at 40K into pellet and supernatant fractions. DNA was extracted from the three fractions and electrophoresed on a 1% agarose gel. After blotting to a nitrocellulose membrane, the DNA was hybridized to pMIA1 probe as previously described (8).

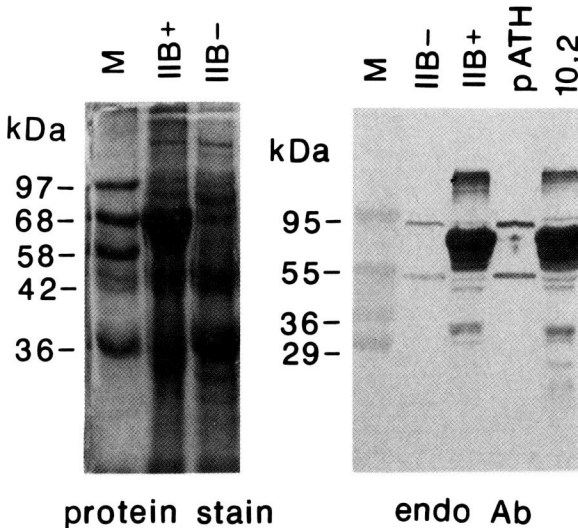

FIG. 5. Expression of the IAP endonuclease in *Escherichia coli*. A region encoding the IAP endonuclease (*Xba*I/*Hind*III fragment from 4.7 to 6.3 on the IAP map in Fig. 1) was cloned in a pATH vector (26) to permit expression as a *trp*E–endonuclease fusion protein. The vector was generously provided by S. Goff. Conditions for growth of the cultures and preparation of lysate have been described (26). The insoluble fraction from 200 μl of bacterial culture was fractionated by sodium dodecyl sulfate/polyacrylamide gel electrophoresis. Lanes on the left were stained with Coomassie blue; those on the right show reaction of the proteins transferred to an Immobilon (Millipore) membrane with antibody to endonuclease. M, Molecular weight markers; IIB, the fragment from a genomic type-IIB element cloned in the sense (+) or antisense (−) orientation downstream of the *trp*E promoter; pATH, vector without an insert; 10.2, the fragment from a type-IΔ1 element cloned in the sense orientation.

same fragment cloned into the vector in the opposite orientation gave no such product.

To determine whether this product resulted from translation in the appropriate reading frame, antibodies were raised in rabbits by immunization with an oligopeptide predicted from the nucleotide sequence of the endonuclease gene in the full-length genomic IAP element. These antibodies were used to test bacterial products blotted to filters. A strong positive reaction was given by the 68-kDa fusion protein, showing that the ORF was the same as that expected for the endonuclease. A construct made by cloning a fragment from a previously sequenced deleted IAP element known to have an ORF capable of encoding endonuclease (clone 10.2) into the same vector served as a positive control.

D. A Protein Associated with Endonuclease Activity Present in MOPC-315 and MOPC-104E Myeloma Cells

The rabbit antibody raised against the endonuclease peptide was also tested for its ability to detect this protein in MOPC-315 cells. When assayed by immunofluorescence (Fig. 6), the cells were brightly stained, with the reaction occurring entirely in the cytoplasm. The antibody detected a protein of 47 kDa in the nuclear supernatant of these cells when this fraction was separated by sodium dodecyl sulfate/polyacrylamide gel electrophoresis and blotted to membrane. The reaction was relatively weak compared with the reaction of an antibody that detects the IAP structural protein, p73, indicating that the endonuclease is present at very low levels. The reaction of the endonuclease antibody with the 47-kDa protein was blocked by the addition of peptide, as was the reaction detected by immunofluorescence, while the peptide had no effect on reaction of the myeloma cells with the control IgA antibody. The relatively strong reaction shown by the cells compared with the relatively weak reaction of the separated protein on the Western blot is probably a function of the need to denature the protein in the blotting procedure. The endonuclease antibody generally required overnight incubation with the Western blot, suggesting that some renaturation had to occur.

The results indicating that an endonuclease was present in the myeloma cells led us to test IAP fractions isolated from MOPC-315 and MOPC-104E cells for endonuclease activity measured by conversion of supercoiled plasmid DNA to relaxed and linear forms. Reaction conditions were those used by Nissen-Meyer and Eikhom (27) to study the putative endonuclease associated with IAPs in MOPC-11 myeloma cells. The enzyme characterized by these investigators was easily solubilized by 0.2% Triton X-100 and was greatly stimulated by ATP.

The components essential for the endonuclease reaction of IAPs were determined (Fig. 7A) using an IAP-enriched fraction isolated from MOPC-315 cells. The particles were treated with 0.2% NP-40 detergent at 0°C to activate the endonuclease and tested for ability to "nick" supercoiled ϕX174 RF DNA. In the presence of Mn^{2+} and ATP, at levels used by Nissen-Meyer and Eikhom, virtually all of the substrate was converted from supercoiled to relaxed form by 50 ng of protein after 30 minutes' incubation at 37°C. This reaction was absolutely dependent on addition of Mn^{2+}. Substitution of Mg^{2+} for Mn^{2+} evoked a topoisomerase-like activity that did not require

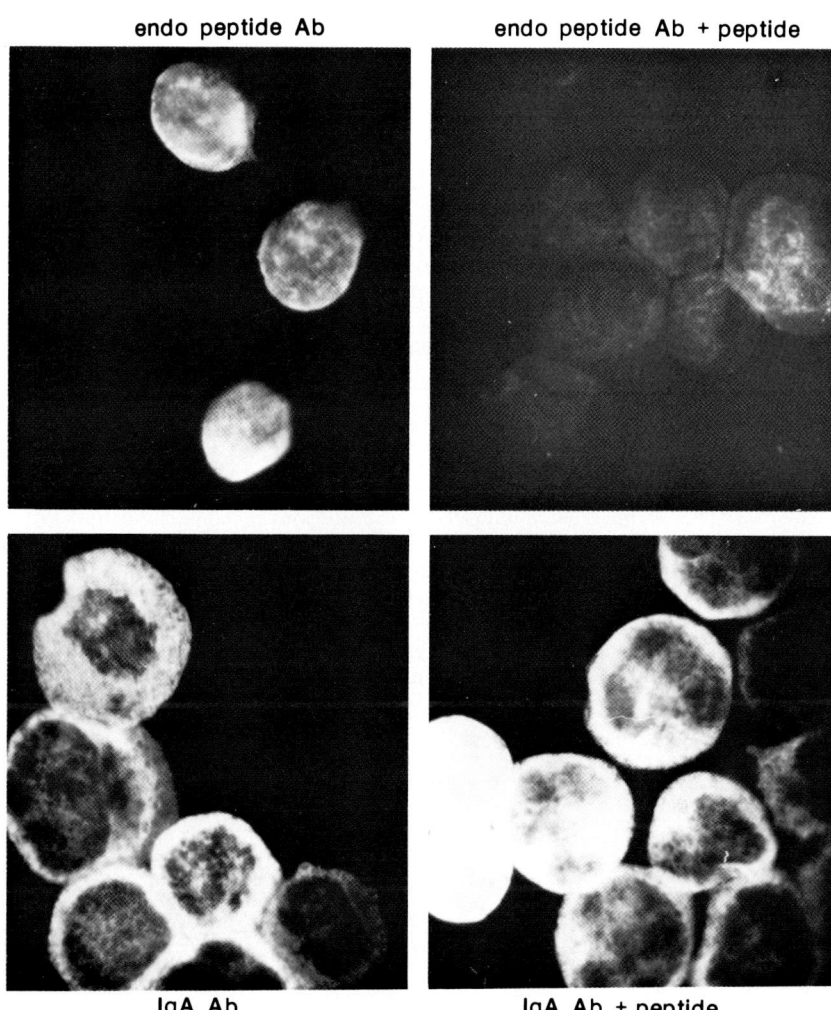

Fig. 6. Reaction of MOPC-315 myeloma cells with endonuclease antibody visualized by immunofluorescence. The cells were spun onto glass slides, fixed with 80% acetone, and treated with endonuclease (endo) peptide antibody (Ab) (top panels) or immunoglobulin A heavy-chain (IgA) antibody as a positive control (bottom panels). Sufficient peptide to block the endonuclease antibody reaction was allowed to react with each antibody before reaction with the cells in the right-hand panels. Reaction of these rabbit antibodies was visualized by addition of fluorescein isothiocyanate-conjugated goat anti-rabbit antibody.

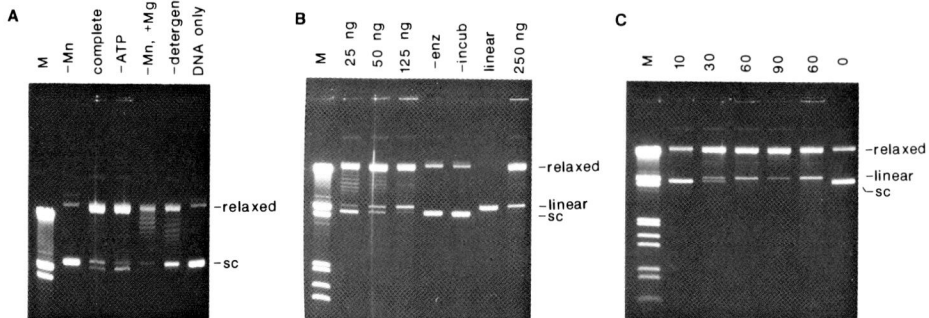

FIG. 7. Characterization of endonuclease activity associated with the IAP fraction of MOPC-315 myeloma cells. The activity was measured by the extent of nicking of supercoiled (sc) φX174 replicative-form DNA to relaxed and linear forms. Each standard 10-μl reaction contained 46 ng of DNA, 20 mM Tris–HCl, pH 7.5, 10 mM 2-mercaptoethanol, 0.1 mM EDTA, 10 mM MnCl$_2$, and an IAP fraction previously treated with NP-40 detergent as indicated. Some reactions contained 2.5 mM ATP. The entire reaction mix was analyzed on a 1% agarose gel, stained with ethidium bromide, and photographed. The M lane in all panels is λ phage DNA cut with EcoRI/HindIII as marker. (A) Shown is the reaction of 50 ng of IAP protein for 30 minutes at 37°C to determine components essential for the reaction. −Mn, Mn^{2+} omitted; complete, standard reaction including ATP; −ATP, ATP omitted; −Mn, +Mg, Mg^{2+} in place of Mn^{2+} with ATP in the reaction; −detergent, no pretreatment of the IAP fraction with NP-40. (B) Protein concentration series reacted for 30 minutes; 25–250 ng of protein was used as shown. φX DNA was cut with PstI to linearize the molecules. (C) Time course of the reaction in minutes with 50 ng of protein. The 60-minute reaction on the right was done with a fraction stored in the refrigerator overnight after NP-40 treatment, showing that the activity is stable.

detergent activation and was stimulated by ATP. The nicking activity did not require added ATP. The activity was proportional to protein concentration (Fig. 7B) and time of incubation (Fig. 7C). In addition to relaxed DNA, the enzyme also produced linear molecules of unit size, as expected from previously characterized retroviral endonuclease (28).

Upon further fractionation of the detergent-activated particles into particulate and soluble fractions, the endonuclease activity remained in particulate form (up to 1% Triton X-100 treatment) while the topoisomerase-like activity was removed. IAPs isolated from MOPC-104E cells by two cycles of sedimentation and isopycnic banding in a sucrose density gradient gave the same results as the MOPC-315 IAP fraction. Therefore, this IAP endonuclease activity is distinct from the previously characterized putative IAP endonuclease (27), and the endonuclease associated with Friend leukemia virus (29). Its properties are also distinct from those of an endonuclease associated with Moloney murine leukemia virus (30) which is easily solubilized with detergent, is stimulated by ATP, and does not appear to be encoded by the *gag* or *pol* genes of the virus. It is likely that the easily solubilized, ATP-dependent nicking activities associated with a variety of retroviral particles are contaminants whose source is unknown.

II. Conclusions

We have examined two myeloma cell lines in which marked amplification of type-II IAP elements has occurred to define factors involved in retrotransposition of these elements. In these cells, there was good correlation between the type of IAP RNA transcribed, the presence of unintegrated proviral elements of the same type, and association of endonuclease activity with IAP fractions. We suggest that the type-IIB elements may be capable of encoding the endonuclease function involved in their own reinsertion into the genomic DNA of these cells.

Although the IIB elements have a large deletion of *gag* and some *pol* sequences, the endonuclease-coding region is intact. Sequencing of a number of type-II insertions (7; K. Lueders, unpublished results) reveals that these insertions introduce a splice-acceptor site. Splicing of an RNA from the 5′ IAP splice-donor site (5) to this site would result in an RNA of approximately 3.5 kb. In the case of type-IIB elements, this RNA would have the potential to encode a protein of 47 kDa, including the endonuclease gene. An RNA of this size is seen with the type-I probe, but not the type-II probe in MOPC-104E cells (not shown), consistent with splicing out of AIIins. The splicing mechanism might result in higher levels of endonuclease than can be achieved by expression of full-sized elements in which a frame-shift is required for translation of the RNA and subsequent cleavage of the polymerase gene product has to occur.

The fact that a randomly selected type-IIB element from embryo DNA had such an open reading frame suggests that this may be a property of most of these elements. Because the deleted elements are incapable of encoding the entire *gag* and the reverse transcriptase region of *pol*, these functions are presumably provided by the products of other expressed IAP elements in *trans*. Therefore, efficient reinsertion of IAP transcripts depends not only on expression of type-IIB elements but also on expression of nondefective undeleted elements. We do not know what factors are involved in specifically activating the type-II elements in some myelomas.

Factors that tend to minimize retrotransposition of IAP elements are discussed in detail elsewhere (*31*).

REFERENCES

1. E. L. Kuff and K. K. Lueders, *Adv. Cancer Res.* **51**, 183 (1988).
2. K. K. Lueders and E. L. Kuff, *JBC* **25**, 5192 (1975).
3. E. L. Kuff, L. A. Smith, and K. K. Lueders, *MCBiol* **1**, 216 (1981).
4. E. L. Kuff and K. K. Lueders, in "Transposition" (A. J. Kingsman, K. F. Charter, and S. M. Kingsman, eds.), p. 247. Cambridge Univ. Press, Cambridge, England, 1987.
5. J. A. Mietz, Z. Grossman, K. K. Lueders, and E. L. Kuff, *J. Virol.* **61**, 3020 (1987).
6. G. L. C. Shen-Ong and M. D. Cole, *J. Virol.* **42**, 411 (1982).
7. K. K. Lueders and J. A. Mietz, *NARes* **14**, 1495 (1986).
8. K. K. Lueders and E. L. Kuff, *PNAS* **77**, 3571 (1980).
9. M. D. Cole, M. Ono, and R. C. C. Huang, *J. Virol.* **42**, 123 (1982).
10. M. Ono, M. D. Cole, A. T. White, and R. C. Huang, *Cell* **21**, 465 (1980).
11. R. G. Hawley, M. J. Shulman, and N. Hozumi, *MCBiol* **4**, 2565 (1984).
12. S. Ymer, W. Q. J. Tucker, H. D. Campbell, and I. G. Young, *NARes* **14**, 5901 (1986).
13. E. L. Kuff, J. A. Mietz, M. L. Trounstine, K. W. Moore, and C. L. Martens, *PNAS* **83**, 6583 (1986).
14. Z. Grossman, J. A. Mietz, and E. L. Kuff, *NARes* **15**, 3823 (1987).
15. K. Lueders, A. Leder, P. Leder, and E. Kuff, *Nature (London)* **295**, 426 (1982).
16. E. Canaani, O. Dreazen, A. Klar, G. Rechavi, D. Ram, J. B. Cohen, and D. Givol, *PNAS* **80**, 7118 (1983).
17. S. Gattoni-Celli, W.-L. W. Hsiao, and I. B. Weinstein, *Nature (London)* **306**, 795 (1983).
18. S. Aota, T. Gojobori, K. Shigesada, H. Ozeki, and T. Ikemura, *Gene* **51**, 1 (1987).
19. D. W. Burt, A. D. Reith, and W. J. Brammar, *NARes* **12**, 8579 (1984).
20. L. Pikó, M. D. Hammons, and K. D. Taylor, *PNAS* **81**, 488 (1984).
21. G. L. C. Shen-Ong and M. D. Cole, *J. Virol.* **49**, 171 (1984).
22. E. L. Kuff and J. W. Fewell, *MCBiol* **5**, 474 (1985).
23. A. T. Panganiban and H. M. Temin, *Cell* **36**, 673 (1984).
24. M. S. Grigoryan, D. A. Kramerov, E. M. Tulchinsky, E. S. Revasova, and E. M. Lukanidin, *EMBO J.* **4**, 2209 (1985).
25. P. O. Brown, B. Bowerman, H. E. Varmus, and J. M. Bishop, *Cell* **49**, 347 (1987).
26. N. Tanese, M. Roth, and S. P. Goff, *PNAS* **82**, 4944 (1985).
27. J. Nissen-Meyer and T. S. Eikhom, *J. Virol.* **40**, 927 (1981).
28. J. Leis, G. Duyk, S. Johnson, M. Longiaru, and A. Skalka, *J. Virol.* **45**, 727 (1983).
29. J. Nissen-Meyer and I. F. Nes, *BBA* **609**, 148 (1980).
30. A. Panet and D. Baltimore, *J. Virol.* **61**, 1756 (1987).
31. E. L. Kuff, *Banbury Rep.* **30**: *Eukaryotic Transposable Elements as Mutagenic Agents*, p. 79 (1988).

Use of Variable Number of Tandem Repeat (VNTR) Sequences for Monitoring Chromosomal Instability

> Paul M. Kraemer[1]
> Robert L. Ratliff
> Marty F. Bartholdi
> Nancy C. Brown and
> Jonathan L. Longmire
>
> *Life Sciences Division, Los Alamos National Laboratory, Los Alamos, New Mexico 87545*

Chromosomal and genetic instability appear to be important, if not essential, characteristics of the neoplastic process (1–5). We have used the spontaneous neoplastic transformation of cultured Chinese hamster cells to study these relationships (6–13). In this experimental model, cells from normal animals sequentially and consistently acquired neoplastic properties when passaged in culture without experimental treatment of any kind. The process was accompanied by ongoing chromosome changes first detectable by conventional cytogenetic analysis during the earliest stages, with further changes thereafter. Certain specific karyotypic changes were selectively favored during the neoplastic evolution of these cells during serial passages *in vitro* as well as during *in vivo* tumorigenesis after inoculation of cells into nude mice. However, repeated cloning and other experiments clearly demonstrated that numerous chromosomal alterations dispersed throughout the genome were being generated continuously in these cell lineages. Thus, neoplastic progression appears to be driven by the continuing selection of variants generated by a pervasive genetic instability inherent to the neoplastic process (2, 13). In this paper, we describe our recent efforts to determine the molecular nature of this genetic instability.

---

[1] Speaker.

I. DNA Fingerprinting

A potential approach to the problem of designing a generic assessment of genetic stability in neoplastic cell populations derives from recent studies of hypervariable sequences—that is, sequences that are extremely polymorphic among individuals. Unlike the typical anonymous dimorphic loci that have been used for restriction-fragment-length-polymorphism (RFLP) analysis, hypervariable sequences show numerous alleles and high heterozygosity among individuals. For instance, the original observation of such sequences reported 15 different alleles of an anonymous locus in a small sample of unrelated individuals (14). Since then, many such polymorphic loci have been reported (15–20). In each case the mechanism of polymorphism has related to the fact that each of these loci includes a short tandemly repeated sequence (9–45 bp). Polymorphism results from variable numbers of the repeated unit within each cluster and hence these loci are referred to as variable number of tandem repeats (VNTRs) (20).

An additional consideration, of importance to the present project, is the discovery that subsets of VNTR loci scattered throughout the genome can have a common core sequence (21). When genomic DNA is restricted with common "4-base cutters," Southern blot analysis with core-sequence probes at moderate stringency yields complex patterns that can be individual-specific. Such DNA "fingerprints" have proven useful for forensic medicine and studies of wildlife populations (22–27). Within the present context, DNA fingerprint methodology offers an opportunity to examine simultaneously the stability of many loci dispersed throughout the genome. We were also encouraged by evidence from Jeffreys' group (28) that although changes were detectable in some tumors, the sequences were somatically stable in normal cells (29).

An example of the fingerprinting technique, as applied to related and unrelated humans, is illustrated in Fig. 1. DNA isolated from peripheral blood leukocytes was digested with *Hae*III and the fragments were separated by electrophoresis in 1% agarose. After Southern transfer, blots were hybridized with probes representing a family of VNTRs discovered by Vassart et al. (30), who unexpectedly found that two clusters of 15-b repeats within the M13 bacteriophage genome could be used as a probe to detect a family of VNTRs in human and animal DNA. Figure 1A illustrates a blot probed with the 282-bp *Hae*III/*Cla*I fragment of the phage that contains one of the repeat clusters. As shown, moderate stringency hybridization detected numerous fragments in the DNA of each individual. Among the seven unrelated individuals, the pattern of the larger, resolvable bands was clearly unique to each individual. While each of these individuals showed between 11 and 20 clearly resolvable bands, there are almost 100 different fragments among the whole group. The patterns reflect both a high degree of heterozygosity in the population being sampled, and the fact that the probe hybridizes with multiple alleles at each of a large number of loci. The patterns for DNAs from a small family show some bands that were obviously alleles, in that all three children had combinations of fragments inherited from one parent or the other. Even here, however, unique fingerprints were obtained, due to the high degree of heterozygosity of the parents.

Figure 1B illustrates the same blot as Fig. 1A, in this case hybridized with a cloned (C1 47) human DNA sequence complementary to the M13 repeat. For this purpose, the M13 fragment was used to screen a chromosome-16-specific Charon-40 library (LA-16NL01) constructed as part of the Los Alamos National Gene Library Project (31). Many of the same bands were detected with the human VNTR probe as were detected with the repeat cluster of the M13 phage. Additional loci were also detected with this probe.

A direct demonstration that this M13-related family of VNTRs is widely distributed throughout the human genome was obtained by hybridizing the M13 probe with spot blots of sorted chromosomes. As illustrated in Fig. 2, all the human chromosomes contain members of this VNTR family; however, the distribution was apparently not uniform. Using Cot_{50} human DNA as a probe to measure the relative amount of DNA on each spot, the relative densities of these VNTR sequences on each chromosome were estimated. The data indicated that several chromosomes (especially 2, 14, and 17) had a large amount, while some (3 and the sex chromosomes) had very little.

The M13 family of VNTRs is also present in Chinese hamster DNA, with characteristics similar to that of human DNA. Individual-specific fingerprints were obtainable from DNA samples from our breeding colony of animals (Fig. 3). In addition, spot blot analyses of sorted chromosomes also were positive throughout the karyotype, with chromosomes 5 and 7 showing enrichment (Fig. 4).

From this evidence, we concluded that the M13–VNTR family discovered by Vassart et al. (30) had appropriate characteristics for use in our project on genomic instability in neoplasia.

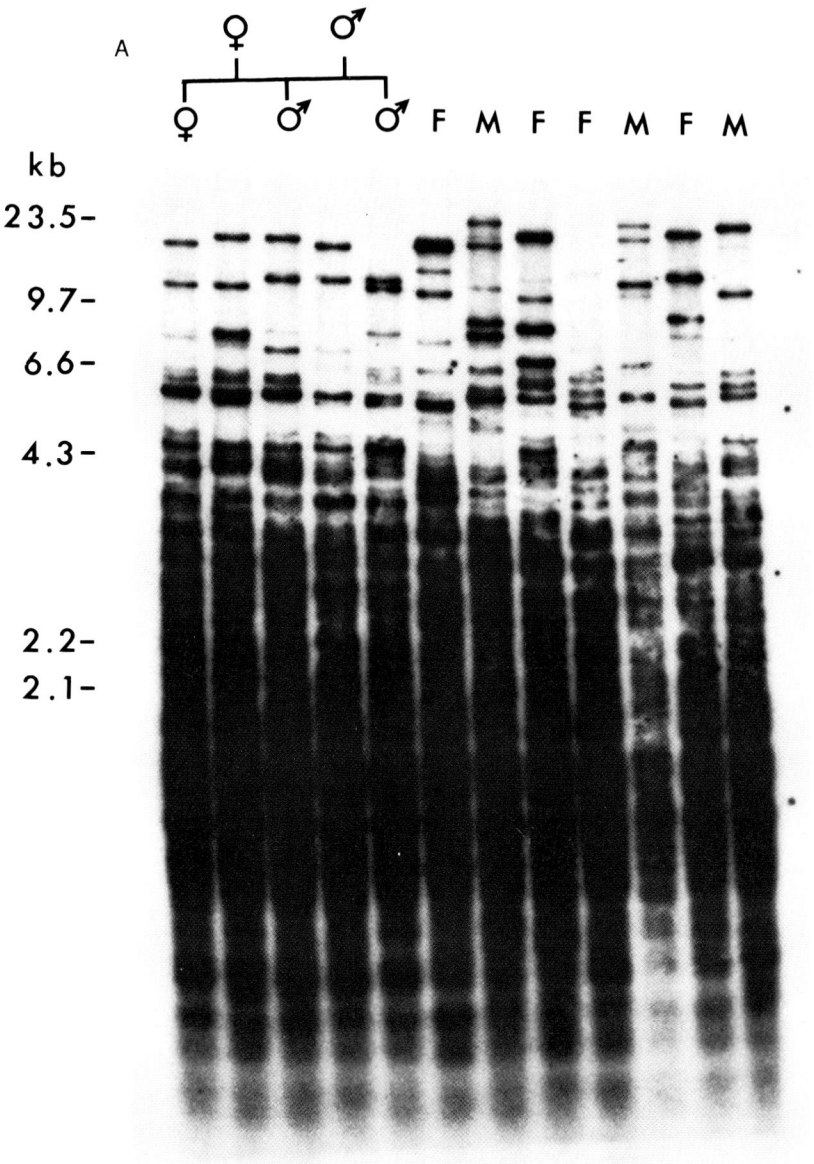

FIG. 1. Restriction patterns detected in human DNA from related and unrelated individuals. HaeIII-digested DNAs (10 µg each) from blood leukocytes were electrophoresed in a 1.0% agarose gel in 40 mM Tris acetate, 1 mM EDTA (pH 7.8) at a constant voltage of 3.5 V/cm for about 36 hours. The gels were soaked in 0.25 M HCl for 15 minutes, then 0.4 N NaOH (2 × 20 min), then transferred overnight with 0.4 N NaOH to nylon filters (Zetabind). After a 2-hour prehybridization, filters were hybridized with probe overnight at 42°C in 40% formamide, 6× SSC (standard saline citrate), 5 mM EDTA, and 0.25% dried skimmed milk. Following hybridization, the filters were

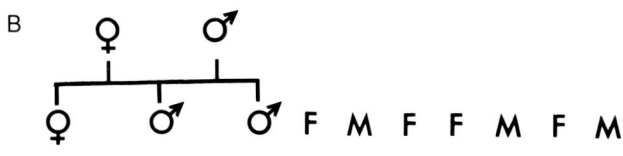

washed twice for 15 minutes in 2× SSC + 0.1% SDS at room temperature; twice in 2× SSC + 0.1% SDS at 65°C for 15 minutes, and twice in 1× SSC + 0.1% SDS for 20 minutes. Filters were exposed to Kodak XAR-5 film at −70°C with intensifying screens. (A) Probe was prepared by primer-extension labeling with [^{32}P]dCTP of a 282-bp fragment (HaeIII/ClaI) of the RF form of M13 phage. (B) Probe was prepared by nick translation of clone 47 isolated from a human chromosome-16-specific library made in Charon 40. F, Female; M, male.

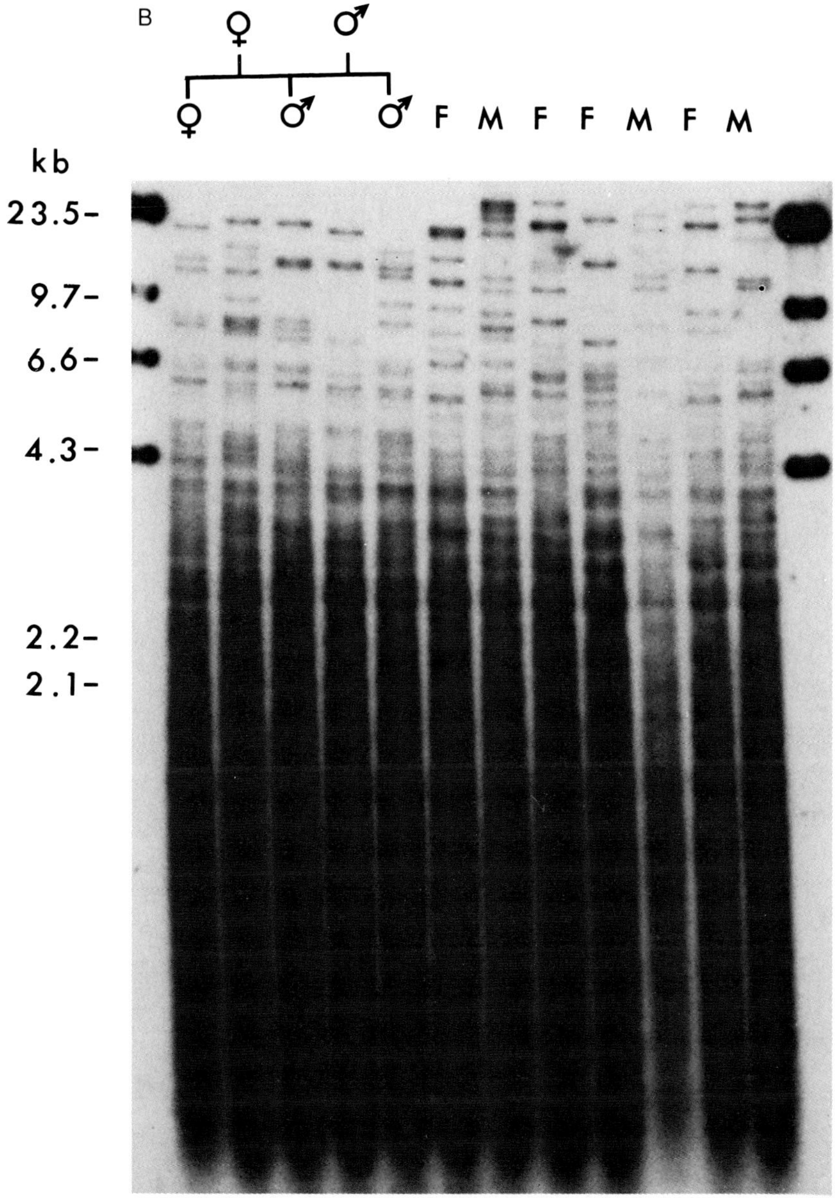

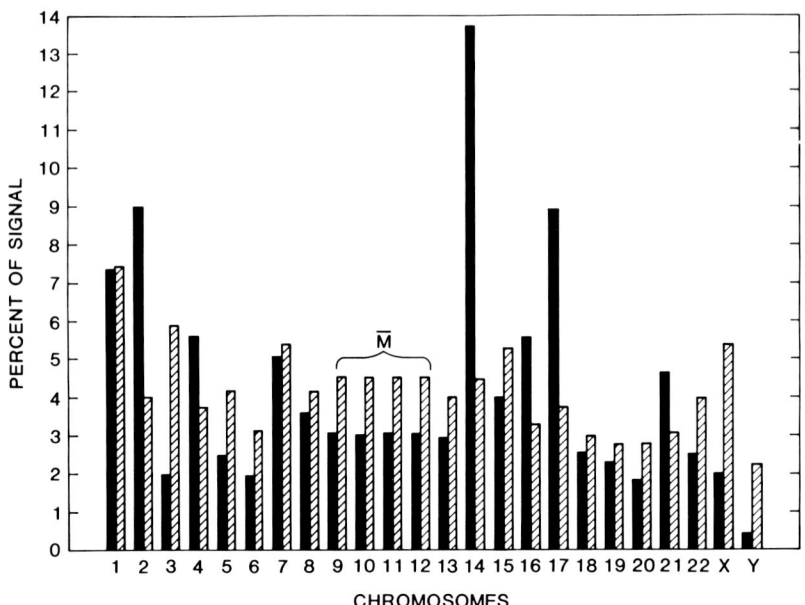

FIG. 2. Densitometry results of spot blots prepared from sorted human chromosomes. Chromosomes were prepared and 30,000 of each type were sorted onto nitrocellulose filters (38). Filters were first hybridized with an M13 probe similar to that in Fig. 1A (solid bars), washed, and exposed to film, then rehybridized with human DNA isolated at Cot_{50} (hatched bars), rewashed, and exposed to another film. Densitometry readings of the latter procedure were used as a measure of the DNA content of each spot. Chromosomes 9–12 were not separated by flow sorting; average values were therefore used.

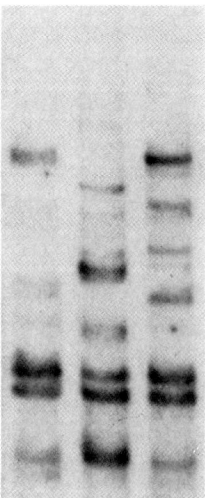

FIG. 3. Restriction patterns of DNA from three individual adult Chinese hamsters from our breeding colony. The methods used were similar to those used in Fig. 1A.

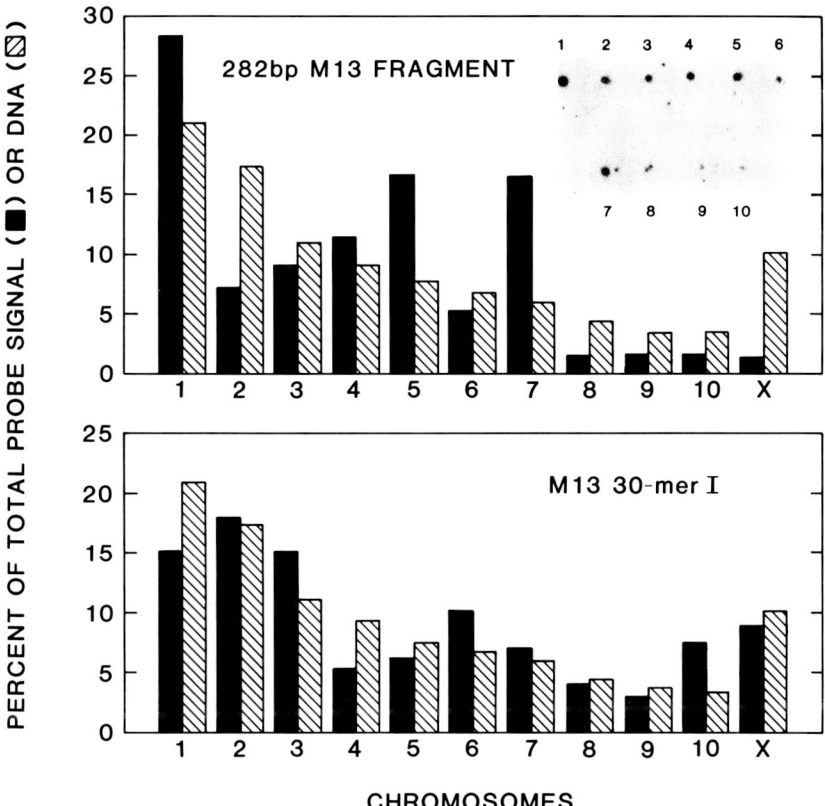

FIG. 4. Densitometry results of spot blots prepared from sorted Chinese hamster chromosomes. Chromosomes were prepared and 30,000 of each type were sorted onto nitrocellulose filters. Filters were hybridized, washed, and exposed to film as described in Fig. 1A. Relative chromosome DNA contents were calculated from published estimates of Chinese hamster chromosome size (39). (Top) Probe was prepared as in Fig. 1A; (bottom) probe was prepared by end-labeling synthetic oligonucleotide 30-mer I: GAG-GGT-GGG-GGT-TCT-GAG-GGT-GGT-GGC-TCT.

II. A Cell-Culture Model of Neoplasia

For studies of the Chinese hamster neoplasia model, cells from single individual fetuses were carried in serial culture. At various passage levels, cells were characterized with respect to the acquisition of *in vitro* indicators of neoplasia, tumorigenicity in nude mice, and cytogenetic analysis. DNA was also isolated from these cell populations. When tumors appeared, they were briefly recultured, and the cells were characterized as before. Table I presents represen-

TABLE I
PHENOTYPIC CHANGES IN THE WCHE/5 LINEAGE

Characteristic	p8	p17	p29	p52	p81	T1	T2	T3
Immortalized	0	+	+	+	+	+	+	+
Karyotype changes	0	+	++	++	++	++	++	++
High saturation density	0	0	+	+	+	+	+	+
Growth in PPP[a]	0	0	+	+	+	+	+	+
Growth in 1% serum	0	0	±	+	+	+	+	+
Morphological transformation	0	0	±	+	+	+	+	+
Tumorigenicity	0	0	Cond.[b]	+	+	+	+	+
Growth in methylcellulose	0	0	0	±	+	+	+	+

[a] PPP: Platelet poor plasma
[b] Cond: Tumorigenic only in implanted sponges (32, 33).

tative biological results for the WCHE/5 lineage. These results clearly illustrate the multistep character of the neoplastic process. In particular, the earliest discernible neoplastic change in this model was the appearance of cells that were serially clonogenic, i.e., immortalized (11).

In the case of the WCHE/5 lineage (one of many studied), clonal evolution within the serially passaged mass culture resulted in most of the cells showing trisomy of chromosome 5 at a very early stage (Table II). Other lineages also showed early chromosome changes, though not necessarily any specific change (8). As shown in Tables I and II, as well as other lineages not reported here, cell-line evolution continued in an ongoing fashion, resulting in the stepwise acquisition of neoplastic characteristics and further karyotypic changes. In some

TABLE II
KARYOTYPIC CHANGE IN THE WCHE/5 LINEAGE

Type	Change
p8	Euploid (67%), no recurrent markers or losses
p17	Trisomy 5 (90%), i(5q), (10%), no recurrent losses
p29	Trisomy 5 (90%), 3q$^+$ (55%), +8q (17%), no recurrent losses
p52	Trisomy 5 (95%), trisomy 8 (50%), +8q (50%), i(3q) (45%), 3q$^+$, +3p, 3 other unknown markers
p81	Trisomy 5 (95%), trisomy 8 (95%)
T1	t(3:5)(3qter→3ql : :5q6→5pter), t(3:x)(3qter→3q5 : :xq3→Xpter), 3 other unknown markers
T2	Trisomy 3, -x, one other unknown marker
T3	Trisomy 10 (70%)

Chinese hamster lineages (for example, the CHO cell line), a relatively stable, homogeneous tumorigenic cell line emerges after many serial passages *in vitro*. Other examples continued to generate new phenotypic and karyotypic variants seemingly without any "climax" cell population. The latter situation, of which WCHE/5 is an example, has been much the more common in our experience.

III. DNA Fingerprinting the Neoplastic Process

High-molecular-weight DNA was prepared from WCHE/5 cells at the various stages of the neoplastic process described above and in Tables I and II. These preparations were digested with *Hae*III or *Hin*fI and the fragments were analyzed by gel electrophoresis, Southern blot transfer, and hybridization at moderate stringency to labeled probes. Initially, the probe consisted of the 282-bp *Hae*III/*Cla*I fragment of the M13 phage that, as shown above, could fingerprint human or Chinese hamster individuals. When this fragment was labeled by primer extension and hybridized to the WCHE/5 DNAs, most of the hybridization signal reflected numerous smaller fragments (<4 kb) that appeared invariant within the entire lineage set. However, the larger fragments included some major and minor invariant bands as well as numerous new bands between 4 and 11 kb that appeared and disappeared in the samples from the various stages of the WCHE/5 lineage (Fig. 5A). These novel bands show characteristic microheterogeneity compared to the invariant bands. The appearance of the new bands that accompanied the neoplastic process was quite similar and reproducible for DNA sets digested with either *Hae*III or *Hin*fI (as shown in Fig. 5A), as well as other 4-base-cutters, such as *Rsa*I, *Alu*I, and *Sau*3AI (data not shown). These observations serve to minimize the likelihood that the bands were artifactual. In addition, comparison of the two sets in Fig. 5A shows that the variant *Hin*fI bands are all similar but smaller (by about 800 bp) than the parallel *Hae*III bands. This suggests that the variant fragments may have a common sequence flanking a common VNTR locus and may therefore all represent a single hypermutable locus in the WCHE/5 lineage.

Both the cytogenetic data (Table II for the WCHE/5 lineage and references 8 and 13 for a more general analysis of Chinese hamster lineages) and the fingerprint data shown in Fig. 5A suggest that new stem lines are constantly being generated throughout the neoplastic process. Any particular DNA sample undoubtedly represents a mix-

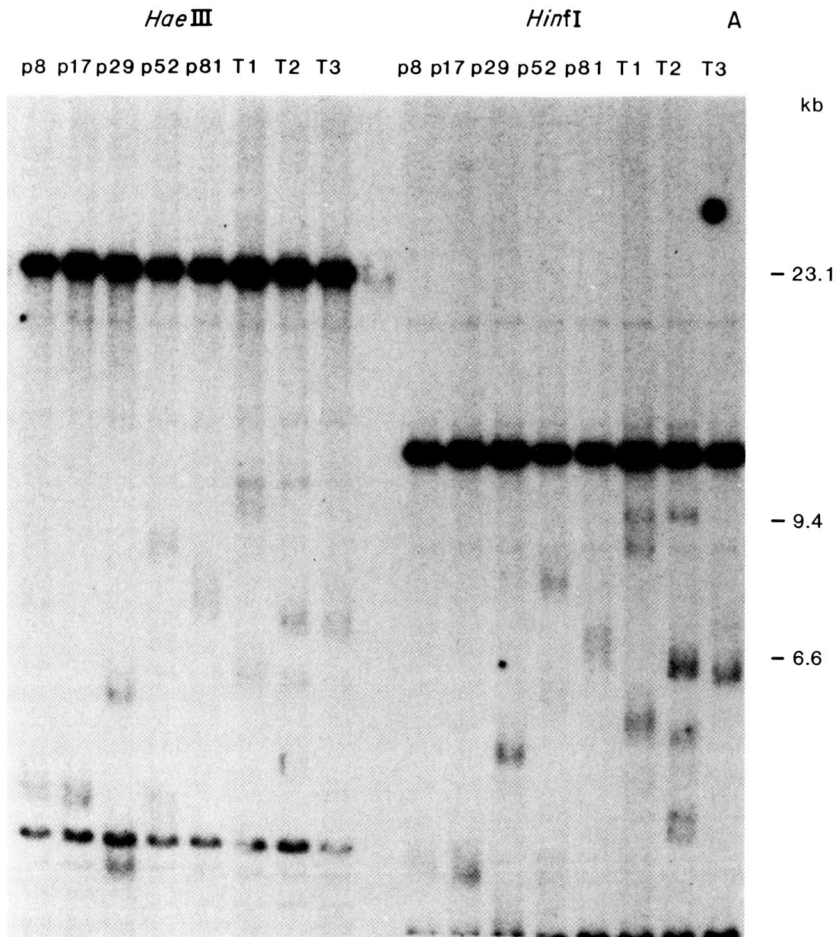

FIG. 5. Restriction patterns detected in Chinese hamster DNA from various passage levels and tumors of the WCHE/5 lineage. Procedures were similar to those used in Fig. 1. (A) Probe was prepared from a 282-bp M13 fragment. (B) Probe was prepared from clone 57 of a human chromosome-16-specific library.

ture of competing stem lines. However, the "fingerprint" methodology appears to detect a form of genetic instability not clearly manifest by the changes in karyotype. The latter, as shown in Table II, are to a great extent consistent with changes in sequence copy number, while the former showed new fragments that may reflect a particularly reactive locus, not necessarily located on any of the chromosomes showing cytogenetic alterations.

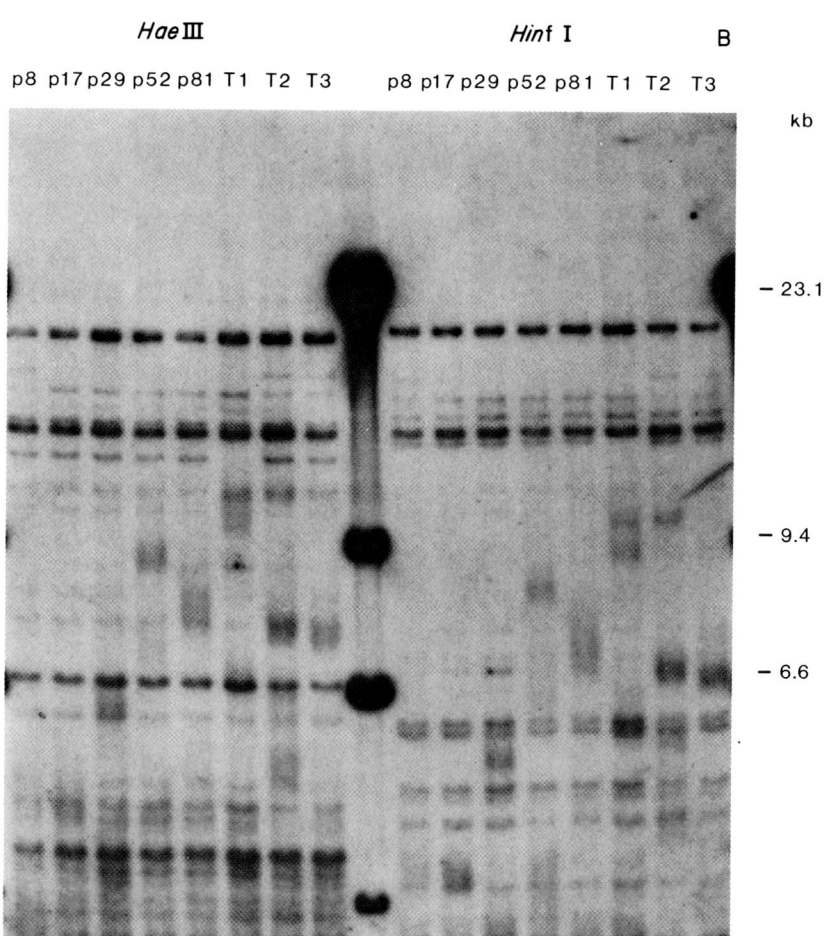

FIG. 5. (continued)

The changes occurring at this locus appear to reflect the generalized genetic instability associated with the neoplastic process rather than being specifically related to the selective alterations in phenotype. The three tumors are especially instructive in this regard, since the bands of each one are different from the others and from any stage of the *in vitro* lineage. T1 and T2 derived from the inoculation of p29 cells into preimplanted sponges in nude mice (32, 33), and both had latent periods in excess of 100 days prior to harvesting and reculturing. T3 derived from a clone isolated from the WCHE/5 lineage during midcrisis (p13). It is clear that different variations of the "fingerprint" changes were all compatible with progression of the cells to the tumorigenic phenotype. We have previously shown that many alternative karyotype changes are also compatible with neoplastic progression, though perhaps not equally so (*13*).

A second approach to fingerprinting with M13-related VNTRs utilized cloned mammalian DNA sequences isolated from libraries by screening with the M13 fragment. As stated above, we have obtained a number of such clones from a human-chromosome-specific Charon-40 library made here at Los Alamos. The blot illustrated in Fig. 5B is the same as in Fig. 5A, stripped and rehybridized with one of these clones (clone 47). As shown, the unstable fragments are readily detected with this clone, but many other differences are seen, such as the absence of signal from the large major allele seen in Fig. 5A. In addition, however, bands present at passage 8 disappeared during *in vitro* passage and may or may not reappear during *in vivo* tumorigenesis. When the disappearance of the band is associated with the enhancement of another band, loss of heterozygosity is suggested, a phenomenon that has been shown to be very common in cultured Chinese hamster cell lines (*34*).

A third approach has utilized synthetic oligonucleotide probes. These were made from variations of the M13 consensus sequence. As illustrated in Fig. 6, the 282-bp *Hae*III/*Cla*I M13 fragment contains 105 bp of a slightly variable tandem repeat with a 15-bp consensus sequence. Synthetic 30-mers were made from these variations and end-labeled with ^{32}P for use as probes of WCHE/5 lineage DNA sets. Each variation yielded unique bands, different from each other and different from the patterns shown in Fig. 5A or B.

An example of these is illustrated in Fig. 7, showing that this variant probe also detected losses of some bands during neoplastic evolution. An even clearer example of transient band loss (Fig. 8) was obtained with an oligonucleotide probe consisting of a "myoglobin" 16-mer from the consensus sequence (GGA-GGT-GGG-CAG-GAA-G) of a VNTR family described by Jeffreys *et al.* (*21*). Evidently, the pair of bands at about 19 kb represents parental alleles present in the fetus from which the WCHE/5 lineage derived. The larger fragment gradually diminished during *in vitro* growth, while the smaller fragment became enhanced. However, tumors resembled either the parental heterozygotic types or the cells at late passage *in vitro*. We think this reflects transient emergence of a stem line that has lost heterozygosity at the specific locus being monitored, perhaps resulting from ongoing rounds of tetraploidization and chromosome reduction by segregation (*34*). We have not determined what chromosome

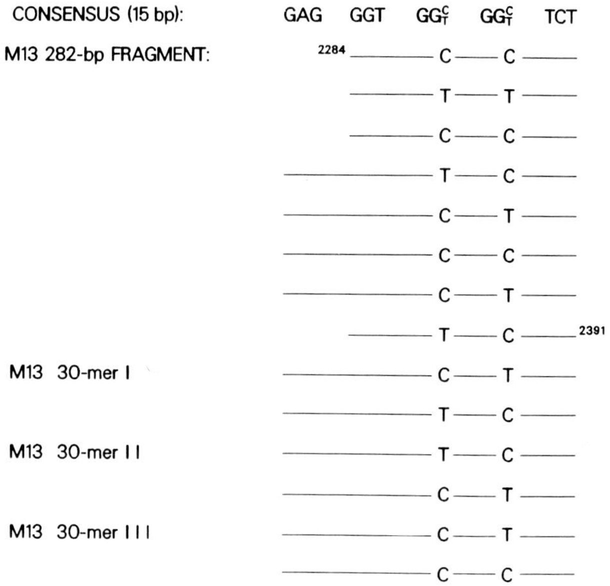

```
CONSENSUS (15 bp):      GAG   GGT   GG^C_T   GG^C_T   TCT
M13 282-bp FRAGMENT:    2284 ———————— C ———— C ————
                             ———————— T ———— T ————
                             ———————— C ———— C ————
                             ———————— T ———— C ————
                             ———————— C ———— T ————
                             ———————— C ———— C ————
                             ———————— C ———— T ————
                             ———————— T ———— C ———— 2391
M13 30-mer I                 ———————— C ———— T ————
                             ———————— T ———— C ————
M13 30-mer II                ———————— T ———— C ————
                             ———————— C ———— T ————
M13 30-mer III               ———————— C ———— T ————
                             ———————— C ———— C ————
```

FIG. 6. Diagram of the M13 phage repetitive sequences and the synthetic 30-mer oligonucleotide variations used as probes.

was involved in the changes illustrated in Fig. 8; in any case, the homozygous pattern detected was not critical to tumorigenesis in the WCHE/5 lineage.

IV. Discussion

Our data suggest that highly dispersed VNTR families may be useful in assaying the genetic stability of cellular lineages, especially neoplastic ones. The possibility that these sequences are themselves involved in rearrangements that drive the neoplastic process should be considered. Such a possibility could explain the persistent generation of translocations, rearrangements, and deletions commonly seen in solid tumors. Some cytogenetic changes (e.g., the trisomy 5 and 8 in the WCHE/5 lineage), which appear to represent nondisjunctive changes in the number of copies of normal chromosomes, are difficult to envision in this way. Our data also suggest that fingerprinting may offer detection of gross chromosomal changes that are not obvious on conventional cytogenetic inspection. Loss of heterozygosity could be extremely important in the neoplastic process, as it could result in the

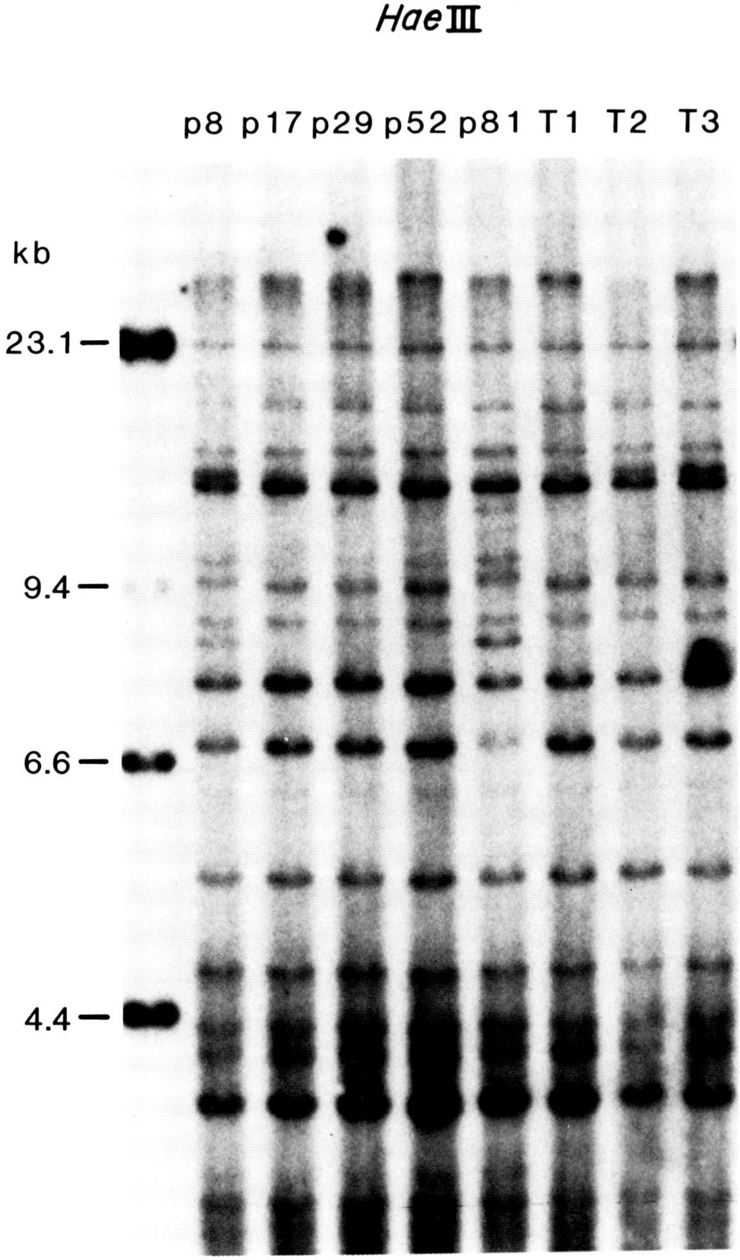

FIG. 7. Restriction patterns detected with the 30-mer II probe in Chinese hamster DNA from various passage levels and tumors of the WCHE/5 lineage.

Hae III

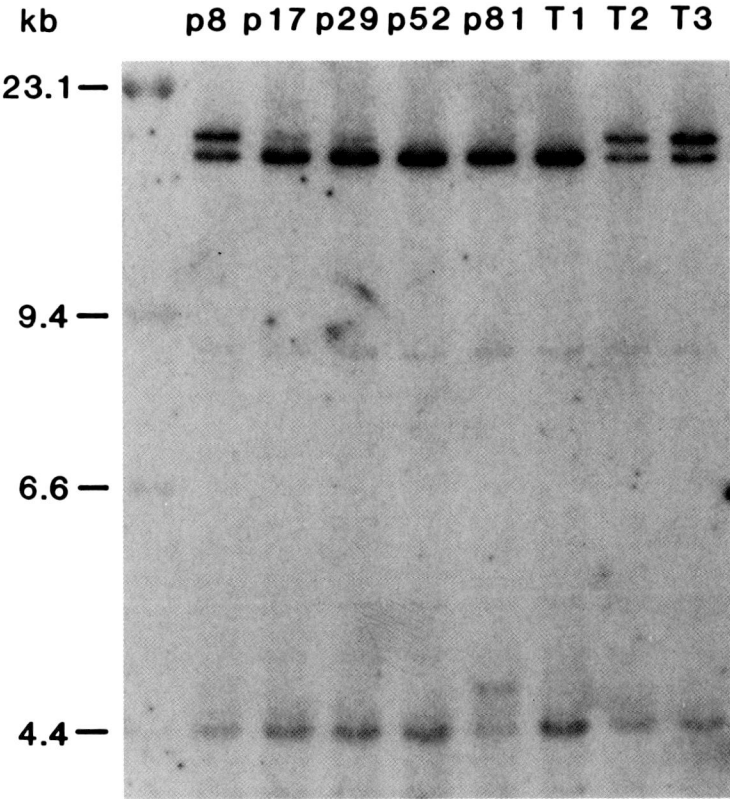

FIG. 8. Restriction patterns detected in the Chinese hamster DNAs following hybridizing to the myoglobin VNTR sequence (21). Southern blot analysis methods were the same as those used in Fig. 1, except an end-labeled 16-mer synthetic oligonucleotide (5' GGA-GGT-GGG-CAG-GAA-G 3') was used as a probe.

expression of recessive genes such as mutated tumor suppressor genes (35). The use of VNTR probes would appear to be an ideal way to follow such phenomena, especially if large chromosomal pieces or whole chromosomes are involved.

The polymorphic nature of VNTR sequences among human or animal populations is thought to be due to unequal exchange during meiotic recombination (21). Normally, these sequences are somatically stable (29), though it seems clear from our data that during neoplasia, flagrant instability at the molecular level can occur. The novel bands that we suspect to be a hypermutable locus in WCHE/5 may be a "sentinel" locus responsive to *trans*-acting abnormal activities that are required for driving the neoplastic process. Under this construction, the hyperactive rate of generation of genetic variants is viewed as a "root" cause of cancer; the selection of variants with activated oncogenes, or mutated tumor-suppressor genes, or other changes in phenotype would follow inevitably from the generalized genetic instability. The mechanism of change in the case of the hypermutable locus in WCHE/5 is, at present, unknown, although somatic recombinational events are one possibility. Jeffreys and co-workers have shown that the germ-line stability of both mouse and human VNTRs is quite variable among different loci (29, 36), but whether *trans*-acting activities are involved in these differences has not been determined.

It must be emphasized that our data on using VNTRs for monitoring genetic instability in cancer are encouraging, but still very preliminary. We believe that further progress and a clear answer to the usefulness of this approach will require considerable work in several areas. First, clones must be isolated and mapped so as to allow the parallel use of common core-sequence hybridization procedures and locus-specific probes representing known allelic systems. Second, it is probable that many VNTR families remain to be discovered (20, 37); presumably, a much greater portion of the genome could be monitored with the use of each new family. Finally, the methods and concepts that we have explored for the WCHE/5 cell-culture model of neoplasia will be applied to other models, including models based upon the transformation of human cells.

Acknowledgment

This work was performed under the auspices of the U.S. Department of Energy.

References

1. J. Cairns, *Nature (London)* **289**, 353 (1981).
2. P. C. Nowell, *Cancer Res.* **46**, 2203 (1986).
3. A. A. Sandberg, C. Turc-Carel, and R. M. Genemell, *Cancer Res.* **48**, 1049 (1988).
4. P. M. Kraemer, L. L. Deaven, H. A. Crissman, and M. A. Van Dilla, *Adv. Cell Mol. Biol.* **2**, 47 (1972).
5. J. German, *in* "Chromosome Mutation and Neoplasia" (J. German, ed.), p. 97. Academic Press, New York, 1983.
6. P. M. Kraemer, G. L. Travis, F. A. Ray, and L. S. Cram, *Cancer Res.* **43**, 4822 (1983).
7. L. S. Cram, M. F. Bartholdi, F. A. Ray, G. L. Travis, and P. M. Kraemer, *Cancer Res.* **43**, 4828 (1983).

8. F. A. Ray, M. F. Bartholdi, P. M. Kraemer, and L. S. Cram, *Cancer Genet. Cytogenet.* **21**, 35 (1986).
9. M. F. Bartholdi, F. A. Ray, L. S. Cram, and P. M. Kramer, *Cytometry* **5**, 534 (1984).
10. E. Wakshull, P. M. Kraemer, and W. Wharton, *Cancer Res.* **45**, 2070 (1985).
11. P. M. Kraemer, F. A. Ray, A. R. Brothman, M. F. Bartholdi, and L. S. Cram, *JNCI* **76**, 703 (1986).
12. M. F. Bartholdi, F. A. Ray, L. S. Cram, and P. M. Kraemer, *Somatic Cell Mol. Genet.* **13**, 1 (1987).
13. P. M. Kraemer, F. A. Ray, M. F. Bartholdi, and L. S. Cram, *Cancer Genet. Cytogenet.* **27**, 273 (1987).
14. A. R. Wyman and R. White, *PNAS* **77**, 6754 (1980).
15. G. I. Bell, M. J. Selby, and W. J. Rutter, *Nature (London)* **295**, 31 (1982).
16. D. R. Higgs, S. E. Y. Goodbourne, J. S. Wainscot, J. B. Clegg, and D. J. Weatherall, *NARes* **9**, 4213 (1981).
17. N. J. Proudfoot, A. Zil, and T. Maniatis, *Cell* **31**, 553 (1982).
18. S. E. Y. Goodbourne, D. R. Higgs, J. B. Clegg, and D. J. Weatherall, *PNAS* **80**, 5022 (1983).
19. D. J. Capon, E. Y. Chen, A. D. Levinson, P. H. Seeburg, and D. V. Goedal, *Nature (London)* **302**, 33 (1983).
20. Y. Nakamura, M. Leppert, P. O'Connell, R. Wolfe, T. Holm, M. Culver, C. Martin, E. Fujimoto, M. Hoff, E. Kumlin, and R. White, *Science* **235**, 1616 (1987).
21. A. J. Jeffreys, V. Wilson, and S. L. Thein, *Nature (London)* **314**, 67 (1985).
22. A. J. Jeffreys, V. Wilson, and S. L. Thein, *Nature (London)* **316**, 76 (1985).
23. P. Gill, A. J. Jeffreys, and D. J. Werrett, *Nature (London)* **318**, 577 (1985).
24. A. J. Jeffreys, V. Wilson, S. L. Thein, D. J. Weatherall, and B. A. J. Ponder, *Am. J. Hum. Genet.* **39**, 11 (1986).
25. J. H. Wetton, R. E. Carter, D. J. Parkin, and D. Walters, *Nature (London)* **327**, 147 (1987).
26. T. Burke and M. W. Buford, *Nature (London)* **327**, 149 (1987).
27. J. L. Longmire, A. K. Lewis, N. C. Brown, J. M. Bucklingham, L. M. Clark, M. D. Jones, L. J. Meinke, J. Meyne, R. L. Ratliff, F. A. Ray, R. P. Wagner, and R. K. Moyzis, *Genomics* **2**, 14 (1988).
28. S. L. Thein, A. J. Jeffreys, H. C. Zooi, F. Cotter, J. Flint, N. T. J. O'Conner, D. J. Weatherall, and J. S. Wainscot, *Br. J. Cancer* **55**, 335 (1987).
29. A. J. Jeffreys, N. J. Royle, V. Wilson, and Z. Wong, *Nature (London)* **322**, 279 (1988).
30. G. Vassart, M. Georges, R. Monsievr, H. Brocas, A. S. Sequarre, and D. Christophe, *Science* **235**, 683 (1987).
31. L. L. Deaven, M. A. Van Dilla, M. F. Bartholdi, A. R. Carrano, L. S. Cram, J. C. Fuscoe, J. W. Gray, C. Hildebrand, R. K. Moyzis, and J. Perlman, *CSHSQB* **51**, 159 (1986).
32. R. S. Wells, E. W. Campbell, D. E. Swartzen Cruber, L. M. Holland, and P. M. Kraemer, *JNCI* **69**, 415 (1982).
33. P. M. Kraemer, G. L. Travis, G. C. Saunders, F. A. Ray, A. P. Stevenson, K. Bame, and L. S. Cram, *in* "Immune Deficient Animals" (B. Sordet, ed.), p. 214. Karger, Basel, Switzerland, 1984.
34. R. G. Worton and S. G. Grant, *in* "Molecular Cell Genetics" (M. Gottesman, ed.), p. 831. Wiley, New York, 1985.
35. G. Klein, *Science* **238**, 1539 (1987).
36. A. J. Jeffreys, V. Wilson, R. Kelly, B. A. Taylor, and G. Bulfield, *NARes* **15**, 2823 (1987).

37. J. W. Schumm, R. C. Knowlton, J. C. Braman, D. F. Barker, D. Botstein, G. Akota, V. A. Brown, T. C. Gravius, C. Helms, K. Hsiao, K. Rediker, J. G. Thurston, and H. Donis-Keller, *Am. J. Hum. Genet.* **42**, 145 (1988).
38. R. V. Lebo, D. R. Tolan, B. D. Bruce, M. C. Cheung, and Y. W. Kan, *Cytometry* **6**, 478 (1985).
39. T. C. Han and M. T. Zenzes, *JNCI* **32**, 857 (1964).

V. Retroviruses

A Retroviral Insertion in the Dilute (*d*) Locus Provides Molecular Access to This Region of Mouse Chromosome 9   207
NANCY A. JENKINS, MARJORIE C. STROBEL, PETER K. SEPERACK, DAVID M. KINGLSEY, KAREN J. MOORE, JOHN A. MERCER, LIANE B. RUSSELL, AND NEAL G. COPELAND

Spontaneous Germ-Line Ecotropic Murine Leukemia Virus Infection: Implications for Retroviral Insertional Mutagenesis and Germ-Line Gene Transfer   221
NEAL G. COPELAND, LESLIE F. LOCK, SALLY E. SPENCE, KAREN J. MOORE, DEBORAH A. SWING, DEBRA J. GILBERT, AND NANCY A. JENKINS

The Specific Consequences of c-*fos* Expression in Transgenic Mice   235
ULRICH RÜTHER AND ERWIN F. WAGNER

Mouse Endogenous Retroviral Long-Terminal-Repeat (LTR) Elements and Environmental Carcinogenesis   247
WEN K. YANG, L.-Y. CH'ANG, C. K. KOH, F. E. MYER, AND M. D. YANG

A Retroviral Insertion in the Dilute (*d*) Locus Provides Molecular Access to This Region of Mouse Chromosome 9

Nancy A. Jenkins[1]
Marjorie C. Strobel
Peter K. Seperack
David M. Kingsley
Karen J. Moore
John A. Mercer
Liane B. Russell* and
Neal G. Copeland

Mammalian Genetics Laboratory, Bionetic Research, Inc.–Basic Research Program, National Cancer Institute–Frederick Cancer Research Facility, Frederick, Maryland 21701, and
* Biology Division, Oak Ridge National Laboratory, Oak Ridge, Tennessee, 37831

I. Genetic Analysis of the *d* Locus

A. Characterization of the First *d* Allele

The original *dilute* (*d*) mutation is an old mutation of the fancy mouse and was among the coat-color variants first studied by mouse geneticists. This recessive mutation, when homozygous, is associated with a single phenotypic effect, a lightening of coat-color. The reduction of pigment in the coat hairs of *d*/*d* mice is not associated with a decrease in melanin synthesis, but is instead correlated with aberrant melanocyte morphology (*1*). Melanocytes isolated from mice wild-type at *d* display many thick dendritic processes extending from the cell, and the melanin granules (melanosomes) are dispersed along these processes. In contrast, melanocytes from mice homozygous for *d*

[1] Speaker.

have few dendritic processes, and most melanosomes clump around the nucleus. The abnormal melanocyte morphology is believed to inhibit proper interaction between the melanocyte and the hair bulb, resulting in irregular distribution of melanosomes within the hair shaft and the appearance of a dilution in coat color (2, 3). Additional studies with this d allele have suggested that the mutation is cell autonomous(4). When differentiating melanoblasts isolated from mice homozygous for d are transplanted to the neutral environment of the anterior chamber of the eye, they give rise to melanocytes with adendritic structures. In contrast, transplanted melanoblasts from mice wild-type at d differentiate into normal, highly dendritic melanocytes.

The first report of linkage between two mouse mutations was described in 1915 by Haldene et al., who noted that albino (c) and pink-eyed dilution (p) were linked (5). This relationship was subsequently named linkage group I (LG-I). The second linkage group (LG-II) to be identified involved the original d mutation and a variant termed *short-ear* (*se*) (6, 7). The recessive *se* mutation causes a marked decrease in the size of the external ear, due to an abnormal cartilage framework. This mutation is also associated with a number of other skeletal and soft-tissue abnormalities (8, 9). These early studies indicated that d and *se* are closely linked; current estimates suggest they reside within 0.16 centiMorgans (cM) of each other. More than 40 years after the linkage relationship between d and *se* was established, LG-II was assigned to mouse chromosome 9 (10, 11). In addition to being used in a variety of linkage studies, the original d mutation was also incorporated into a number of inbred strains. In fact, the first inbred strain to be derived, DBA, was selected to carry three coat-color mutations: *dilute, brown,* and *non-agouti*. This d allele has also been incorporated into a large number of inbred and recombinant inbred strains, as well as many linkage testing stocks.

B. Spontaneous Forward Mutations to *d*

In large breeding populations of mice, it is possible to detect forward mutations to dilute; it has been estimated that they occur at a rate of 1.2×10^{-5} events per gamete examined (12). These new alleles are usually detected in F_1 hybrid progeny from the cross of two inbred strains, one which carries d and is usually the DBA strain, and the other which is wild-type at d and is usually C57BL/6. The vast majority of these spontaneous d mutations are classified as *dilute–lethal* (d^l) alleles. Mice of the d^l/d^l genotype are indistinguishable in color from d/d mice, but they die about 14–21 days postpartum with a neuromuscular disorder resembling opisthotonus, which is characterized by a convulsive arching upward of the head and back. The affected cell types associated with this neurological lethality have not yet been clearly described. The postnatal lethality of d^l is itself recessive, since d^l/d mice do not exhibit the neuromuscular abnormalities but are dilute in color. Occasionally, mutations at the d locus are recovered that display, in addition to a dilution of color, a less severe neurological disorder. These alleles are generally viable, since the mice either recover from the neurological difficulties as they age, or the neurological abnormalities are very mild and persist.

C. Induced Forward Mutations to d

An important step in the analysis of the $d-se$ region of chromosome 9 was the incorporation of these mutations into tester stocks that were used in mutagenesis screens (13). In these screens, wild-type mice are treated with a potential mutagen (radiation or chemical) and are then crossed to a multiple-recessive tester stock homozygous for a number of phenotypic mutations, including d and se. In the first descendant generation, forward mutations at any of these loci can be identified. About 300 mutations at d and/or se have been recovered in these studies.

Some of the recovered mutations were phenotypically similar to the original d and se mutations carried in the tester stock. Another, probably heterogeneous, class of alleles, termed *dark–dilute* (d^x), determines a coat color intermediate between dilute and wild-type. Likewise, these experiments generated a class of se alleles (se^x) that specify an ear length intermediate between short ear and wild type. Similar to results of spontaneous mutation studies, the vast majority of d mutations recovered in this screen are d^l-like alleles and have been designated *dilute–opisthotonic* (d^{op}). Several unique classes of mutants also emerged from these radiation mutagenesis experiments. These include *dilute prenatal lethal* mutations (d^{pl}), which are indistinguishable from d in the genotype d^{pl}/d but are prenatally lethal when homozygous, and recessive *short-ear–lethal* (se^l) mutations, which are indistinguishable from se/se mice in the se^l/se genotype, but se^l/se^l embryos die before birth. Finally, a class of animals simultaneously mutant in d and se was identified. This last class of mutations, designated *Df* (*dse*), are also lethal when homozygous and are thought to be large deficiencies that remove the d and se loci as well as at least the chromosomal region between them (14).

The genetic complexity of the $d-se$ region was demonstrated by extensive complementation analysis, originally undertaken to obtain information about the chromosomal nature of mutations induced by different types of radiation treatment (14). More than 800 pairwise combinations of mutants were examined for their ability to complement the dilute, short-ear, neuromuscular, and prenatal lethality phenotypes. At least six factors responsible for prenatal lethality (*pl-1* through *pl-6*), two factors for juvenile survival (*nl-1*, *nl-2*), one factor for dilute (*d*), one factor for the opisthotonic phenotype (*op*), one factor for short ear (*se*), and one factor for Snell's waltzer (*sv*, a mutation specifying another type of neurologic disorder that maps 2 cM distally to *se*) were identified in these studies. With few exceptions, these factors can be placed on a linear complementation map (14; Fig. 1). It is important to note that the *d* and *op* phenotypes, where they coexist, cannot be separated by complementation analysis. They are shown as two functional units because mutations such as the original *d* mutation, which displays no neurological phenotype, exist. Thus, these complementation analyses have described an extensive panel of

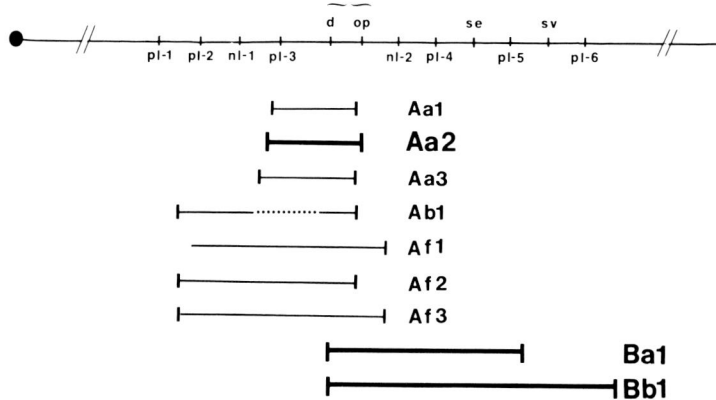

FIG. 1. This complementation map of *d–se* region of mouse chromosome 9 is drawn linearly, with the centromere (●) on the left. No correlation with physical distance is implied. Functional units include prenatal lethal factors 1 through 6 (*pl-1* through *pl-6*); neonatal lethal factors 1 and 2 (*nl-1* and *nl-2*); dilute (*d*); opisthotonus (*op*); short-ear (*se*); and Snell's waltzer (*sv*). Certain *d–se* region mutations, which have end points in *d*, are shown (14). The A series mutants correspond to d^{pl} alleles; the B series mutations are *Df (dse)* alleles. Members of the same complementation group are indicated by the same second-letter designation and individual members of a complementation group are identified by a numerical designation (24). Horizontal bars represent the extent of the presumed deficiency for mutants in each complementation group. Dotted lines represent regions "skipped" by a deficiency (14). Broader lines represent mutations for which the deletion breakpoint fusion fragment has been identified.

overlapping deficiency mutations that, in turn, define the location of several distinct genes that control diverse developmental pathways in the mouse.

D. Reverse Mutations at the *d* Locus

In large breeding populations of mice, it is possible to detect spontaneous reversions of the original *d* allele. The revertant (d^+) phenotype is dominant and is identified as an intensely colored mouse. The germinal reversion rate is estimated to be 4.5×10^{-6} events per gamete screened (15), a rate much higher than that for other recessive coat-color mutations that have been tested (12). Only one somatic reversion of *d* has been reported (15). It was identified as a mouse with a mottled coat, containing relatively equal numbers of mutant and revertant coat hairs. The frequency of somatic reversion of the original *d* mutation is quite low, 9×10^{-7} events per mouse screened. No reversion of any other *d* allele has been reported, although it is possible that too few mice carrying other *d* alleles are bred to detect a reversion event.

II. Molecular Analysis of the *d* Locus

A. Mechanism of Reversion of the Original *d* Allele

The unusually high germinal reversion rate of the original *d* mutation can be explained by the observation that this allele was generated by the integration of a mobile element into the *d* gene (16, 17). The element has been identified as a retrovirus, an ecotropic murine leukemia virus provirus termed *Emv-3*. The original *d* allele has thus been renamed d^v (dilute virally induced). Using virus-specific sequences as a probe, it was possible to clone molecularly the provirus and flanking cellular DNA sequences. Several unique regions of cellular DNA were identified, one of which is referred to as p0.3 (Fig. 2). Southern blot analysis of DNA prepared from mice homozygous for d^v, d^+, or the wild-type allele at the *d* locus revealed that reversion from d^v to d^+ resulted from excision of most, but not all, of *Emv-3* (17, 18). The d^+ chromosome retains a single copy of the viral long-terminal-repeat (LTR). The presence of a single LTR in revertant chromosomes was further confirmed by molecular cloning and DNA sequence analysis of two independent revertant sites (18). That a single LTR remains in each revertant d^+ chromosome examined to date suggests that proviral excision proceeds by homologous recombination mediated by the direct repeats (LTRs) flanking the

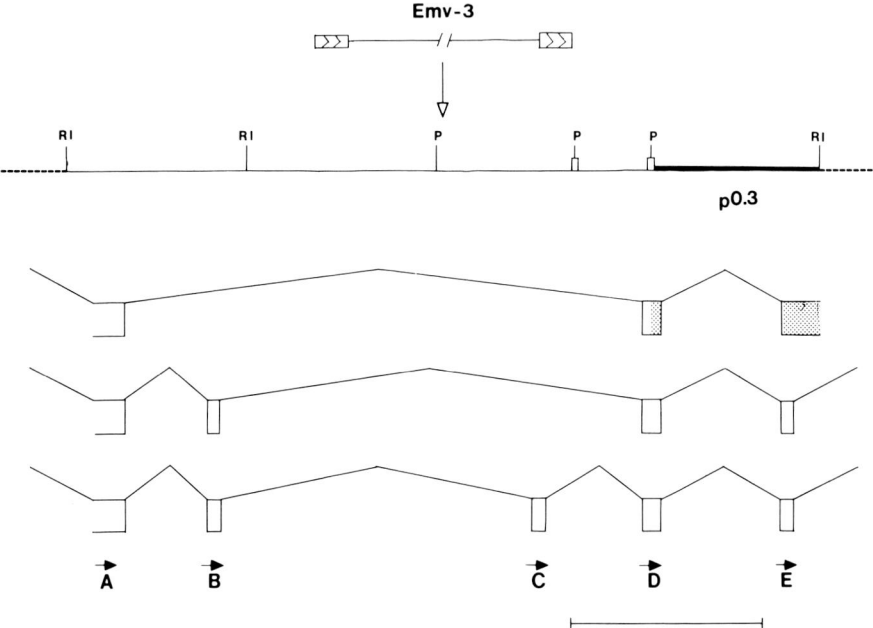

FIG. 2. Shown are splicing patterns identified at the d locus. The top line represents the *Emv-3* provirus, bounded by the LTR sequences. The direction of provirus transcription is from left to right. The location of the provirus in genomic DNA of the d^v allele is indicated by an arrow. A partial restriction map of the d gene is shown. Restriction sites shown include *Eco*RI (RI) and *Pst*I (P). The unique sequence genomic probe, p0.3, used in several studies described in the text, is indicated by a broader line. This probe is a 2.6 kb *Pst*I–*Eco*RI restriction fragment (17). Dotted lines represent d gene sequences not shown. The three splicing patterns identified from the analysis of six d locus cDNA clones is also shown. Coding exons are represented as nonstippled boxes; noncoding sequences are indicated by stippled boxes. The five exons that have been identified thus far are arbitrarily labeled A through E. The direction of d-gene transcription is from left to right, as indicated by the dark arrows. The deletion of exon D in the d^{120J} allele is shown (⊢——⊣). No correlation with physical size is implied.

viral structural genes. Additionally, Southern blot analysis of the DNA of the one somatic revertant of d^v indicated that, in this case, reversion proceeded by an intrachromosomal homologous recombination event (15).

B. Location of *Emv-3* in the *d* Gene

Since revertant mice are wild-type in coat color yet carry an LTR at the d locus, it was hypothesized initially that the *Emv-3* provirus induces the dilute mutation by inserting into noncoding sequences of the *d* gene. This was supported by DNA sequence analysis of the area immediately flanking *Emv-3*; no significant open reading frames were detected within 600 base-pairs (bp) of the provirus (*18*). Recent evidence confirms this hypothesis and demonstrates that the provirus is located within an intron of the *d* gene.

Initially, the genomic *d* locus probe, p0.3, was used in Northern blot analysis of RNA derived from the B16 melanoma cell line, which is wild-type at *d*. A melanoma cell line was used, since the *d* locus would be expected to be expressed in melanocyte derivatives. Transcripts were identified and a cDNA library was then prepared from the B16 cell line. The library was screened with p0.3 and one cDNA, 2.5 kilobases (kb) in length, was isolated. When this cDNA was used as a probe in Northern blot analysis of B16 RNA, three principal transcripts of 11, 9, and 7 kb were detected. While the developmental and tissue-specific expression pattern of the *d* gene is currently being determined, initial studies revealed that the adult brain is a rich source of *d*-gene transcripts. As shown in Fig. 3, Northern blot analysis of brain RNA from mice wild-type at *d* revealed the 11-, 9-, and 7-kb transcripts previously detected in B16 RNA.

Using specific regions of the initial cDNA as probes, five additional cDNA clones have now been isolated from the B16 cDNA library. The six cDNA clones can be aligned in a colinear fashion and collectively span about 6,000 bp. Nucleotide sequence analysis of certain regions of the cDNA clones and the corresponding genomic region around the viral integration site has been performed and the results can be summarized as follows (Fig. 2). The direction of transcription of the *d* gene is the same as that established for *Emv-3*, and three splicing patterns have been identified. In all cases, there is an open reading frame 5' to the virus (designated exon A in Fig. 2). In splicing pattern 1, exon A is spliced to a 199-bp exon that is 3' to the virus (exon D). The first 39 bp of exon D are in-frame; the remaining exon sequences presumably represent 3' noncoding sequences. In splicing pattern 2, exon A is spliced to a small 101-bp exon (exon B) that is 5' to *Emv-3*. The presence of exon B shifts the reading frame of exon D. This exon A, B, D transcript is also spliced to a small 75-bp exon (exon E); the entire sequence is coding. The third splicing pattern is similar to the second, except that an additional 81-bp exon (exon C), which maintains the reading frame, is included (Fig. 2). In all three cases, the *Emv-3* provirus is located within intron sequences. It remains to be determined precisely how the provirus induces the d^v mutation. However, the evidence presented below suggests that the provirus affects *d* gene transcription.

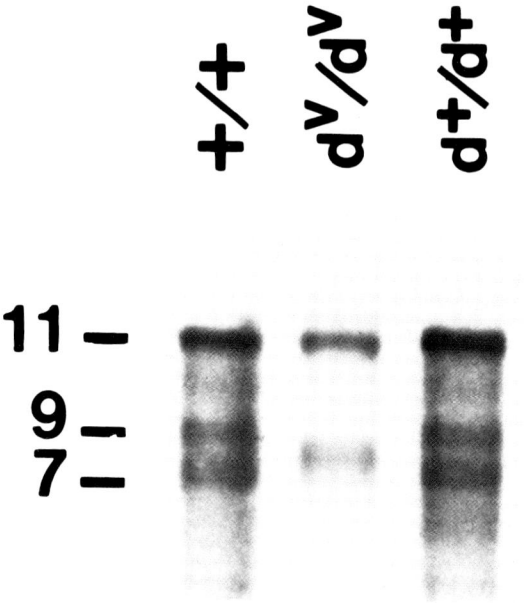

FIG. 3. Northern blot analysis of RNA derived from mice wild-type at d or homozygous for the d^v or d^+ alleles. Poly (A)+ mRNA was prepared from adult brain of +/+ (C57BL/6J), d^v/d^v (DBA/2J), and $d+$ (DBA/2J$-d^{+l8J}/d^{+l8J}$) mice and subjected to Northern blot analysis. The probe was the 2.5-kb cDNA described in the text (Section II). Transcript sizes were estimated relative to an RNA marker ladder.

C. Expression and Genomic Characterization of the *d* Locus

As mentioned above, three transcripts, 11, 9, and 7 kb in size, have been detected in RNA from mice and cell lines wild-type at d. Northern blot analysis of RNA from d^v/d^v mice revealed only two transcripts, of 11 and 8 kb (Fig. 3). The structure of these transcripts, as well as the mRNAs detected in mice wild-type at d, is currently being determined. However, the 11-kb RNA of d^v/d^v mice appears to be similar in size and abundance to that detected in wild-type animals (Fig. 3). The 8-kb transcript has not been detected in mice of other d-locus genotypes. These results suggest that the presence of the virus results in abolition of the normal 9- and 7-kb transcripts and the generation of an 8-kb RNA. Excision of the provirus in d^+ mice results in a wild-type phenotype that is correlated with the restoration of the expected 7- and 9-kb transcripts (Fig. 3). From these studies, we hypothesize that the product(s) of the 7- and/or 9-kb transcript(s) is necessary for normal melanocyte morphology, while the product of the 11-kb transcript is required for normal neurological function. This notion is supported further by several other observations. The 11-kb transcript is the predominant RNA detected in the PC12 cell line (data not shown), a rat adrenal pheochromocytoma cell line that responds to nerve growth factor by generating neuronlike processes (19). Preliminary *in situ* hybridization data suggest that, within the brain, the *d* gene is expressed primarily in neurons. Finally, characterization of a *dilute–lethal* allele (d^{l20J}), homozygotes for which display both the coat-color and neurological abnormalities, revealed an altered transcription pattern.

The spontaneous d^{l20J} mutation is the result of a simple, discrete deletion that begins 2 kb 3′ of the proviral integration site in the d^v allele and extends for approximately 3.6 kb, removing most of p0.3 (data not shown). The deletion results in aberrantly sized *d* transcripts that are present in greatly reduced levels. The only exon removed by the deletion is exon D (data not shown; Fig. 2). These observations suggest that the d^{l20J} mutation is functionally null. The findings that the d^{l20J} allele is caused by a 3.6-kb deletion in the *d* gene and that homozygotes for this allele display both the dilute coat-color and neurological abnormalities strongly suggest that both phenotypes are the consequence of mutation at a single locus (the *d* locus) rather than two closely linked genes. This is consistent with the aforementioned observation that when the *d* and *op* phenotypes coexist, they cannot be separated by complementation analysis.

In summary, we conclude that the *d* gene is quite large; the existing 6 kb of cDNA corresponds to more than 70 kb of genomic sequence. Since the largest mRNA detected to date is about 11 kb, the genomic locus may be on the order of 150 kb in size. Multiple transcripts are encoded by this locus; two may be required for normal melanocyte function, while one may function primarily in neuronal tissue. Mutation in various regions of the *d* gene can result in only a melanocyte effect (such as the d^v allele) or a melanocyte as well as a neural effect of varying severity. The true null mutation is represented by the *dilute–lethal* allele, d^{l20J}.

The nature of the *d*-gene product has not yet been determined. The protein predicted from the nucleotide sequence was not identical to any protein sequence in the available data bases. However, computer analysis revealed that the predicted protein contains many α-helices and may generate a rodlike protein similar to many cytoskeletal proteins. That the product of the d gene may be a component of the cytoskeleton is consistent with all observations made about this locus. A continuation of the combined molecular and genetic approach to the study of the d gene will allow a precise description of this locus, provide insight into how it is developmentally and temporally regulated, and reveal the nature of the d-gene product.

III. Dilute Suppressor

A gene previously reported to be recessive that is unlinked to d and can suppress the mutant phenotype of d^v/d^v animals has been identified (20). This mutation has been designated *dilute–suppressor* (*dsu*). Eukaryotic recessive suppressors have been identified in several different organisms, including *Drosophila* and yeast, and many of these suppressors have been shown to act specifically on retrotransposon-induced mutations. However, it had not been resolved whether *dsu* acts specifically on retrotransposon-induced mutations. Previous studies showed that *dsu* suppresses the coat-color phenotype but not the postnatal lethality of a spontaneous d^l allele; however, the DNA alteration in this mutation is not known (20). We have now extended the analysis of *dsu* by examining its action on other alleles of d and on genes that closely mimic the dilute phenotype.

To further evaluate the mechanism of suppression by *dsu*, we determined whether *dsu* can suppress the d^{l20J} mutation, which carries a small deletion in the d gene. Animals of the genotype a/a, d^{l20J}/d^{l20J}, *dsu*/*dsu* were generated and examined by visual inspection. The results indicated that *dsu* suppresses the coat-color phenotype but not the neurological defect and postnatal lethality of the d^{l20J} deletion allele (21). Similar findings with *dsu* have been reported for considerably longer d locus deletions made homozygous by combining d^{pl} mutations that complement for prenatal lethality (22). Another characteristic of the *dsu* gene, which distinguishes it from the eukaryotic recessive suppressor genes mentioned previously, is its mode of inheritance. Although *dsu* was originally reported to be recessive, our results show that it acts semidominantly; zero, one, and two doses of *dsu* can be identified phenotypically by their differential effects on d^v (21).

There are two other recessive mutations that resemble d: *ashen* (*ash*), which is closely linked (1 cM) to d on chromosome 9, and *leaden* (*ln*), which is located on chromosome 1, as is *dsu*. The genes *ash* and *ln* cause a dilution of coat color that is associated with an altered melanocyte morphology, identical to d^v. Genetic analysis revealed that both *ash* and *ln* are suppressed by *dsu* (23). In contrast, *dsu* did not suppress the dilution of coat color associated with 11 other recessive mutations that do not specifically affect melanocyte morphology.

The simplest interpretation of the experimental results obtained in these studies is that *dsu* produces a mutant protein that can substitute for the lack of the *d*-, *ash*-, or *ln*-gene product in melanocytes (but not neural tissue in the case of *d*) or, alternatively, that suppression results from the abnormal temporal or developmental expression of an otherwise normal *dsu*-gene product. Either model is compatible with the observation that *dsu* acts in a dosage-dependent manner and that the product of *dsu* on a wild-type genetic background has no effect on coat color. Consistent with either hypothesis, *dsu* does not appear to affect the level or sizes of *d* transcripts identified in tissues of d^{l20J} homozygous mice. The compensatory model of action of *dsu* suggests that *d*, *ash*, *ln*, and possibly *dsu* may be evolutionarily and functionally related. If so, probes obtained from various *d*-locus transcripts may cross-hybridize to *ash*, *ln*, and/or *dsu* transcripts, facilitating the identification and molecular characterization of these loci and our understanding of the nature of *dsu* suppression at the biochemical level.

IV. Strategies for Identifying and Characterizing Other Functional Units of the *d–se* Complex

A. Deletion Mapping Utilizing the Radiation-Induced Forward Mutations of the *d–se* Complex

The *d* alleles generated by radiation and thoroughly characterized by complementation analysis have been, and will continue to be, an important resource for studying the *d–se* region. In initial studies, genomic probes isolated from the *d* region surrounding the *Emv-3* integration site (including p0.3) were used in Southern blot analysis of DNAs prepared from four d^{op}, 11 d^{pl}, and 18 Df (*dse*) mutations (24). Five of 11 d^{pl} and all of 18 Df (*dse*) mutations examined were deleted for sequences 3' to *Emv-3*. This was the first direct physical proof that at least some of the *d–se* radiation-induced mutations were indeed deletions. However, five d^{pl} and all four d^{op} mutations carried no rearrangements or deletions detectable with the probes used in this analysis. More recently, using the cDNA probes in Southern blot analysis, it has been possible to detect additional rearrangements. These results are in agreement with the observation that the d locus is physically quite large. As probes representative of more of the locus are identified, it will be possible to detect additional structural alterations in the radiation-induced mutations.

Another important observation from the initial studies was the identification of a deletion breakpoint fusion fragment in a d^{pl} mutant referred to as Aa2 (24) (Figs. 1 and 4). The fusion fragment was identified in Southern-blot analysis as a unique restriction fragment. One end of the deletion breakpoint was in p0.3. The other end of the fusion fragment, referred to as p94.1, was subsequently cloned and used in Southern blot analysis of the radiation-induced mutations. The distribution of all of the original cloned probes (p94.1 and those 3' to the Emv-3 integration site) across the panel of d–se mutations allowed the placement of cloned DNAs into specific intervals on the complementation map. This allowed us to orient the emerging physical map with the functional map (24).

B. Future Directions

The identification of a deletion breakpoint fusion fragment is quite important for another reason, as well. That is, by cloning the end that is not in the d locus, it is possible to "jump" across the deficiency into regions of the d–se complex previously defined only at the level of complementation analysis. An example of the usefulness of this approach is described for the Aa2 mutation.

The Aa2 mutation has one end between nl-1 and pl-3 and the other end in the d locus (Figs. 1 and 4). As described earlier, the proximal end lies 1 kb to the right of probe p94.1, and the distal end lies within p0.3 (24). The p94.1 sequence can now be used in a series of experiments designed to recover sequences from nl-1 and pl-3. By "chromosome-walking" starting at p94.1, additional unique sequence genomic clones can be identified and characterized at a molecular genetic level. The cloned probes can also be sequenced and used in Northern blot analysis. Recent sequence analysis of p94.1 has indicated that there is an open reading frame associated with this DNA sequence. In Northern blot analysis, these sequences detect several mRNAs that are transcribed in the same direction as d (Fig. 4). These results suggest the p94.1 sequences are part of a gene, perhaps pl-3. Further molecular, genetic, and developmental studies will resolve this issue. Such work will be aided by the observation that there are two other members (Aa1 and Aa3) of the Aa2 complementation group available for analysis (Fig. 1).

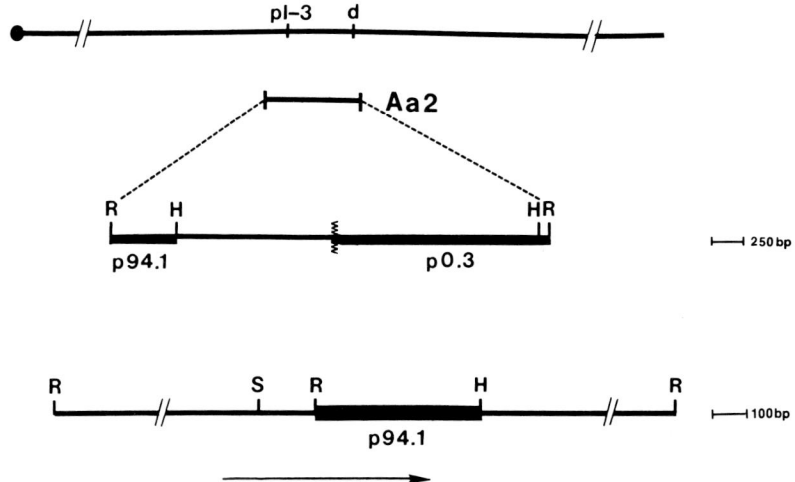

FIG. 4. The complementation map (Fig. 1) is again shown to emphasize that the Aa2 d^{pl} deficiency includes two functional units, pl-3 and d(op). A partial restriction map of the Aa2 deletion breakpoint fusion fragment is shown (24). The restriction sites include EcoRI (R) and HindIII (H). The locations of the unique sequence genomic probes, p94.1 and p0.3, are shown by darkened lines. The genomic sequences identified by p94.1 are shown in the bottom line. The restriction sites include EcoRI and HindIII, as described previously, as well as SalI (S). The direction of transcription is indicated by the arrow.

In principle, similar studies can be performed with all deficiency mutations that have end points in d. These include, but are not limited to, members of four other complementation groups: the d^{pl} groups include the Ab1 mutation and the Af1, Af2, and Af3 mutations. The Df (dse) groups include the Ba1 and Bb1 mutations (Fig. 1). To date, the deletion breakpoint fusion fragments associated with Ba1 and Bb1 have been identified and molecularly cloned. Such sequences should be very useful for identifying and characterizing pl-5 and pl-6 as well as Snell's waltzer (sv). Overall, these types of studies should eventually provide a precise description of genes of the d–se complex involved in normal development of the embryo, maintenance of neuromuscular function, and control of melanocyte, cartilage, and skeletal morphologies, as well as provide insight into how these genes are expressed in both time and space. Moreover, continued analysis of spontaneous and mutagen-induced d mutations will allow for a detailed study of the mutagenic effects of mobile genetic elements, radiation, and chemicals on the mammalian genome.

Acknowledgments

This research was sponsored by the National Cancer Institute, Department of Health and Human Services, under contract No. N01-CO-74101 with Bionetics Research, Inc., and the Office of Health and Environmental Research, U.S. Department of Energy, under contract DE-AC05-84OR21400 with the Martin Marietta Energy Systems, Inc. (L.B.R.). M.C.S. is the recipient of a postdoctoral fellowship from the National Institutes of Health (HD 07014). J.A.M. is a Leukemia Society of America fellow. We thank Linda Brubaker for typing the manuscript.

References

1. W. K. Silvers, in "The Coat-Colors of Mice," p. 83. Springer, New York, 1979.
2. E. S. Russell, *Genetics* **33**, 228 (1948).
3. C. L. Market and W. K. Silvers, *Genetics* **41**, 429 (1956).
4. C. L. Market and W. K. Silvers, in "Pigment Cell Biology" (M. Gordon, ed.), p. 241. Academic Press, New York, 1959.
5. J. B. S. Haldane, A. D. Sprunt, and N. M. Haldane, *J. Genet.* **5**, 133 (1915).
6. W. H. Gates, *Genetics* **13**, 170 (1928).
7. G. D. Snell, *PNAS* **14**, 926 (1928).
8. C. J. Lync, *Am. Nat.* **55**, 421 (1921).
9. M. C. Green, in "Genetic Variants and Strains of the Laboratory Mouse," p. 218. Gustav Fischer Verlag, New York, 1981.
10. O. J. Miller, D. A. Miller, R. E. Kouri, P. W. Allderdice, V. G. Dev, M. S. Grewal, and J. J. Hutton, *PNAS* **68**, 1530 (1971).
11. M. Nesbitt and U. Francke, *Science* **174**, 60 (1971).
12. G. Schlager and M. M. Dickie, *Mutat. Res.* **11**, 89 (1971).
13. W. L. Russell, *CSHSQB* **16**, 327 (1951).
14. L. B. Russell, *Mutat. Res.* **11**, 107 (1971).
15. P. K. Seperack, M. C. Strobel, D. J. Corrow, N. A. Jenkins, and N. G. Copeland, *PNAS* **85**, 189 (1988).
16. N. A. Jenkins, N. G. Copeland, B. A. Taylor, and B. K. Lee, *Nature* **293**, 370 (1981).
17. N. G. Copeland, K. W. Hutchison, and N. A. Jenkins, *Cell* **33**, 379 (1983).
18. K. W. Hutchison, N. G. Copeland, and N. A. Jenkins, *MCBiol* **4**, 2899 (1984).
19. L. A. Greene and A. Tischler, *PNAS* **73**, 2424 (1976).
20. H. O. Sweet, *J. Hered.* **74**, 305 (1983).
21. K. J. Moore, P. K. Seperack, M. C. Strobel, D. A. Swing, N. G. Copeland, and N. A. Jenkins, *PNAS* **85**, 8131 (1988).
22. L. B. Russell and C. S. Montgomery, *Genetics* **116**, s7 (1987).
23. K. J. Moore, D. A. Swing, E. M. Rinchik, M. L. Mucenski, A. M. Buchberg, N. G. Copeland, and N. A. Jenkins, *Genetics* **119**, 933 (1988).
24. E. M. Rinchik, L. B. Russell, N. G. Copeland, and N. A. Jenkins, *Genetics* **112**, 321 (1986).

Spontaneous Germ-Line Ecotropic Murine Leukemia Virus Infection: Implications for Retroviral Insertional Mutagenesis and Germ-Line Gene Transfer

> Neal G. Copeland[1]
> Leslie F. Lock
> Sally E. Spence
> Karen J. Moore
> Deborah A. Swing
> Debra J. Gilbert and
> Nancy A. Jenkins
>
> *Mammalian Genetics Laboratory, Bionetics Research, Inc.–Basic Research Program, National Cancer Institute–Frederick Cancer Research Facility, Frederick, Maryland 21701*

    Ecotropic murine leukemia viruses (MuLVs) provide important tools for the study of mammalian development. Not only can they act as germ-line insertional mutagens, facilitating the identification and cloning of genes important in mammalian development, but they can also serve as vectors for the efficient delivery of genes into mammalian cells. Many experimental approaches have been taken for incorporating these viruses into the mouse germ line, including infection of preimplantation embryos *in vitro*, postimplantation embryos *in vivo*, or pluripotent embryonic stem cell lines *in vitro* followed by their reintroduction into the mouse germ line via blastocyst injection (1). Although these procedures work, they are technically difficult and labor-intensive, limiting the usefulness of these approaches for viral mutagenesis studies.

    The development of inbred and/or hybrid strains of mice that spontaneously acquire new germ-line proviruses at high frequencies

---

[1] Speaker.

would provide a technically simple experimental system for viral insertional mutagenesis and may ultimately allow for selected mutagenesis similar to what is now being done in *Drosophila melanogaster* with P elements. The development of such a system has, however, been hindered by the low frequency at which ecotropic proviruses are spontaneously acquired in the mouse germ line and by our lack of knowledge concerning the mechanism and developmental stage of spontaneous germ-line provirus acquisition. Our identification of a hybrid mouse strain combination, SWR/J-RF/J, that has higher frequencies of spontaneous provirus acquisition than previously reported (2), has enabled the execution of experiments designed to delineate the mechanism and developmental stage of provirus acquisition and to evaluate the potential of this approach for retroviral insertional mutagenesis. The results of these experiments, described below, also have important implications for the use of retroviral vectors to introduce foreign genes into the germ line of mice and, perhaps, other mammalian species.

I. Organization, Distribution, and Stability of Endogenous Ecotropic Proviruses in Inbred Strains of Mice

The ability to discriminate endogenous ecotropic proviruses from the other related families of endogenous MuLVs carried by inbred mouse stains was made possible when Chattopadhyay *et al.* (3) identified and subcloned a small 400-base-pair (bp) fragment from the ecotropic virus envelope (*env*) gene that is ecotropic virus-specific (pEco). Using this probe, we characterized the endogenous ecotropic MuLV DNA content of 54 inbred strains and substrains of mice (4). In these studies, high-molecular-weight DNA was isolated from spleens, digested to completion with one of a number of different restriction endonucleases, and subjected to Southern blot analysis using probe pEco. One restriction endonuclease that has proven particularly informative is *Pvu*II, which cleaves twice within each ecotropic provirus, producing a single detectable 3' proviral DNA–cellular DNA junction fragment for each provirus. Among the strains analyzed, 12 were ecotropic provirus-negative. The other 42 strains carried from one to six proviral loci (Fig. 1; 4).

Analysis of the ecotropic proviral DNA content of a number of wild mouse species indicated that these proviruses are limited to one Asian species, *Mus musculus molossinus* (5). This suggests that these viruses entered the *Mus* germ line via Asian mice. These data also

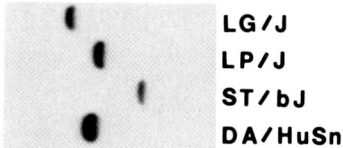
LG/J
LP/J
ST/bJ
DA/HuSn

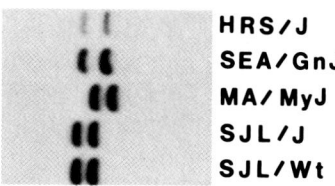
HRS/J
SEA/GnJ
MA/MyJ
SJL/J
SJL/Wt

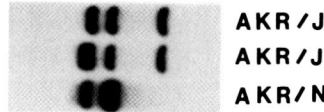
AKR/J
AKR/J
AKR/N

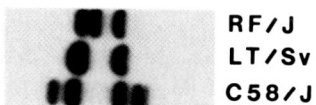
RF/J
LT/Sv
C58/J

I/LnJ
PL/J

FIG. 1. Ecotropic proviral DNA content of selected inbred strains and substrains of mice. High-molecular-weight DNA was isolated from spleens of the indicated strains. DNAs were digested with *Pvu*II and submitted to Southern blot analysis using probe pEco (ecotropic virus-specific envelope gene sequence).

support the conclusion that ecotropic proviruses carried by laboratory mouse strains were inherited from Asian mice, which are known progenitors of standard laboratory strains. This limited geographical distribution is also consistent with virus structural data suggesting that these proviruses were only recently introduced into the mouse germ line (6).

The expression of endogenous ecotropic MuLVs varies considerably among inbred strains: some strains express ecotropic virus early in life, while other strains express ecotropic virus only late in life or never at all (7). The variability in MuLV expression is the consequence of both *trans*- and *cis*-acting regulatory factors.

Important to the experiments described below, the ecotropic proviral DNA content of low viremic strains is relatively stable, while that of high viremic strains, such as AKR, is not stable (4). For example, compare the DNAs from the two AKR/J mice and the one AKR/N mouse shown in Fig. 1. Not only are differences observed in the ecotropic proviral DNA content of AKR/J and AKR/N mice, differences in the proviral content of different AKR/J mice are also apparent (the fastest migrating proviral fragment represented in the DNA of the second AKR/J mouse shown in Fig. 1 is a doublet). New ecotropic proviral loci are observed only in the progeny of viremic female mice, suggesting that provirus acquisition results from germ-line reinfection of virus produced in the female parent (8). Although it is difficult to obtain an accurate estimate of the provirus acquisition frequency in these studies, estimates have ranged from one new provirus fixed every 15 to 50 years of inbreeding (4). This frequency is sufficiently low to have precluded the execution of experiments to determine the mechanism and developmental stage of provirus acquisition, to identify factors that influence the frequency of proviral acquisition, or to institute large-scale retroviral insertional mutagenesis studies.

II. SWR/J-RF/J Hybrid Mice Spontaneously Acquire New Germ-line Ecotropic Proviruses at High Frequency

In 1985, Jenkins and Copeland (2) described a hybrid mouse strain combination that spontaneously acquires new germ-line ecotropic proviruses at frequencies considerably higher than those previously reported. This suggested that it might be possible to develop strains of mice that spontaneously acquire new germ-line ecotropic proviruses at frequencies sufficiently high to be useful for large-scale viral insertional mutagenesis studies. When viremic female SWR/J-RF/J hybrid mice carrying two closely linked RF/J ecotropic proviruses, *Emv-16* and *Emv-17*, were bred to ecotropic virus-negative SWR/J mice, as many as 19% of the progeny spontaneously acquired new proviral loci. The frequency of provirus acquisition was 0.35 proviruses per mouse. The ecotropic proviral loci acquired in these crosses have been designated *Srev* loci (SWR/J-RF/J ecotropic viral loci). The *Srev* content of some of these mice is shown in Fig. 2 (lanes 1–10). Similar numbers of *Srev* loci were acquired regardless of whether or not an animal carried *Emv-16* and *Emv-17*. Many mice carrying new proviral loci acquired more than one provirus (i.e., the mouse analyzed in lane 2 carried at least 10 *Srev* loci). The hybridization intensity of these loci relative to *Emv-16* and *Emv-17*, which are present at one copy per cell, suggested that they were present at less than one copy per cell.

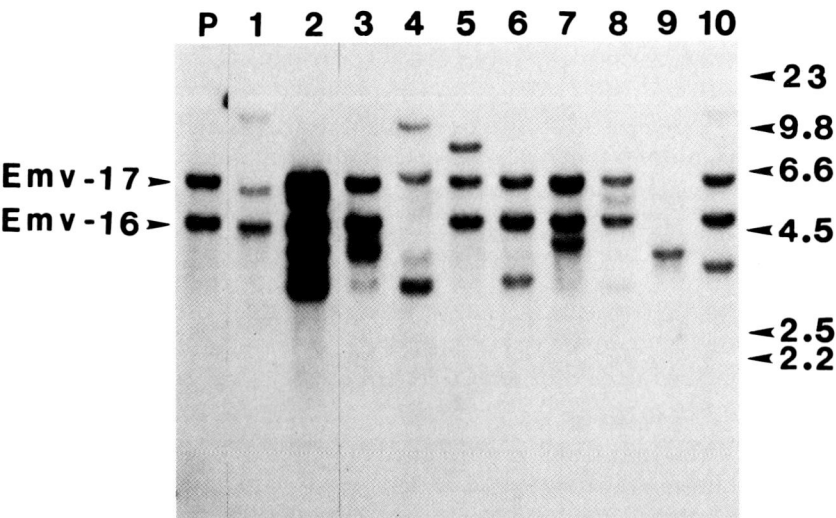

FIG. 2. Multiple *Srev* loci are detected in the progeny of SWR/J-RF/J hybrid mice. High-molecular-weight DNA was isolated from spleens of 10 SWR/J-RF/J hybrid mice carrying multiple newly acquired *Srev* loci (lanes 1–10) and an N_2 SWR/J-RF/J hybrid female parent carrying *Emv-16* and *Emv-17* (lane P). DNAs were digested with *Pvu*II and submitted to Southern blot analysis using probe pEco. The sizes in kilobases (kb) of ^{32}P-labeled *Hind*III-digested λ DNA marker fragments electrophoresed in parallel lanes of the same gels are indicated.

A. Srev Loci Equally Represented in All Somatic Tissues

Srev loci were originally identified in spleen DNA (Fig. 2). To measure the relative contribution of these loci to other somatic tissues, we screened DNAs from tissues representing all three germ layers for their Srev content. Representative results for DNAs of three mice are shown in Fig. 3. The number and relative contribution of Srev loci to each somatic lineage were virtually identical. The proviral copy number of new Srev loci measured by densitometric analysis of mosiac founder mice ranged from less than 0.04 to 0.74 proviral copies per cell; the average proviral copy number was 0.33 copies per cell (9). This suggests that these proviruses are acquired early enough in development that they subsequently become equally represented in all somatic tissues.

B. Srev Loci Transmitted through the Germ Line

The relative contribution of each Srev locus to the germ line was estimated by its transmission frequency. The majority (78%) of the proviruses identified in somatic tissue were transmitted through the germ line (Fig. 4). Transmission frequencies ranged from 6% to 46%; the average transmission frequency was 19% (9, 10). This frequency agrees well with the average somatic proviral copy number of 0.33 proviruses per cell. All proviruses transmitted through the germ line were detectable in somatic tissues. The somatic and germ-line content of new Srev loci calculated in these studies suggests that the majority of viral integration events giving rise to new Srev loci occur either after DNA replication in the zygote or before DNA replication in the two-cell embryo, and that integration precedes the divergence of the somatic and germ-line lineages.

C. Srev Loci Acquired by Extracellular Infection of Oocytes

Ovary transplantation studies have been used to demonstrate that Srev loci are acquired by extracellular virus infection and to define the developmental stage of provirus acquisition (11). In the first series of experiments, ovaries from SWR/J mice congenic for two wild-type coat-color alleles (A, agouti, and C, albino) donated by C57BL/6J mice (SWR.B6-A,C mice) were transplanted into the ovarian bursa of SWR/J mice congenic for Emv-16 and Emv-17 (SWR.RF-Emv-16,-17 mice). These mice were chosen for several reasons. First, SWR.RF-Emv-16,-17 mice spontaneously acquire new germ-line ecotropic

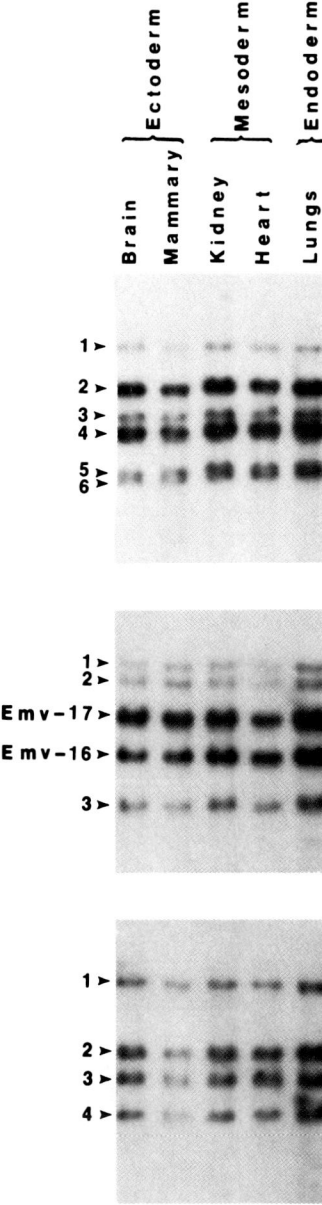

FIG. 3. *Srev* loci are equally represented in all somatic lineages. DNAs were isolated from five tissues representing each of the three possible germ layers from three mosaic SWR/J-RF/J hybrid mice carrying multiple new *Srev* loci. DNAs were digested with *Pvu*II and submitted to Southern blot analysis using probe pEco. The number of *Srev* loci observed for each animal is indicated to the left of each panel. The mouse whose DNAs are shown in the middle panel also inherited *Emv-16* and *Emv-17*, as indicated.

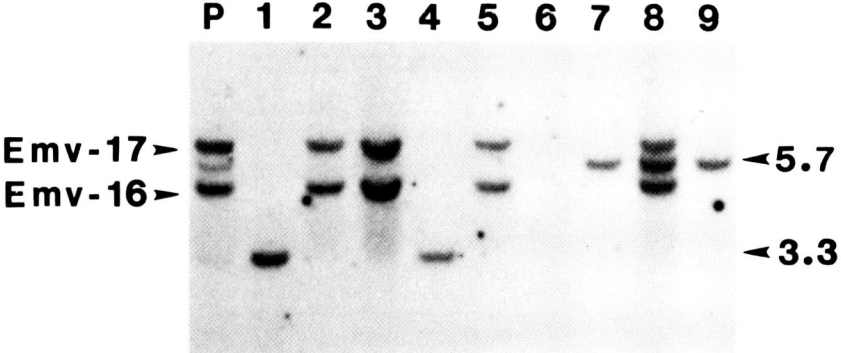

FIG. 4. Spontaneously acquired ecotropic proviral loci are transmitted through the germ line. DNAs from nine progeny (lanes 1–9) derived from a single mosaic *Srev* founder parent (lane P) were analyzed by Southern blot analysis for their *Srev* proviral DNA content. Fragments representing *Emv-16* and *Emv-17* are indicated at the left of the panel. The molecular weights (in kilobases) of the *Pvu*II-digested proviral DNA–cellular DNA junction fragments representing the two *Srev* loci segregating in this cross are indicated to the right of the panel.

proviruses at high frequency [0.11 new proviruses per mouse (9)]. Second, these two congenic strains should be identical at most, if not all, major and minor histocompatibility loci, thus allowing successful ovary transplantation. Third, discrimination of progeny derived from donor or residual host ovary can be made by visual inspection of coat color. Transplant recipients were mated to SWR/J male mice and the progeny were scored for their ecotropic proviral DNA content. If new proviral loci are detected in the progeny of the donor ovary, the proviruses must have resulted from extracellular virus infection, since neither the donor ovary nor the SWR/J male carry any endogenous ecotropic proviral loci. Of 10 donor ovary-derived progeny analyzed, four acquired new *Srev* loci (Table I; *11*). A total of 12 new proviruses were acquired, for a frequency of 1.2 new proviruses acquired per animal. This experiment provides definitive proof that *Srev* loci are acquired by extracellular virus infection.

Since the majority of viral integration events giving rise to new *Srev* loci appear to take place in the zygote or two-cell embryo (judged by proviral copy number in mosaic *Srev* founder mice), and the fertilized zygote is not thought to be susceptible to virus infection (*12*; R. Jaenisch, personal communication), it is likely that *Srev* loci are acquired by infection of oocytes either before or shortly after ovulation. To determine if the oocyte is a target for virus infection,

TABLE I
S<small>REV</small> L<small>OCI</small> A<small>CQUIRED BY</small> E<small>XTRACELLULAR</small> I<small>NFECTION OF</small> O<small>OCYTES</small>

Transplant	No. of donor ovary-derived progeny	No. of progeny with new proviruses	No. of new proviruses	No. of new proviruses/ No. of Donor Ovary-Derived Progeny
SWR.B6-*A,C*→SWR.RF-*Emv-16,-17*	10	4	12	1.2
SWR.RF-*Emv-16,-17*→C57BL/6NCr-*nu/nu*	39	2	6	0.15

SWR.RF-*Emv-16,-17* ovaries were transplanted into the ovarian bursa of C57BL/6NCr-*nu/nu* mice. These mice were chosen for several reasons. First, the immunodeficiency conferred by the nude (*nu*) mutation should allow for successful ovary transplantation. Second, these two strains differ at the *A* and *C* coat-color loci, allowing the discrimination of progeny derived from donor oocytes. Third, C57BL/6NCr-*nu/nu* mice do not carry any replication-competent *Emv* loci that could give rise to spontaneous germ-line infection events. Fourth, C57BL/6NCr-*nu/nu* mice carry a restrictive allele at the *Fv-1* locus (*Fv-1^b*), a dominant *trans*-acting locus that restricts replication of *Emv-16*, *Emv-17*, and related viruses. This restrictive allele should suppress the spread of virus from the donor ovary into the host oviduct. Transplant recipients were mated to SWR/J male mice and the progeny were analyzed for their ecotropic proviral DNA content. Since the only source of infectious virus is the donor ovary, and virus spread should be restricted in this host, the presence of new proviral loci in progeny would strongly indicate that germ-line virus infection occurred within, or very near, the donor ovary.

Of 39 donor oocyte-derived progeny analyzed, two carried new proviral loci (Table I). The frequency of provirus acquisition was 0.15 new proviruses per mouse. To confirm that virus spread from the donor ovary to the host oviduct does not occur, oviducts from a transplant recipient that had produced progeny carrying new proviral loci were analyzed by *in situ* hybridization for ecotropic virus expression. Ecotropic viral RNA was not detected above background levels in oviductal tissue, indicating that virus spread from the donor ovary had not occurred (L. Lock, unpublished results). These transplantation studies strongly indicate that the target of virus infection is the oocyte and that virus infection occurs within or very near the ovary. Since the frequency of provirus acquisition in donor oocyte-derived progeny was similar to that occurring spontaneously in SWR.RF-*Emv-16,-17* mice, it appears that the majority of new *Srev* loci arise by oocyte infection. However, this does not rule out the possibility that some *Srev* loci arise by embryo infection.

If infection does take place in oocytes, why are newly acquired proviral loci present at less than one copy per cell? If viral integration *in vivo* requires DNA synthesis, as has been proposed (*13, 14*), viral integration would be restricted in the oocyte, since it is arrested in the first meiotic prophase. Unintegrated proviral DNA may be synthesized in infected oocytes; however, integration would proceed only following the DNA synthetic phases that occur in the early cleavage stages of embryogenesis. Alternatively, infection may take place just before or after ovulation, but the time required for viral DNA synthesis and integration to occur is sufficiently long that integration occurs only after DNA synthesis in the zygote.

D. Ecotropic Viral RNA Expressed in the SWR.RF-*Emv-16,-17* Female Genital Tract

Ecotropic virus must be expressed within the SWR.RF-*Emv-16,-17* ovary to provide a virus source for oocyte infection. *In situ* hybridization studies have confirmed that this is indeed the case (*11*). Sections of SWR.RF-*Emv-16,-17* ovaries were hybridized with a ^{35}S-labeled antisense ecotropic virus-specific RNA probe. Virus expression was detected in the theca, a specialized epithelial cell layer that surrounds the growing follicle. Viral RNA was detected in lower amounts in some cells of the corpora lutea and the ovarian stroma. Viral RNA was not detected in the granulosa cells. The theca cells are spatially separated from the oocyte by the granulosa cell layer and its associated basal lamina. Thus, the exact route of virus infection of the oocyte is unclear.

As expected, viral RNA was not detected in the ovaries of ecotropic virus-negative SWR/J mice. Virus expression was also not detected in the ovaries of RF/J mice, the parental donor of *Emv-16* and *Emv-17*. RF/J mice carry genetic as well as nongenetic factors that inhibit ecotropic virus expression (*15*). This inhibition of virus expression could be responsible for the low frequency of spontaneous germ-line virus infection observed in RF/J mice. Presumably, genetic transfer of *Emv-16* and *Emv-17* onto the permissive SWR/J strain background relieves this inhibition, resulting in high-frequency germ-line virus infection.

III. *Srev* Proviruses Acting as Insertional Mutagens

Spontaneous integration of ecotropic MuLVs into the mouse germ line has already been shown to be mutagenic. The recessive dilute (*d*) coat-color mutation carried by DBA/2J mice results from spontaneous germ-line viral integration (*16*). Among all of the *Srev* loci screened to date for visible mutations, no visible phenotypic mutations have been observed. However, only a limited number of *Srev* loci have been screened and the frequency of visible mutations predicted from studies of transposon-induced mutations in other species is quite low (*17*). In contrast, the frequency of recessive lethal mutations produced by transposon insertion can be quite high. For example, the frequency of recessive lethal mutations resulting from P-element insertion in *Drosophila melanogaster* is approximately 10% (*17*). To determine the frequency of recessive lethal mutations associated with *Srev* loci, we have intercrossed heterozygous *Srev* mice representing each of 18 different *Srev* loci (*9*). Failure to generate homozygous *Srev* progeny, as measured by proviral hybridization intensity, was used as the screen for prenatal lethal recessive mutations.

Of 18 *Srev* lines analyzed, one (*Srev-5*) carried a recessive lethal mutation (*9*). Preliminary studies indicate that homozygous *Srev-5* progeny die between 8.5 and 13.5 days of gestation (S. Spence, unpublished results). Similar frequencies (5%) of recessive lethal mutations induced by viral integration following preimplantation embryo infection *in vitro* have been observed (*18*; R. Jaenisch, personal communication). This high frequency of recessive lethal mutations induced by spontaneous ecotropic viral integration indicates that the approach to viral insertional mutagenesis described here is a viable strategy for inducing developmental mutations in the mouse.

IV. Gene Susceptibility Determinants Affecting the Frequency of Germ-Line Virus Infection

The frequency of spontaneous germ-line ecotropic virus infection in CBA/CaJ-RF/J hybrid mice is at least 20-fold lower than in SWR/J-RF/J hybrid mice (*10*). This is surprising since CBA/CaJ, like SWR/J, is ecotropic virus-negative and appears to be highly susceptible to exogenous ecotropic virus infection. This differential frequency of spontaneous germ-line infection suggests that host genetic factors can affect the frequency of spontaneous germ-line proviral acquisition.

Virally encoded determinants also appear to affect profoundly the proviral acquisition frequency. For example, Moloney-MuLV (Mo-MuLV) has recently been introduced into the SWR/J germ line by preimplantation infection of SWR/J or SWR/J hybrid embryos (*19*; R. Jaenisch, personal communication). Among the progeny of several SWR/J-derived *Mov* lines subsequently screened, no spontaneous germ-line infection events have been detected. Likewise, very few spontaneous germ-line virus infection events have been detected in many thousands of BALB/c- or C57BL/6-*Mov* lines analyzed (R. Janenisch, personal communication). Many differences exist among *Emv* and Mo-MuLV isolates. *Emv-16* and *Emv-17* are known germ-line parasites (i.e., they already have demonstrated their ability to infect the mouse germ line spontaneously), whereas Mo-MuLV is a laboratory isolate. The tissue-tropism of these viruses also differs. *Emv*-encoded viruses are apathogenic and have a broad tissue-tropism. In contrast, Mo-MuLV is thymotropic and rapidly induces T-cell lymphomas in infected hosts. Elucidation of the viral as well as host genetic determinants that influence the frequency of provirus acquisition may make it possible to significantly increase the frequency of spontaneous germ-line provirus acquisition, thus facilitating this approach for the study of mammalian development.

V. Spontaneous Germ-Line Virus Infection following Subcutaneous Infection of Newborn SWR/J Mice

The results described above suggest that the SWR/J oocyte is sensitive to ecotropic virus infection. To investigate whether exogenously administered virus can also infect the SWR/J germ line, newborn SWR/J mice were injected subcutaneously with high-titer virus stocks. Viruses used for inoculation included *Emv-16*- and *Emv-17*-encoded virus as well as AKV623-encoded virus. AKV623 is similar in structure to *Emv-16* and *Emv-17*, except that it has a duplication in the enhancer located in the viral long-terminal-repeat that slightly increases the titers of infectious virus produced *in vitro* and *in vivo*.

In situ hybridization of ovaries and oviducts of 6- to 9-week-old females infected subcutaneously as newborns demonstrated that ecotropic viral RNA is expressed in the genital tract in a pattern indistinguishable from that observed in SWR.RF-*Emv-16,-17* mice

(11). When viremic females were mated to SWR/J males, the resultant progeny acquired new germ-line proviruses at high frequency (11, 20). In our study, the frequency of germ-line provirus acquisition in AKV623-infected mice was similar to the spontaneous frequency observed in SWR.RF-*Emv-16,-17* mice, whereas the frequency of provirus acquisition in *Emv-16,-17*-virus-infected mice was about a fourth of that. While it is still too early to know whether the higher titers resulting from AKV623 infection will extrapolate into higher frequencies of germ-line infection, these studies demonstrate that the SWR/J oocyte can be infected efficiently following subcutaneous virus infection.

VI. Future Prospects

The ability to infect the mouse germ line by subcutaneous infection of newborn mice should greatly simplify experiments to identify host and viral genetic determinants affecting the frequency of provirus acquisition. The use of genetically marked retroviruses that induce visible phenotypic changes in the host, such as a change in coat color, would obviate the need for Southern blot analysis to identify germ-line infection events. This ability, combined with an increase in the frequency of germ-line infection, should make this approach a method of choice for inducing new developmental mutations in the mouse. Derivation of replication-competent *Emv*-related retroviral vectors should facilitate the introduction of selected DNA sequences into the mouse germ line by subcutaneous virus infection. Helper-free packaging cell lines expressing *Emv*-based defective retroviral vectors may, following their introduction into the ovarian bursa of female SWR/J mice, also facilitate the introduction of defective retroviral vectors into the mouse germ line.

Finally, we are in the process, in collaboration with John Eppig (The Jackson Laboratory, Bar Harbor, Maine), of determining whether mouse oocytes in culture are susceptible to exogenous virus infection. If oocytes are susceptible, these experiments should make it possible to define the stage(s) of oocyte growth and maturation susceptible to virus infection. This approach may also facilitate the introduction of foreign DNA into the germ line of other mammalian species.

The introduction of foreign DNA into the germ line of large mammals by microinjection is often restricted by the low number of zygotes available for microinjection. Further, the egg cytoplasm, which often obscures the pronuclei from view, also makes microinjection difficult in many mammalian species. Conversely, the genomes of all large mammals analyzed in any detail have been found to contain large numbers of retrovirus-related sequences. Presumably, at least some of these sequences were acquired by oocyte infection. Since large numbers of oocytes of many developmental stages can be obtained from a single animal, the introduction of foreign DNA sequences into the mammalian germ line by oocyte infection may ultimately simplify germ-line gene transfer in many mammalian species.

Acknowledgments

This research was supported by the National Cancer Institute, Department of Health and Human Services, under Contract N01-CO-74101 with Bionetics Research, Inc. L.F.L. is supported by Damon Runyon–Walter Winchell Cancer Fund Fellowship DRG-918. We thank Robin Handley for typing this manuscript.

References

1. A. Gossler, T. Doetschman, R. Korn, E. Serfling, and R. Kimler, *PNAS* **83**, 9065 (1986).
2. N. A. Jenkins and N. G. Copeland, *Cell* **43**, 811 (1985).
3. S. K. Chattopadhyay, M. R. Lander, and W. P. Rowe, *PNAS* **77**, 5774 (1980).
4. N. A. Jenkins, N. G. Copeland, B. A. Taylor, and B. K. Lee, *J. Virol.* **43**, 26 (1982).
5. C. A. Kozak and R. R. O'Neill, *J. Virol.* **61**, 3082 (1987).
6. S. King, J. Horowitz, and R. Risser, *Virology* **157**, 542 (1987).
7. J. Coffin, in "RNA Tumor Viruses: Molecular Biology of Tumor Viruses" (R. Weiss, N. Teich, H. Varmus, and J. Coffin, eds.), p. 410. CSHLab, Cold Spring Harbor, New York, 1982.
8. W. P. Rowe and C. A. Kozak, *PNAS* **77**, 4871 (1980).
9. S. E. Spence, D. J. Gilbert, D. A. Swing, N. G. Copeland, and N. A. Jenkins, *Mol. Cell Biol.* (in press) (1989).
10. V. Bautch, *J. Virol.* **34**, 373 (1986).
11. L. F. Lock, E. Keshet, D. J. Gilbert, N. A. Jenkins, and N. G. Copeland. *EMBO J.* **7**, 4169 (1988).
12. W. Baranska, W. Sawicki, and H. Koproski, *Nature (London)* **230**, 591 (1971).
13. H. E. Varmus, T. Padget, S. Heasly, G. Simon, and J. M. Bishop, *Cell* **11**, 307 (1977).
14. H. E. Varmus, P. R. Shank, S. E. Hughes, H.-J. Kung, S. Heasly, J. Majors, P. K. Vogt, and J. M. Bishop, *CSHSQB* **43**, 851 (1979).
15. A. Mayer, F. Duran Struuck, M. L. Duran-Reynals, and F. Lilly, *Cell* **19**, 431 (1980).
16. N. A. Jenkins, N. G. Copeland, B. A. Taylor, and B. K. Lee, *Nature (London)* **293**, 370 (1981).
17. L. Cooley, R. Kelley, and A. Spradling, *Science* **239**, 1121 (1988).
18. P. Soriano, T. Gridley, and R. Jaenisch, *Genes Dev.* **1**, 366 (1987).
19. P. Soriano and R. Jaenisch, *Cell* **46**, 19 (1986).
20. J. Panthier, H. Condamnie, and F. Jacob, *PNAS* **85**, 1156 (1988).

The Specific Consequences of c-*fos* Expression in Transgenic Mice

ULRICH RÜTHER[1] AND
ERWIN F. WAGNER

European Molecular Biology
Laboratory, 6900 Heidelberg,
Federal Republic of Germany

Proto-oncogene products are thought to play an important role in cellular differentiation and development (for reviews, see *1* and *2*). The proto-oncogene c-*fos* has been the subject of extensive analysis and the results suggest that the c-*fos* gene product exerts different functions in different biological processes (*3, 4*).

Expression of c-*fos* during development has been demonstrated in extraembryonic tissues in mice and humans (*3, 4*). In addition, *in situ* hybridization experiments show c-*fos* expression in the growth regions of fetal bone of day-17 mouse embryos (*5*). In adults, only hemopoietic cells like macrophages and neutrophils express c-*fos* mRNA at high levels (*6, 7*). Although a large number of *in vitro* studies demonstrate that *fos* expression can be induced by various growth factors (*8, 9*) and is required for cell proliferation (*10, 11*), the function of c-*fos* in development is still unknown.

Recent studies have shown that the *fos* protein, localized in the nucleus, can function as a transcription factor in yeast, and is part of a transcription complex in an adipocyte differentiation system (*12, 13*). The *fos* protein complex recognizes the binding site for the transcription factor AP-1, which is one of the major transcription factors in mammalian cells (*14–17*). These observations suggest that the *fos* protein can modulate the expression of certain genes. To study the role of c-*fos* in development and to define the genes regulated by *fos*, we have generated transgenic mice expressing different c-*fos* gene constructs in a wide range of tissues. In this review, we summarize our data on c-*fos* expression in transgenic mice and discuss the influence of c-*fos* on the development of certain tissues.

[1] Speaker.

I. Metallothionein (MT)-c-*fos* Mice

In the first series of experiments, we introduced the murine c-*fos* gene under the control of human MT promoter into fertilized eggs. Seven independent transgenic lines were analyzed for exogenous c-*fos* expression. Transgene expression was undetectable, except for a low level of expression in testis and pancreas in two lines (Table I). Injection of mice with cadmium chloride ($CdCl_2$) had no effect on the expression of exogenous c-*fos* mRNA. However, when transgenic mice were injected with cycloheximide, a protein synthesis inhibitor known to stabilize c-*fos* expression *in vitro*, significant levels of exogenous c-*fos* transcripts were found in most tissues analyzed, with highest levels observed in testis and pancreas (18). This result suggests that the transgene is transcribed in several tissues and that an accumulation of c-*fos* mRNA is prevented by posttranscriptional regulatory mechanisms. Since the mRNA destabilizing sequence seems to be positioned in the 3' nontranslated region of c-*fos* (19), we next used a DNA construct in which this region was exchanged with a retroviral (LTR) (20). This construct, p76/21, can induce morphological transformation in NIH/3T3 and rat 208F cells and can therefore be defined as a transforming gene.

II. MT-c-*fos*LTR Mice

Five independent transgenic mouse lines were established from eggs injected with p76/21. High levels of exogenous c-*fos* RNA were detected in two of these lines, in particular, in the pancreas, kidney, heart, brain, muscle, and bone (Table I). Upon induction with $CdCl_2$, the steady-state RNA level increased nearly tenfold in most tissues. We have never detected expression of exogenous c-*fos* RNA in the liver, spleen, or thymus of these mice. The transgenic mice expressing c-*fos* developed characteristic swellings on the long bones of the legs as early as 2–3 weeks after birth, predominantly in males (18). A histopathological analysis revealed marked disturbances of bone genesis, characterized by bone marrow fibrosis and enhanced formation of new bone. The coexistence of lamellar old bone and osteoblast-deposited osteoid and irregular woven bone suggested also that bone remodelling occurred in an unregulated way, with a dominance of bone formation. The specific bone swellings did not increase in size beyond the fourth week after birth; however, in older animals, characteristic bone tumors, mainly chondrosarcomas, were observed (Fig. 1). On the average, 15% of the transgenic animals developed

TABLE I
EXPRESSION OF DIFFERENT CONSTRUCTS IN TRANSGENIC MICE[a]

Construct	No.[b]	S	P	L	T	H	Lg	G	K	M	B	Sg	Bo	Tu
MT-c-fos	7	−	±	−	−	−	−	±	−	−	−	−	ND	
MT-c-fosLTR	2	−	++	−	−	+++	+	+	++	++	+++	+	++	+++
H2-c-fosLTR	2	+	−	±	+	+++	+	−	++	++	+++	+	++	+++
H2-c-fos	5	+++	±	+	+++	+	+++	+	±	±	−	+++	+	

[a] Shown are the relative abundances of exogenous c-fos RNA expression. S, spleen; P, pancreas; L, liver; T, thymus; H, heart; Lg, lungs; G, gonads; K, kindeys; M, skeletal muscle; B, brain; Sg, salivary glands; Bo, bone; Tu, bone tumor; ND, not determined.
[b] Number of independent mouse lines (integration sites). For the H2-c-fosLTR construct, only founder animals could be tested.

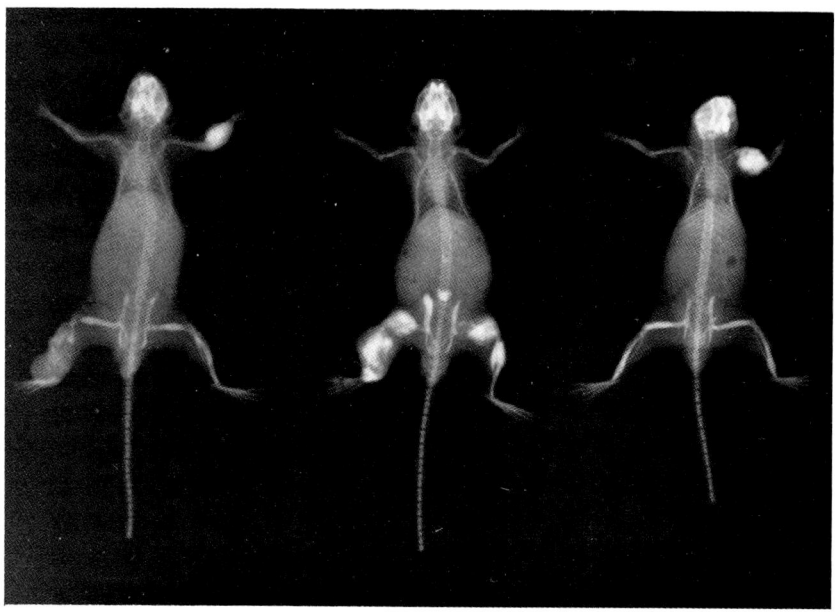

FIG. 1. X-rays of three MT-c-*fos*LTR mice. Shown are transgenic mice which have developed multiple bone tumors by the age of 8–12 months.

tumors when the mice were more than 9 months old. A clear sex preference was again detected, as 34 of 43 animals were males.

It is worth noting that none of the other tissues in the transgenic mice revealed any abnormalities when histologically analyzed. Thus, the bone cells appear to be the only target for *fos*-induced disturbances in MT-c-*fos*LTR mice.

III. H2-c-*fos*LTR Mice

We were interested to know whether the specificity of the *fos* effect in the MT-c-*fos*LTR mice was due to the altered *fos* construct or due to the promoter specificity, and therefore exchanged the human MT promoter against the murine H2-K^b class I MHC promoter. This H2 promoter is active in a broad range of cell types in transgenic mice (*21*).

Using this construct, only two founder animals were obtained; both failed to produce transgenic offspring. However, both animals exhibited a more pronounced phenotype of bone lesions; in particular, in animal No. 288-5, these lesions occurred in almost all bones in the body (Fig. 2). The histological examination revealed extensive bone alterations with numerous preneoplastic lesions, but all other tissues were normal. Both founder animals developed chondrosarcomas with characteristics similar to those observed in the MT-c-*fos*LTR mice. We also analyzed the expression pattern of the transgene in these mice and found stable expression of exogenous c-*fos* in almost all tissues analyzed (Table I). These findings suggest that the specificity of the *fos* effects on bone development is independent from one particular construct.

IV. H2-c-*fos* Mice

In the MT-c-*fos* mice, we obtained expression of *fos* only when the protein synthesis inhibitor cycloheximide had been injected. Since the expression profiles of the MT and H2 genes are distinctly different, we wanted to find out whether in tissues expressing H2 the accumulation of c-*fos* mRNA is also prevented by a posttranscriptional mechanism. If this is not the case, we could investigate the consequences of *fos* in hemopoietic tissues, in which we never found

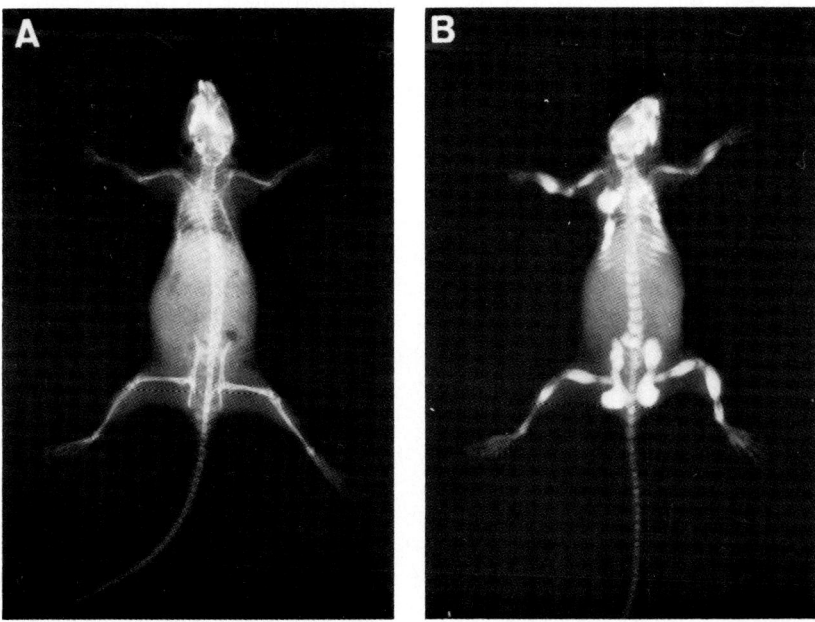

FIG. 2. X-rays of a H2-c-*fos*LTR transgenic mouse. (A) Control mouse; (B) transgenic mouse No. 288-5, with drastic bone alterations.

expression of *fos* in the MT-c-*fos*LTR mice. Therefore, we generated five independent transgenic mouse lines with a H2-c-*fos* construct without the alteration on the 3' nontranslated part of the c-*fos* gene (22). All of these lines expressed the exogenous c-*fos* gene at high levels in hemopoietic organs such as the spleen and thymus, and at low levels in some other organs like the liver, heart, and kidney (Table I). These mice have enlarged spleens and hyperplastic thymuses with an increased number of thymic epithelial cells. The development of T cells in these thymuses is affected, which is indicated by an increased fraction of mature thymocytes. However, this altered distribution of T-cell subsets is not a direct effect of c-*fos* expression within the T-cell lineage. Transfer of transgenic bone marrow into a normal donor mouse resulted in normal differentiation of transgenic thymocytes (22). It appears that c-*fos* specifically stimulates the proliferation of thymic epithelial cells and affects T-cell development indirectly. In the spleen and lymph nodes, no changes in the proportion of hemopoietic cell lineages are detectable and none of these mice have yet developed a lymphoid malignancy.

Histological analysis of all tissues of these mice revealed that the thymus is the only organ with a change of its morphology (22). It appears that c-*fos* expression specifically influences the proliferation of thymic epithelial cells, which has distinct consequences for the immune system. B- and T-cell function is impaired and the H2-c-*fos* mice are immune-deficient as indicated by reduced antibody titers.

V. AP-1/c-*jun* Expression

Although we have general several transgenic mouse lines expressing exogenous c-*fos* in about 12 different tissues (Table I), only two tissues (the bone and thymic epithelium) seem to be affected. These findings suggest that c-*fos* may need additional factor(s) for its action, and these might be present only in these two cell types. For the adipocyte differentiation system, c-*fos* and AP-1 are part of a transcription complex that binds to an identical DNA sequence (13). This suggests that both proteins might interact in the regulation of gene expression. Recently, several groups demonstrated that the c-*jun* proto-oncogene encodes the transcription factor AP-1 (23, 24). We analyzed the expression pattern of c-*jun* in mice. As shown in Fig. 3, c-*jun* is expressed in every organ of the mouse except the pancreas. Two transcripts of 2.7 and 3.2 kb are present as described (24), with the highest level of expression in the lungs. This result suggests that AP-1/c-*jun* is not the limiting factor for the action of *fos*.

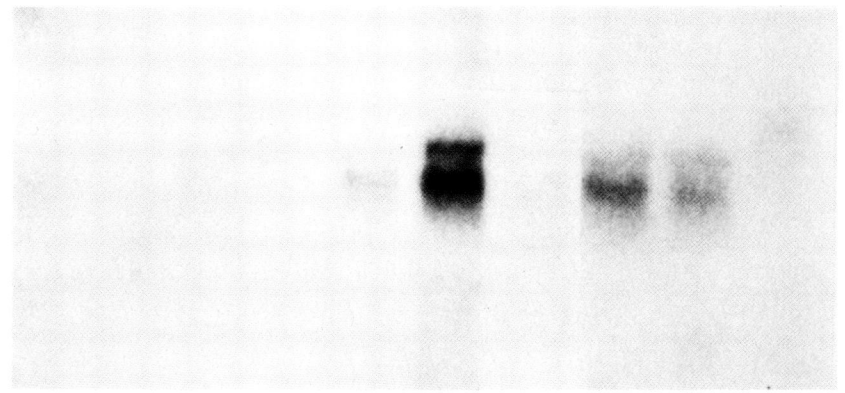

FIG. 3. c-*jun* expression in various tissues of the mouse. RNA was isolated from an adult mouse and probed with a *c-jun*-specific sequence. S, spleen; P, pancreas; L, liver; T, thymus; H, heart; Lg, lungs, G, gonads; K, kidneys; B, brain; Sg, salivary glands.

VI. Discussion

In this review, we have shown that overexpression of c-*fos* from different constructs in transgenic mice generates distinct phenotypes that can be described at the molecular or cell biological level. First, stable expression from the human MT promoter is only possible when destabilizing sequences of the 3' noncoding part of the mRNA are removed. Second, stable expression of exogenous c-*fos* in several tissues seems to result in induction of proliferation of either bone or thymic epithelial cells. The specificity of the induction is controlled by the construct used. Each construct that contained the 3' LTR in place of the destabilizing sequences affected bone development. Third, specific tumors arise only in mice where bone cell proliferation is influenced.

A. The 3' LTR of c-*fos* Transgenes Seems to Define Its Expression Specificity

In Table I, we have summarized the relative abundances of c-*fos* RNA expression from the different constructs. The expression profile of H2-c-*fos* resembles the expression of endogenous H2 class I genes. Highest expression is detected in the spleen, thymus, lungs, and salivary glands. However, expression of c-*fos* constructs containing the 3' LTR with the same H2 promoter results in a pattern of expression very similar to that of the MT-c-*fos* LTR construct. Highest expression of RNA is seen in the heart and brain; intermediate levels are found in the kidney, muscle, and bone. In addition, only with these two constructs (MT-c-*fos*LTR and H2-c-*fos*LTR) have we observed the effects on bone development and their progression to bone tumors in which high levels of c-*fos* RNA could also be detected. These results imply that the 3' LTR in these two constructs may be responsible for the specificity of expression and for the observed effects. There is good evidence that retroviral LTRs introduced into the germ line of mice become irreversibly inactivated (25). In our analysis, we have indeed never found a transcript initiated from the 3' LTR, but our data suggest that the enhancer of the LTR might function, thereby influencing the expression specificity from the H2 and MT promoter.

In transgenic mice in which the c-*fos* gene is fused to the human MT promoter, stable expression was obtained only when the 3' end of the gene was modified. However, in H2-c-*fos* mice, the transgene is highly expressed without manipulation of the 3' end. This discrepancy could be attributed to the fact that a different cell type expresses the c-*fos* gene in the same organs. Alternatively, the different 5' end of the RNA and the promoter choice could influence the expression levels.

B. Cell-specific Induction of Proliferation by *fos*

Our investigations of the consequences of deregulated c-*fos* expression in transgenic mice have identified the bone-forming cells and the thymic epithelium as target cells where *fos* seems to induce proliferation. In older animals, a progression to specific bone tumors was observed. Recently, it has been shown that *fos* is part of a protein complex which binds to the AP-1 consensus sequence (13). This AP-1 binding site is identical to the TPA-responsive element (TRE) defined as the mediator of gene activation through the tumor promoter TPA (26). The human collagenase gene contains such TRE/AP-1 binding site, and it was shown that *fos* is sufficient and necessary for the activation of the collagenase gene (27).

Based on these data, we propose the following interpretation for our observations in the transgenic mice. Overexpression of *fos* results in activation of TPA-inducible genes containing TRE elements. Due to the differentiation stage of the cell, these activated subsets of genes might vary between several cell types. Activation of TPA-inducible genes does not necessarily lead to a mitogenic response, as shown in several studies (for a review, see 28). However, this induced state is comitogenic, and a fully mitogenic response can be achieved through so-called progression factors like epidermal growth factor (29). Assuming that these factors are cell-type specific, we can postulate that the tissue-specific enhanced proliferation by *fos* is due to the cooperativity of *fos* acting as a competence factor like TPA with cell-type-specific progression factors (Fig. 4). In cell types in which high levels of *fos* are detectable but no influence on proliferation is seen, the specific progression factor is missing. We can exclude that AP-1 is solely responsible for the tissue specificity and the effect on proliferation since AP-1, which possibly forms a transcription complex with *fos*, is present in almost every tissue, as shown in this report.

Thymus and bone are very different in their growth characteristics in the adult stage of an animal. The thymus is no longer a proliferating tissue, and decreases in size in an adult mouse. On the other hand, there is, of course, no additional growth of bones in an adult; however, a permanent bone remodeling with osteoclast-mediated bone resorption and osteoblast-mediated bone formation takes place. The balance of this process seems to be influenced by *fos* expression, since we seem to observe an induction of proliferation of bone-synthesizing cells (hyperplastic bones). If we assume that the effect of *fos* on growth control predisposes cells to tumor formation, than the bone-forming cells, active during the whole life of an animal, should become malignant after a latency period (in contrast to the quiescent thymus epithelial cells). Indeed, in 15% of the MT-c-*fos*LTR mice, we

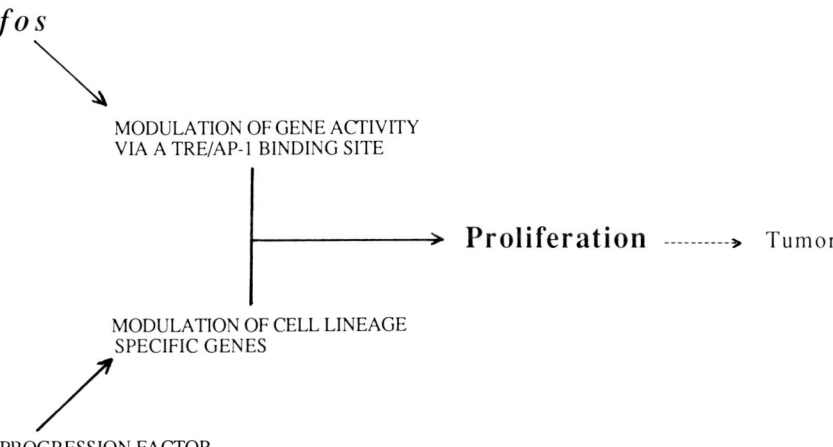

FIG. 4. Cell-type-specific proliferation by *fos*. Shown is a model in which the influence of *fos* protein on proliferation is illustrated.

have seen the development of tumors that were always associated with bone tissue, and we never observed a thymoma in the H2-c-*fos* mice. The bone tumors are reminiscent of the osteosarcoma-like tumors induced in newborn mice by the v-*fos* transducing retroviruses FBJ-MSV and FBR-MSV (3). The tumors develop much more quickly here than in the transgenic mice, in particular, upon infection with the FBR-MSV virus. The set of genes that becomes *trans*-activated by the v-*fos* protein might be different and thereby the effect on growth control is more pronounced. It is remarkable that the target tissue for tumor development for transgenic c-*fos* or retroviral v-*fos* is the same.

A final statement concerning the specificity of *fos* action in transgenic mice and a conclusion about its role in development certainly require additional experiments. To further ascertain that every cell type in a mouse produces *fos* protein we are generating new c-*fos* constructs with strong housekeeping and inducible promoters that should already be active during development. In parallel, we are interested in obtaining mouse lines with mutant *fos* proteins (30) as well as with an inactivated *fos* gene produced via homologous recombination. These approaches should help us to delinate the role *fos* protein has in development, as we should be able to screen the genes that are either activated or repressed by *fos* and so define the molecular mechanism at certain developmental stages.

Ackowledgments

We would like to acknowledge the valuable contribution of our colleagues Christa Garber, Dymitri Komitowski, Rolf Müller, Werner Müller, Thomas Pohl, Klaus Rajewsky, Takayuki Sumida, and Takeshi Tokuhisa in this work and we thank Jacqueline Reid for typing the manuscript.

References

1. P. Kahn and T. Graf (eds.), "Oncogenes and Growth Control." Springer-Verlag, Berlin, 1986.
2. E. D. Adamson, *Development* **99**, 449 (1987).
3. R. Müller, *BBA* **823**, 207 (1986).
4. I. M. Verma, *Trends Genet.* **2**, 93 (1986).
5. C. Dony and P. Gruss, *Nature (London)* **328**, 711 (1987).
6. T. J. Gonda and D. Metcalf, *Nature (London)* **310**, 249 (1984).
7. H. Kreipe, H. I. Radzun, K. Heidorn, C. Marder, and M. R. Parwaresch, *J. Histochem. Cytochem.* **35**, 837 (1987).
8. I. M. Verma and P. Sassone-Corsi, *Cell* **51**, 513 (1987).
9. R. Treisman, *Cell* **46**, 567 (1986).
10. I. T. Holt, T. Venkat Gopal, A. D. Moulton, and A. W. Nienhuis, *PNAS* **83**, 4794 (1986).

11. K. Nishikura and J. M. Murray, *MCBiol* **7**, 639 (1987).
12. K. Lech, K. Anderson, and R. Brent, *Cell* **52**, 179 (1988).
13. F. J. Rauscher III, L. C. Sambucetti, T. Curran, R. J. Distel, and B. M. Spiegelman, *Cell* **52**, 471 (1988).
14. C. Setoyama, R. Frunzio, G. Liau, M. Mudry, and B. de Crombrugghe, *PNAS* **83**, 3213 (1986).
15. W. Lee, A. Haslinger, M. Karin, and R. Tjian, *Nature (London)* **325**, 368 (1987).
16. B. R. Franza, F. J. Rauscher III, S. F. Josephs, and T. Curran, *Science* **239**, 1150 (1988).
17. F. C. Lucibello, M. Neuberg, J. B. Hunter, T. Jenuwein, R. Wallich, B. Stein, A. Schönthal, P. Herrlich, and R. Müller, *Oncogene* **3**, 43 (1988).
18. U. Rüther, C. Garber, D. Komitowski, R. Müller, and E. F. Wagner, *Nature (London)* **325**, 412 (1987).
19. A. D. Miller, T. Curran, and I. M. Verma, *Cell* **36**, 51 (1984).
20. U. Rüther, E. F. Wagner, and R. Müller, *EMBO J* **4**, 1775 (1985).
21. D. Morello, G. Moore, A. M. Salmon, M. Yaniv, and C. Babinet, *EMBO J*. **5**, 1877 (1986).
22. U. Rüther, W. Müller, T. Sumida, T. Tokuhisa, K. Rajewsky, and E. F. Wagner, *Cell* **53**, 847 (1988).
23. D. Bohmann, T. J. Bos, A. Adamson, T. Nishimura, P. K. Vogt, and R. Tjian, *Science* **238**, 1386 (1987).
24. P. Angel, E. Allegretto, S. Okino, K. Hattari, W. J. Boyle, T. Hunter, and M. Karin, *Nature (London)* **332**, 166 (1988).
25. D. Jähner, H. Stuhlmann, C. L. Stewart, K. Harbers, J. Löhler, I. Simon, and R. Jaenisch, *Nature (London)* **298**, 623 (1982)
26. P. Angel, M. Imaguwa, R. Chiu, B. Stein, R. J. Imbra, H. J. Rahmsdorf, C. Jonat, P. Herrlich, and M. Karin, *Cell* **49**, 729 (1987).
27. A. Schönthal, P. Herrlich, H. J. Rahmsdorf, and H. Ponta, *Cell* **54**, 325 (1988).
28. I. M. Verma, R. L. Mitchell, W. Kruijer, C. Van Beveren, L. Zorkas, T. Hunter, and J. A. Cooper, *Cancer Cells* **3**, 275 (1985).
29. C. D. Scher, R. C. Shepard, H. N. Antoniades, and C. D. Stiles, *BBA* **560**, 217 (1979).
30. T. Jenuwein and R. Müller, *Cell* **48**, 647 (1987).

Mouse Endogenous Retroviral Long-Terminal-Repeat (LTR) Elements and Environmental Carcinogenesis

Wen K. Yang[1]
L.-Y. Ch'ang*
C. K. Koh
F. E. Myer and
M. D. Yang

Biology Division, Oak Ridge National Laboratory, Oak Ridge, Tennessee 37831, and
* Department of Microbiology, University of Tennessee, Knoxville, Tennessee 37971

I. Perspective

For the past several years, our working hypothesis has been that chromosomal retrovirus-related gene elements play important roles in gene rearrangement and gene activation events of carcinogenesis induced by environmental agents. This working hypothesis is based on the concept of transposable genes as well as the recent understanding of retroviruses (RNA tumor viruses) in relation to carcinogenesis.

Activation of transposable gene elements was discussed by McClintock (1) from the viewpoint of unprogrammed genomic changes in response to unanticipated genomic shocks. She was also said (2) to apply this view in considering the possibility of transposable gene elements being involved in genetic changes in carcinogenesis. In this regard, this concept is similar to the perspectives of RNA tumor viruses (3, 4), the oncogene–virogene hypothesis (5), and the provirus hypothesis (6) on the basis that retroviruses replicate through DNA forms that carry LTR sequences resembling the insertion sequences (IS) or prokaryotic transposons (7, 8).

The findings of oncogene *myc* activation in leukemogenesis induced by avian leukosis virus (9) and proviral insertion in the dilute locus mutation in mice (10) point to the functional similarity between

[1] Speaker.

retroviruses and transposable genes. The fact that viral oncogenes are actually transduced cellular genes under regulatory control by the LTR (*11*) also demonstrates the potential of retroviruses to undergo genetic recombination and rearrangement with genes important for regulating cell growth and/or differentiation. Furthermore, production of type-C retroviruses is induced in some mouse embryo culture cells following exposure to such chemicals as halogenated deoxyuridines (*12, 13*), indicating that proviruses of these retroviruses may be present in "silent" but activatable form in the mouse genome. Many chemical and physical agents that can activate the mouse endogenous type-C retroviruses are carcinogenic (*14*).

On the other hand, no authentic cancer-causing retrovirus has been isolated from human cancers. A careful evaluation of the information collected in the 1970s, did not establish a direct cause–effect relationship between the known endogenous retrovirus isolates and carcinomas that occur naturally or from environmental factors (*15*). The value of virus-isolation approaches for solving the human cancer problem has also been questioned (*16, 17*). However, with the recent development of the "retrotransposon" concept (*18*), it appears that, while retroviruses represent a special case of retrotransposon evolution, many mobile gene elements in the eukaryotic genome are likely to go through the replicative transposition cycle of transcription–reverse transcription–insertion without a virus stage. Thus, the involvement of retrotransposable gene elements in chromosomes in certain stages of the multistep and multifactorial mechanism of carcinogenesis remains a valid working hypothesis, although very little is known about these elements in mammalian genomes.

In our experimental approach, we have placed major emphasis on retroviral LTRs, as they contain both the enhancer-promoter–polyadenylation signal sequences for gene transcription and the short inverted-repeat sequences for gene insertion. A chromosomal element carrying two LTRs and adjacent priming sequences would be capable of retrotransposition, if provided with the functions of RNA polymerase, reverse transcriptase, insertase, and other *trans*-acting factors for transposition. The mouse genome was chosen for our LTR studies as considerable knowledge has been accumulated concerning murine leukemia viruses in relation to leukemogenesis. With our own research experiences (e.g., *19*), it is possible to investigate LTR-containing gene elements in the mouse germ line that are related but not expressed in an extracellular virus form. Also, various inbred mouse strains have been developed for genetic studies of carcinogenesis and mutagenesis.

It has long been noted that cultured mouse cells show rapid karyotypic changes in parallel with "immortal" growth *in vitro*. The instability of chromosome structure presumably reflects a particular sensitivity of the mouse cell to environmental changes *in vitro*. Whether this is similar to the cancer-prone anomalies of chromosome instability in humans would be worth studying. In this regard, activation of LTR-containing retrotransposable gene elements with subsequent insertion into new chromosomal sites could potentially cause genomic instability. This possibility may be examined experimentally by the isolation and functional characterization of "insertase" protein encoded in the 3' portion of the *pol* gene of these elements, with subsequent tests for the ability of this insertase protein to cause chromosome breakage in the cell.

II. Molecular Cloning and Sequence Characterization of Retroviral LTR-Containing Gene Families

We have previously elucidated the LTR structures of a naturally occurring murine leukemia virus (MuLV) isolate from BALB/c mice (20) and a chemically induced MuLV isolate from RFM/Un mouse cells (21). With MuLV LTR sequences as probes, we isolated by molecular cloning many mouse chromosomal DNA fragments that harbor LTR-homologous sequences (22–24). Sequence characterization of these molecular clones has established the presence of two other distinct retrovirus-like gene families in addition to the well-known murine leukemia provirus family in the mouse genome (Fig. 1). Distinct features of the latter two retroviral LTR gene families are briefly described in the following section.

A. Mu-LV-Related Proviral Sequences

Most of the 30–50 dispersed sequences per haploid mouse genome detectable by molecular probes of MuLV are not the proviruses of ecotropic MuLVs (25, 26) or xenotropic MuLVs (27), but rather belong to a distinct family of proviral sequences in the mouse germ line. Other than being apparently incapable of producing viruses directly, these MuLV-related proviral sequences can be distinguished from ecotropic and xenotropic MuLV proviruses mainly by four characteristic sequence features.

First, their *env* genes are of the mink-cytopathic-focus (MCF) virus specificity (24, 27, 28). This is significant, as MCF viruses have been considered to be intimately associated with lymphomatogenesis

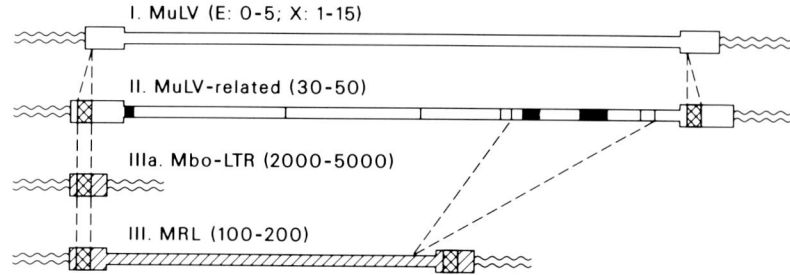

FIG. 1. Sequence relationship of LTR and proviral structures among three retrotransposable gene families in the mouse genome. The approximate copy numbers per haploid mouse genome are indicated in parentheses. E, Ecotropic; X, xenotropic.

mechanisms in the mouse (29). MCF viruses and another class of oncogenic retroviruses, spleen-focus-forming viruses, are presumably generated by recombination between ecotropic MuLV and the MuLV-related proviral sequences, with the latter providing the oncogenic potential.

Second, there is an Fv-1 B-tropism determinant in the gag-p30 region of all MuLV-related proviral clones that we examined (30). Nearly all ecotropic MuLV proviruses in mouse chromosomes are N-tropic and hence may be prevented from inducing leukemogenesis by the $Fv-1^b$ host genetic factor in certain laboratory mouse strains (19). However, if these ecotropic MuLVs gain B-tropism by recombination with a MuLV-related proviral sequence, the host Fv-1 restriction against leukemogenesis will become ineffective.

Third, the primer binding site for initiation of reverse transcription in MuLV-related proviral sequences is complementary (22, 23) to the 3' 18-nucleotide sequences of the major glutamine tRNA (31) or minor UAG suppressor glutamine tRNA (32). As shown in Fig. 2, this is distinctly different from most horizontally infectious retroviruses, such as MuLV and feline leukemia viruses, which utilize proline tRNA as the primer for reverse transcription (33, 34). MuLV-related proviral sequences thus appear to represent a distinct lineage of the retroviral reverse transcription mechanism.

Fourth, because of an extra 190- to 200-nucleotide segment inserted between the promoter area and the putative enhancer area in the U3 region (22, 35), the LTRs of MuLV-related proviral sequences are larger than the approximately 530-nucleotide basic LTR structures of ecotropic and xenotropic LTRs.

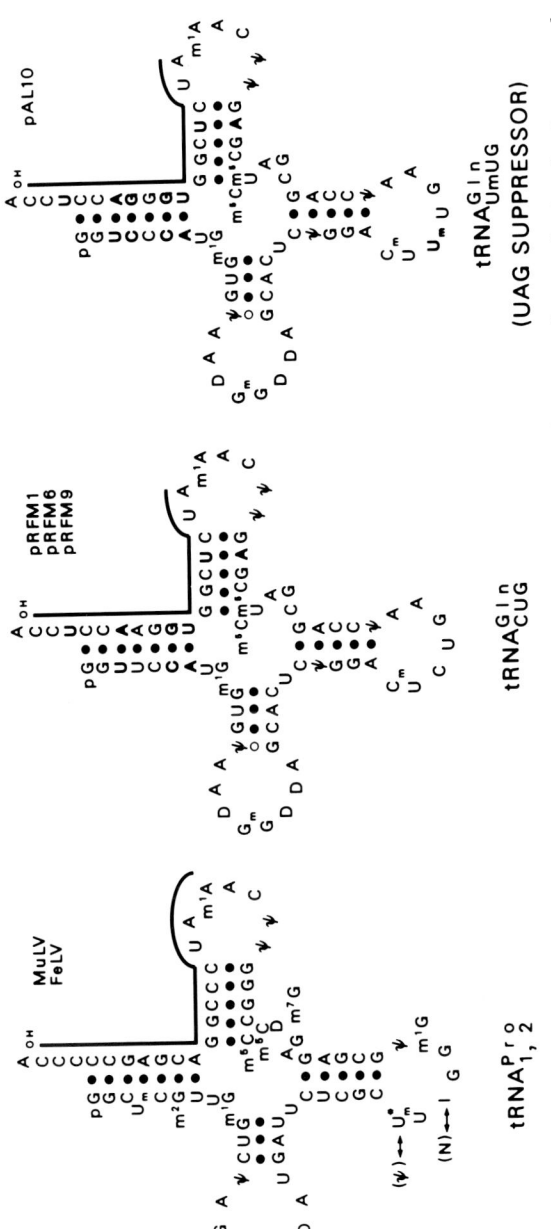

FIG. 2. Shown are three primer tRNA structures. Proline tRNA$_{1,2}$ is the primer for most infectious leukemia viruses, such as murine (MuLV) and feline (FeLV) leukemia viruses. Glutamine tRNA$_{CUG}$ and UAG suppressor glutamine tRNA$_{UmUG}$ (32) are implicated to be the primer for endogenous MuLV-related proviral clones, pRFM1, 6, and 9 and pAL10, respectively, whose primer binding site sequences are complementary to the 3' 18 nucleotides of the tRNAs (indicated by lines).

Stoye and Coffin (28) recognized that chromosomal MuLV-related proviral sequences can be further divided into two classes, called polytropic and modified polytropic, which can be distinguished by specific *env* gene sequences and different restriction enzyme sites in the proviral structure. We have found that the LTR members of the polytropic class of proviral structures have ~700-nucleotide LTRs, whereas the LTRs of the modified polytropic class have ~750 nucleotides. The latter contain an extra 37-nucleotide segment at the immediate 3' side of the enhancer area. As the distinct sequence features are at and surrounding the enhancer area, regulation of gene expression by the MuLV-related LTRs is likely to differ significantly from that by ecotropic or xenotropic viral LTRs.

Restriction-fragment-length polymorphism of chromosomal DNA with respect to the integrated MuLV-related proviral sequences has been noted among inbred strains of laboratory mice (36, 37). With unique cellular flanking sequences of isolated MuLV-related proviral clones as molecular probes, we observed proviral integration at these specific chromosomal DNA sites in a few, but not all, inbred strains of laboratory mice (22, 24). These observations suggest that germ-line integration of MuLV-related proviral sequences occurred rather recently in the evolution of the mouse genome. This is significant since some ecotropic MuLV proviruses in unique chromosomal loci of AKR and C58 strain mice are considered to be a predisposing factor for high leukemia incidences in these mice (39, 40). Different strains of laboratory mice carry MuLV-related proviral sequences at distinct chromosomal locations. The modified polytropic class of MuLV-related proviral sequences, like ecotropic and xenotropic MuLV proviruses, has been detected only in genomes of *Mus musculus domesticus* and other subspecies, but not in genomes of other *Mus* species. The polytropic class MuLV-related proviral sequences, while found predominantly in the *Mus musculus* genomes, is also present in *Mus spretus*.

B. *Mbo*l Repeat LTR (MRL) Elements

As indicated in Fig. 1, the approximately 200-nucleotide-sequence segment found in the MuLV-related LTR is also present in another distinct family of LTRs, numbering at least 2000 to 5000 dispersed copies in the genome of laboratory mice. About 95% of these LTRs are solitary, while the remaining 5% are associated with proviral structures. As these LTRs commonly contain two internal *Mbo*I restriction enzyme recognition sites, one within the 5' inverted repeat and the other close to the 3' border in the U3 region, the generic name "MboI-repeat LTR" or MRL, is used to include both the 2000–5000 copies of the solitary LTR and the 100–200 copies of the proviral structure. It should be mentioned that two other names, "LTR-IS" and MuRR, have been given, respectively, to the solitary LTR and the proviral structure (41, 42).

The nucleotide sequence of MboI-repeat LTRs is enclosed by TGAAAAGATCC/GGAGTCTTTCA inverted repeats. Its 5' half comprises mainly the distinct 200-nucleotide segment found also in the MuLV-related LTR; its 3' half, containing the "TATA" box, the mRNA cap site, and the polyadenylation signal, shows little homology to the MuLV and related LTRs. Two sizes of MboI-repeat LTRs are found in the mouse genome, one ~500 and the other ~600 nucleotides long. The size difference is mainly due to the 5' portion enclosed by the two MboI sites.

The nucleotide sequences of a few isolated MRL proviral clones have been determined (42, and our unpublished data). Although they obviously differ from the known murine leukemia virus genomes, discernible low degrees of sequence homology suggest that these MRL proviral structures contain the gag and pol genes but not the env gene. Also, several apparent base deletions within the gag–pol sequences caused shifts of the reading frame and appearance of the termination codons, implicating the defective nature of thes isolated MRL proviral structures. However, use of the gag–pol sequences of these MRL clones as molecular probes shows that the MRL proviral structures in the mouse genome are highly polymorphic. It is thus possible that, while most of the MRL proviral structures are defective, some may be functionally intact.

Certain MRL proviral structures are present only in the chromosomal DNA prepared from male mice. The male-specific MRL elements are detected as two EcoRI DNA fragments of about 14 kilobase-pair (kbp) and 3.1 kbp in mouse strains AKR, NFS, and RFM, while they appear as two EcoRI DNA fragments of 8.5 kbp and 3.1 kbp in mouse strains BALB/c, C57BL/6, and C3H. The restriction enzyme digestion patterns indicate that they differ from other retroviral genes previously reported to be located on the mouse Y chromosome (reviewed in 40). Also, the hybridization intensities of the 14- and 8.5-kbp EcoRI fragments suggest that these elements, as well as their flanking sequences, are amplified at least severalfold in the chromosomal DNA of adult males. Some of these amplified male-specific proviral sequences have been isolated by molecular cloning in our laboratory. Unique nucleotide sequences exist at the reverse transcriptase–

endonuclease gene junction within the *pol* gene and also in the U3 region of the LTR (L.-Y. Ch'ang *et al.*, unpublished results).

Although most abundant in the genomes of laboratory mice (*Mus musculus domesticus*), both the LTR and the retroviral structural gene components of MRL are also present in the genomes of other *Mus* species (e.g., *M. spretus*, *M. cervicolor*, and *M. caroli*). However, they have not been detected in *Coelomys pahari* (formerly *Mus pahari*), other rodents, or other mammals. This suggests that MRL elements were integrated into the ancestral germ line of the *Mus* species. However, there are also indications that MRL elements have continued to proliferate in the germ line of laboratory mice. If each solitary LTR represents a proviral integration and subsequent exclusion of all the structure genes and one LTR by homologous recombination between the two LTRs, this has occurred at least 2000 times during the evolution of laboratory mice. Another bit of supportive evidence comes from our study on the integration site of an MRL proviral sequence cloned from a BALB/c mouse. Of eight other mouse strains examined, only A/J shows the same MRL integration; the remaining seven contain no integrated provirus nor solitary LTR at the same chromosomal DNA site. These observations suggest that the integration of this MRL provirus occurred as recently as the segregation of the common parental lineage of BALB/c and A/J strains.

III. Tissue-Specific Expression of Different LTR Elements

Since DNA copies of MuLV-related proviral sequences and MRL elements are present in the chromosomes of every cell in the mouse, the potential for their expression is always present. It has long been noted that normal hemopoietic and lymphoid organs may contain antigenic components as well as infectious virion particles of leukemia viruses (5); genetic evidence has established that they are expressed from the ecotropic and/or xenotropic MuLV proviruses (39, 40). Also, the male reproductive tract expresses leukemia virus-like antigenic components (43), although the proviral origin of these antigens is not precisely known. Recent studies reveal that the induction of leukemia in specific tissues by retroviruses, such as thymomas by the Moloney virus and erythroleukemia by the Friend virus, is determined by specific DNA sequences within the LTR (44–46). From the common and distinct LTR structural features shown by our nucleotide sequence analyses, it can be inferred that endogenous MuLV-related proviral sequences and MRL proviral elements may be expressed with different tissue specificities in normal mice. This has been verified experimentally.

A. Detection of MuLV-Related Transcripts in the Liver, Kidney, and Endometrium

To determine whether or not MuLV-related proviral sequences are expressed, we used normal mice of RFM/Un and NFS/N strains. The RFM/Un mouse genome has one ecotropic MuLV and a few xenotropic MuLV proviruses, while the NFS/N mouse genome contains no ecotropic MuLV and only one copy of a xenotropic MuLV provirus. Both contain at least 30 copies of MuLV-related proviral sequences. When poly(A) RNA preparations from virus organs of these mice were examined by electrophoretic gel-blot hybridization analysis using an LTR sequence probe or a 3' *pol* sequence probe (Fig. 3), various positive RNA species were detected, predominantly in the spleen, thymus, liver, and kidney of RFM/Un mice and in the liver and kidney of NFS/N mice.

To distinguish transcripts of MuLV proviruses from those of MuLV-related proviral sequences, specific molecular probes were used to detect distinct *env* gene sequences and the respective absence or presence of the 200-nucleotide LTR "insert" sequence. As expected, both ecotropic and xenotropic MuLV specificities were found in most of the retroviral RNAs in the spleen and thymus of RFM/Un mice. In contrast, the 17-S, 24-S, 28-S, 30-S, 32-S, and 35-S RNAs in the liver and kidney of both RFM/Un and NFS/N mice (Fig. 3) all showed the content of the "inserted" sequence and other MuLV-related proviral specificities. These findings are consistent with the results obtained by others (43, 47).

These observations raise two questions. First, are only a few or all 30–50 copies of the MuLV-related proviral sequences in the mouse genome responsible for the RNA transcripts detected in the liver and kidney? Second, which cells in these two organs are expressing the MuLV-related RNA transcripts? We find that most of the liver and kidney MuLV-related DNA sequences are methylated and that only a minor population of them are relatively sensitive to exogenous nuclease digestion. The latter have hypomethylated LTRs, suggesting active transcription of a few selected MuLV-related proviral sequences in the liver and kidney cell nuclei.

To answer the second question, an *in situ* hybridization method was employed. MuLV-related RNA transcripts were located mainly in hepatocytes and epithelial cells of renal tubules and not as much in other liver and kidney cells. Moderate amounts were also present in

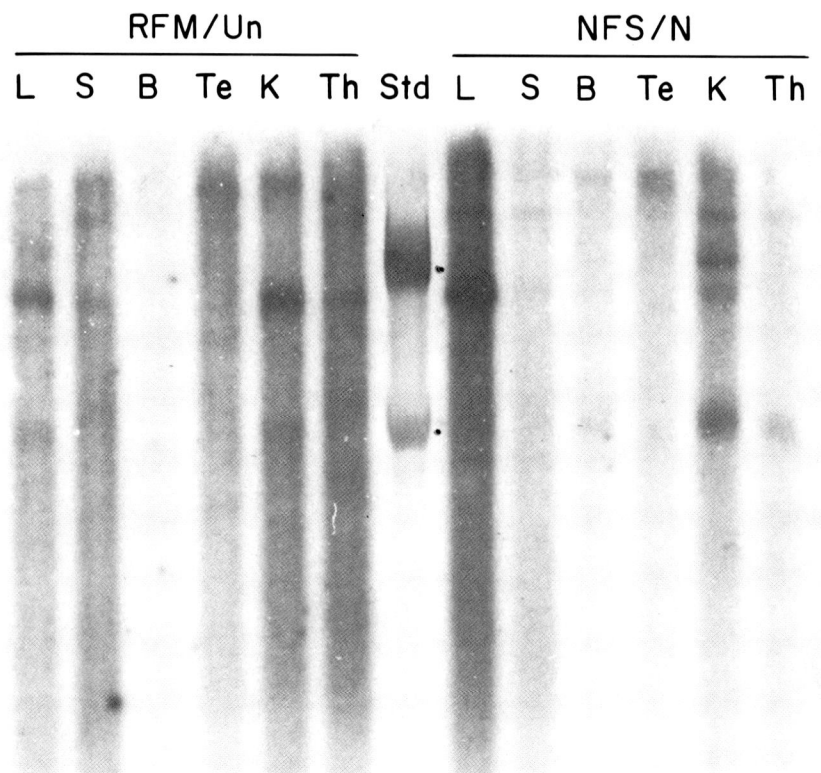

FIG. 3. Autoradiograms of RNA gel blots showing poly(A) RNA transcripts of MuLV and related sequences in the liver (L), spleen (S), brain (B), testis (Te), kidney (K), and thymus (Th) of RFM/Un and NFS/N mice. Each gel lane contained 1 μg of poly(A) RNA preparation. The molecular probe used was a DNA fragment representing the 3' *pol* region. A ^{14}C-labeled ribosomal RNA preparation of mouse cells (Std) was run in parallel to provide molecular-weight markers.

the germinal centers of the spleen and in the cortex–medulla junction of the thymus. However, a surprising finding was that the endometrium of the uterus showed the strongest hybridization with MuLV-related specific sequence probes. Subsequent analyses indicated that the uterine endometrium of the adult mouse at the estrous phase contains high levels of other endogenous retroviral RNAs as well. All of these results imply that various retroviral LTR-containing gene elements of the mouse genome are under a complex mechanism of regulation for their expression in the animal.

B. A Negative Regulatory Element in the MuLV-Related LTRs

To assess the transcriptional activity, LTR sequences of various chromosomal DNA-derived MuLV-related proviral clones were linked to the bacterial chloramphenicol acetyltransferase (CAT) gene and introduced by DNA transfection into mammalian cultured cells, which were then assayed for the appearance of CAT enzyme. As illustrated in Fig. 4, the LTR of an ecotropic MuLV was highly active, whereas the LTRs of all MuLV-related proviral clones showed very little or negligible activity of CAT enzyme, particularly in mouse NIH/3T3 cultured cells. Some CAT activity was detected with these MuLV-related LTR-CAT recombinant DNA in mink CCL64 cells, but it was only slightly higher than that of the promoter-less pCAT-3M control. The MuLV-related LTRs were therefore divided into three segments, which contained the promoter TATA and CAAT boxes, the 200-nucleotide "inserted" sequences, and the enhancer elements as well as the adjacent 5' sequences, respectively. The promoter segment, in combination with a known active enhancer sequence, supported high levels of CAT expression. The "insert" segment was a potential enhancer, especially in combination with SV40 promoter and by examination in CCL64 cells. In contrast, the segment containing the enhancer sequences, or the "protein-binding motifs," and the upstream region of the LTR were not only inactive but also inhibited CAT expression by other active promoter-enhancer sequences. The negative effect was exerted apparently by sequences located within the 5' portion of MuLV-related LTRs. This finding is significant, since it is likely that the "silent" majority of the multiple MuLV-related proviral sequences in the mouse genome are not inert or defective, as generally believed, but rather carry a negative regulatory element to control their expression in the cell.

C. Preferential Expression of MRL Elements in Mature Mouse Gonads

By RNA gel-blot analysis, we have found that adult testis and ovary contain large quantities of MRL proviral transcripts, while other somatic organs are essentially negative. The testis transcripts gave a diffuse electrophoretic pattern, spanning from the 4-S to the 35-S region of the gel lane, while the ovary MRL transcripts were of discrete sizes of 32 S, 28 S, and 17 S. The amount of MRL transcripts in the ovary appeared to fluctuate, being highest during the estrous phase. Very few MRL transcripts were detected in the testis of mice

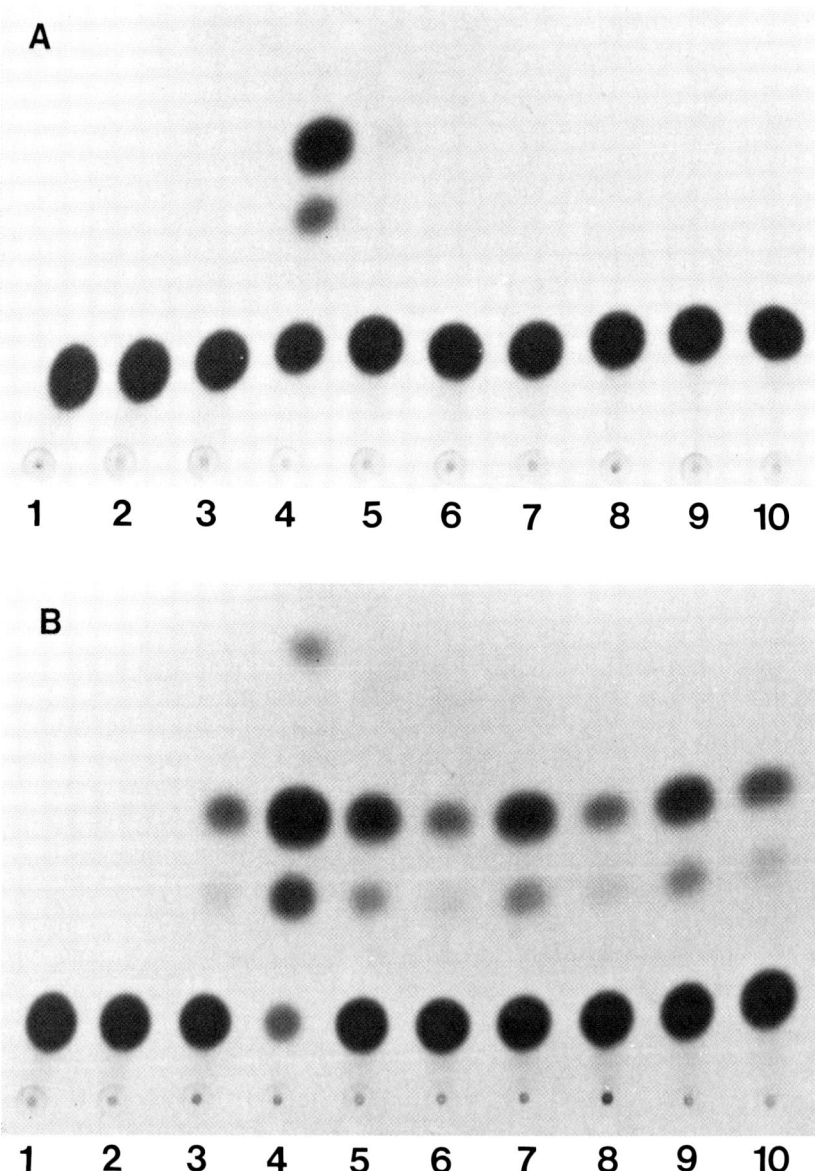

FIG. 4. Thin-layer chromatographic patterns showing chlorampehnicol acetyltransferase (CAT) gene expression in mouse NIH/3T3 cells (A) and mink CCL64 cells (B) 48 hours after DNA transfection with various LTR-CAT recombinant DNA clones. [^{14}C]Chloramphenicol acetylation reactions were performed with control untransfected cell lysate (lane 1), lysate of cells transfected with control plasmid

less than 14 days of age, implying that MRL expression is associated with maturation of late-stage spermatogonia cells. These results of RNA gel-blot analysis have been confirmed by our subsequent screening of a cDNA library from an adult mouse testis. Since the first step for the genomic retrotransposal gene elements to initiate the transposition mechanism is the synthesis of RNA transcripts, our finding of preferential expression of MRL RNAs in the adult testis and ovary is consistent with the idea that MRL elements are propagating actively in the germ-line genome of the mouse.

IV. Stimulation of Endogenous LTR Elements by Environmental Insults

Although chromosomal LTR-containing gene elements in somatic cells are believed to be generally inactive, there are also indications that they may respond to genomic challenges from internal or external stimuli. Our *in situ* hybridization study detected MuLV-related RNA transcripts in germinal centers of the spleen, indicating that the ecotropic and/or xenotropic MuLV proviruses are activated in lymphoid cells during immune responses. Also, production of murine leukemia virus is induced in lymphoid cells undergoing blastogenesis following stimulation by lipopolysaccharides (48). The liver is the major organ of the body for metabolizing exogenous chemicals from food sources, while the kidney is involved mainly in the excretion of metabolic wastes. Expression of endogenous MuLV-related proviral sequences in specific cells of these organs may be associated with these possible chemical challenges. While various environmental chemical and physical agents capable of causing carcinogenesis and/or mutagenesis induce the expression of MuLV proviruses (14), interpretation of these results is complicated by two facts. One is that the extent of viral induction does not correlate with the carcinogenic or mutagenic potency of the agents. The other is that this virus-induction phenomenon is limited to only a few mouse embryo culture cell lines, particularly those from the AKR mouse strain. Some mouse

---

p41LTR-NEO (lane 2), promoter-less pCAT-3M (lane 3), p41LTR-CAT (lane 4), pAL10LTR-CAT (lane 5), pRFM1LTR-CAT (lane 6), pRFM6LTR-CAT (lane 7), pRFM9LTR-CAT (lane 8), pRFM1LTR-CAT (lane 9), and pRMF17LTR-CAT (lane 10). p41 represents the ecotropic MuLV LTR, while the remaining six lanes (5–10), the endogenous MuLV-related LTRs. Radioactive spots of each ascending chromatographic lane are (from the bottom up): origin, chloramphenicol, the two monoacetylated chloramphenicols, and diacetylated chloramphenicol.

strains, such as SWR/J, do not harbor ecotropic or xenotropic MuLV proviruses in their genomes, yet they are susceptible to the carcinogenic and/or mutagenic effects. It is, therefore, important to perform similar studies on the LTR-containing elements found in all mouse genomes, such as MuLV-related proviral sequences and MRL elements, that have now been identified. Our work on this aspect is still ongoing. Two preliminary studies are described here.

A. Carbon Tetrachloride-Induced Liver Damage and Regeneration

Carbon tetrachloride, a common industrial solvent, has been judged a potential human carcinogen (49). A cocarcinogenic effect was clearly demonstrated by the observation that the incidence of neutron-induced hepatoma in LAF_1 mice increased from 14% to 92% if the mice were treated with CCl_4 1 day prior to the radiation exposure (50). Using Moloney MuLV sequences as a molecular probe in RNA gel-blot analysis, a marked stimulation of retroviral RNA expression was detected in the liver of B6C3F1 mice 2 days following CCl_4 treatment (51). We have investigated that particular CCl_4 effect. We find the following: (a) the increased RNA transcripts show sequence specificities of the "MCF" env gene type as well as the LTR "inserted" segment; hence the transcripts are evidently derived from the endogenous MuLV-related proviral sequence family; (b) CCl_4 treatment of other mouse strains including RFM/Un, C3H/Anf, C57Bl/6, NFS/N, and BALB/c gives similar responses; (c) the responses of male C3H/Anf mice are more marked than female C3H/Anf mice of the same age; (d) in situ hybridization analysis (Fig. 5) demonstrated that high levels of MuLV-related RNAs accumulated in hepatocytes of the viable and regenerating areas but not in the areas showing cell death; (e) immunofluorescence analysis reveals corresponding increases in viral antigens in the regenerating areas of the liver; (f) preliminary results of transcription "run-on" experiments using isolated nuclei suggest that the accumulated MuLV-related RNAs are due partially to an increased transcription rate and partially to a decrease in degradation; and (g) a transient appearance of an MuLV-related DNA molecule of less-than-genomic size was detected in some mice 2 days after CCl_4 treatment. Whether or not this molecule is the reverse-transcribed DNA intermediate of the stimulated retroviral RNA transcripts is being determined.

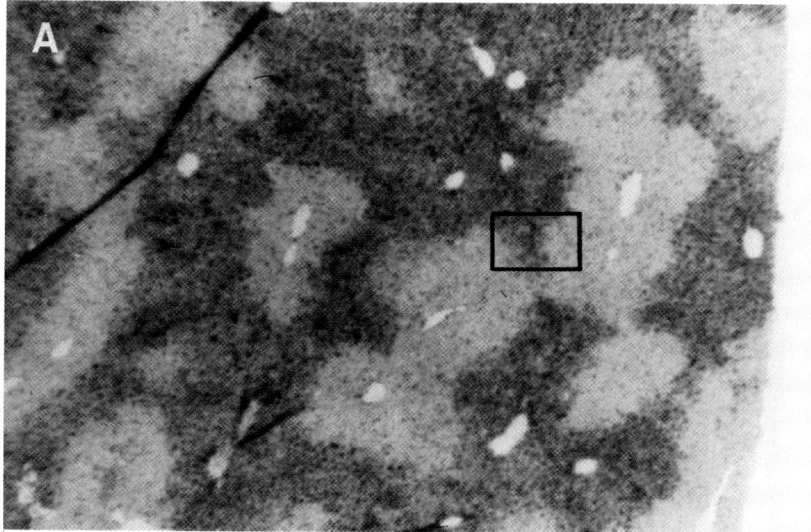

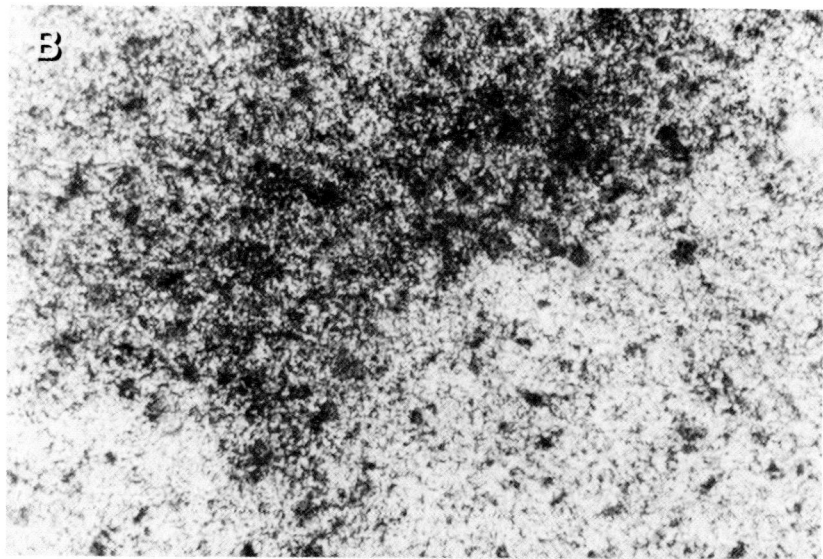

FIG. 5. *In situ* hybridization of a liver section from a female RFM/Un mouse 48 hours after treatment with CCl$_4$. (A) A low-power magnification, showing viable and regenerating areas with darker hematoxylin-staining and autoradiographic density and areas of cell death with lighter staining. (B) A high-power magnification of the squared area indicated in A, showing clusters of silver grains over hepatocytes in the viable regenerating areas. The molecular probe used for *in situ* hybridization was a ^{35}S-labeled MCF sequence specific for MuLV-related transcripts.

B. Azacytidine-Induced NIH/3T3 Cell Transformation

Azacytidine is a potent carcinogenic agent in mice (52). It is also the most active chemical for inducing the expression of endogenous ecotropic murine leukemia viruses. In mouse 10T½ cell cultures, a brief azacytidine treatment causes rapid neoplastic transformation (53). Concomitant endogenous retroviral gene expression with the emergence of a neoplastic phenotype was oberved (54).

We have found that NIH/3T3 cells can also be readily transformed by a brief minimally toxic azacytidine treatment. Since NIH/3T3 cells harbor no ecotropic MuLV provirus and only a single copy of xenotropic MuLV provirus in their genome, the neoplastic transformation is evidently associated with a genetic mechanism more general than the induction of ecotropic MuLV. Six individual azacytidine-transformed NIH/3T3 cell clones were isolated. In all six of them there was no DNA methylation at the LTR-SmaI site in a few specific MuLV-related proviral sequences, although they were methylated in the untransformed NIH/43T3 cells. Whether or not the apparent activation of these specific MuLV-related proviral sequences is associated with neoplastic phenotype expression remains to be studied.

V. Concluding Remarks

The two most important tenets of our working hypothesis are that the mouse genome contains many potentially transposable regulatory gene elements and that these elements may be induced to initiate the molecular processes of transposition upon exposure of cells to environmental carcinogenic agents. As summarized in Table I, our experiments have essentially established the first tenet. The most relevant biological feature is the apparent tissue specificity of these three endogenous retroviral LTR families: the MuLV proviruses for the lympho-hemopoietic system; the MuLV-related proviral sequences for the liver and kidney; and the MRL elements for the mature gonads. In the case of MuLV proviruses, there appears to be a significant correlation between expression in the normal tissues and the associated leukemogenic processes. For example, radiation-induced and naturally occurring neoplasms, such as myeloid leukemias and reticulum cell sarcomas, of RFM/Un mice usually contain somatically acquired ecotropic MuLV proviruses in the chromosomal DNA, suggesting that the single ecotropic MuLV on chromosome 5 of this mouse strain became active in transposition during leukemogenesis (55). It would be important to determine whether mouse

TABLE I
MURINE LEUKEMIA VIRUS (MuLV) AND RELATED GENE FAMILIES IN THE GERM LINE OF LABORATORY MICE

	MuLV		MuLV-related		MRL	
	Ecotropic	Xenotropic	Polytropic	Modified polytropic	Solo-LTR	Proviral
Number/haploid mouse genome	0–5	1–15	15–25	15–25	2000–5000	100–200
LTR size (nt)	~530	~530	~700[a]	~750[a]	~500, ~600	~500, ~600[b]
Primer tRNA	Proline	Proline	Glutamine	Glutamine	—	Proline
Fv-1 tropism in gag-p30	N	?	B	B	—	?
env specificity	eco	xeno	MCF	MCF	—	Deleted
Virus production	+	+	—?	—?	—	—?
Major organs expressing mRNA[c]	Lympho-hematopoietic	Lympho-hematopoietic	Liver, kidney	Liver, kidney	Testis?, ovary?	Testis, ovary
Inducing or stimulating agents	Azacytidine, halogenated deoxyuridine, lymphocyte mitogens	Protein synthesis inhibitors, lymphocyte mitogens	CCl$_4$ and other hepatotoxic agents?	CCl$_4$ and other hepatotoxic agents?	?	?
Related neoplasms	Lymphomas, leukemias	Lymphomas?, leukemias?	Hepatomas?	Hepatoma?	?	Gonadal tumors?

[a] These show considerable sequence homology with the MuLV LTR, except for insertion of a 190- to 200-nucleotide segment between enhancer and promoter domains. N, N-tropic; B, B-tropic; eco, ecotropic; MCF, mink-cell focus-forming.

[b] The 190- to 200-nucleotide segment found in the MuLV-related LTRs occupies most of the 5' half of the 500-nt LTRs; additional small-sequence insertions are present in the 600-nt LTRs. Our recent findings suggested marked LTR polymorphisms in certain MRL proviral structures, particularly the male-specific elements.

[c] All three gene families are expressed in high levels in the endometrium, particularly during the estrous phase.

hepatomas, renal tumors, and gonadal neoplasms can acquire additional integrated copies of MuLV-related proviral sequences and MRL elements, respectively.

More work is required to establish the second tenet of our working hypothesis, although the results of our CCl_4 studies are consistent with the idea that the already active LTR gene elements can respond to cocarcinogenic environmental insults. One particular aspect in question is whether or not, upon carcinogenic insult, the already expressed retroviral RNA molecules would become the enzymatic substrate of reverse transcriptase and thus generate DNA intermediates for subsequent integration into novel chromosomal sites. Since carcinogenesis by environmental carcinogens is generally considered to be a multifactorial and multistep genetic process, the induction of the retrotransposition mechanism may represent one of these steps. Alternatively, since integration of LTR elements is nonspecific, if not random, with respect to the chromosomal target site, multiple integration events could occur to affect multiple cellular genes in the same cell.

In terms of cellular genomic effects, the two most important components of retroviral and retrotransposable genes are the LTRs and the "insertase" gene. Effects of an LTR on cellular genes at the chromosomal site of integration would be dictated by the specific regulatory gene sequences carried by this LTR. To elucidate the functional properties of specific regulatory sequences in different LTR element families that we have isolated from the mouse genome, it will be necessary to construct recombinant DNA clones of these LTRs with a reporter gene for subsequent biological assessment in transfected cells as well as in transgenic mice. The "insertase" gene, on the other hand, may be equally important since this protein or enzyme can potentially cause chromosomal DNA alterations by its endonucleolytic cleavage activity. Thus, the expression of this retrotransposable gene function without actual LTR insertion may be sufficient to cause genomic instability and lead to the neoplastic state. Experiments are currently in progress to test this idea.

Acknowledgment

Earlier studies on the molecular cloning and sequence characterization of mouse endogenous retroviral genes involved collaborative efforts also from L. R. Boone, M. J. Gardner, N. B. Kuemmerle, K. N. Nikbakht, and C. Y. Ou. We thank Neva Hair for typing the manuscript. Funding support is from the National Institute of Environmental Health Sciences (Y01-ES-4-0118), the Office of Health and Environmental Research, and the U.S. Department of Energy under Contract DE-AC05-840R21400 with Martin Marietta Energy Systems, Inc.

References

1. B. McClintock, *Science* **226**, 792 (1984).
2. R. Sager, *Nature (London)* **282**, 447 (1979).
3. P. Rous, *J. Exp. Med.* **12**, 696 (1910).
4. L. Gross, "Oncogenic Viruses," 2nd Ed. Pergamon, New York, 1970.
5. R. J. Huebner and G. J. Todaro, *PNAS* **64**, 1087 (1969).
6. H. M. Temin, *Annu. Rev. Microbiol.* **25**, 609 (1971).
7. J. M. Taylor, *Curr. Top. Microbiol. Immunol.* **87**, 23 (1979).
8. H. E. Varmus, *Science* **216**, 812 (1982).
9. W. S. Hayward, B. G. Neel, and S. M. Astrin, *Nature (London)* **290**, 475 (1981).
10. N. A. Jenkins, N. G. Copeland, B. A. Taylor, and B. K. Lee, *Nature (London)* **293**, 370 (1981).
11. J. M. Bishop, *ARB* **46**, 35 (1978).
12. D. R. Lowy, W. P. Rowe, N. Teich, and J. W. Hartley, *Science* **174**, 155 (1971).
13. S. A. Aaronson, G. J. Todaro, and E. M. Scolnick, *Science* **174**, 157 (1971).
14. R. W. Tennant and R. J. Rascatti, *in* "Modifiers of Chemical Carcinogenesis" (T. J. Slaga, ed.), Vol. 5, p. 185. Raven, New York, 1980.
15. R. A. Weinberg, *Cell* **22**, 643 (1980).
16. H. M. Temin, *Science* **190**, 1075 (1976).
17. D. Baltimore, *Science* **192**, 632 (1976).
18. D. Baltimore, *Cell* **40**, 481 (1985).
19. W. K. Yang, L. R. Boone, R. W. Tennant, and A. Brown, This Series **29**, 175 (1983).
20. L. R. Boone, F. E. Myer, D. M. Yang, J. O. Kiggans, C. K. Koh, R. W. Tennant, and W. K. Yang, *J. Virol.* **45**, 484 (1983).
21. R. S. Liou, L. R. Boone, J. O. Kiggans, D. M. Yang, T. W. Wang, R. W. Tennant, and W. K. Yang, *J. Virol.* **45**, 288 (1983).
22. C. Y. Ou, L. R. Boone, and W. K. Yang, *NARes Res* **16**, 5603 (1983).
23. K. N. Nikbakht, C. Y. Ou, L. R. Boone, P. L. Glover, and W. K. Yang, *J. Virol.* **54**, 889 (1985).
24. K. N. Nikbakht, L. R. Boone, P. L. Glover, F. E. Myer, and W. K. Yang, *J. Gen. Virol.* **68**, 683 (1987).
25. H. W. Chan, T. Bryan, J. L. Moore, S. P. Staal, W. P. Rowe, and M. A. Martin, *PNAS* **77**, 5779 (1980).
26. S. K. Chattopadhayay, M. R. Lander, E. Rands, and D. R. Lowy, *PNAS* **77**, 5774 (1980).
27. R. R. O'Neill, A. S. Khan, M. D. Hoggan, J. W. Hartley, M. A. Martin, and R. Repaske, *J. Virol.* **58**, 358 (1986).
28. J. P. Stoye and J. M. Coffin, *J. Virol.* **61**, 2659 (1987).
29. J. W. Hartley, N. K. Wolford, L. J. Old, and W. P. Rowe, *PNAS* **74**, 789 (1977).
30. L. R. Boone, P. L. Glover, C. L. Innes, L. A. Niver, M. C. Bondurant, and W. K. Yang, *J. Virol.* **62**, 2644 (1988).
31. J. A. Yang, L. W. Tai, P. F. Agris, C. W. Gehrke, and T. W. Wong, *NARes* **11**, 1991 (1983).
32. Y. Kuchino, H. Beier, N. Akita, and S. Nishimura, *PNAS* **84**, 2665 (1987).
33. L. C. Waters and B. C. Mullin, This Series **20**, 131 (1977).
34. F. Harada, G. G. Peters, and J. E. Dahlberg, *JBC* **254**, 10979 (1979).
35. A. S. Khan and M. A. Martin, *PNAS* **80**, 2699 (1983).
36. J. C. Wejman, B. A. Taylor, N. A. Jenkins, and N. G. Copeland, *J. Virol.* **50**, 237 (1984).

37. M. D. Hoggan, R. R. O'Neill, and C. A. Kozak, *J. Virol.* **60**, 980 (1986).
38. C. A. Kozak and R. R. O'Neill, *J. Virol.* **61**, 3082 (1987).
39. W. P. Rowe, *Harvey Lect.* **71**, 173 (1978).
40. C. A. Kozak, *Adv. Cancer Res.* **44**, 295 (1985).
41. T. Wirth, K. Gloggler, T. Baumruker, and I. Horak, *PNAS* **80**, 3327 (1983).
42. T. Wirth, M. Schmidt, T. Baumruker, and I. Horak, *NARes* **12**, 3603 (1984).
43. D. E. Levy, R. A. Lerner, and M. C. Wilson, *J. Virol.* **56**, 691 (1985).
44. D. Celander and W. A. Hazeltine, *Nature (London)* **312**, 159 (1984).
45. P. A. Chatis, C. A. Holland, J. E. Silver, T. N. Frederickson, N. Hopkins, and J. W. Hartley, *J. Virol.* **52**, 248 (1984).
46. L. Des Groseillers, E. Rassart, and P. Jolicoeur, *PNAS* **80**, 4203 (1983).
47. A. S. Khan, F. Laigret, and C. P. Rodi, *J. Virol.* **61**, 876 (1987).
48. J. P. Stoye and C. Moroni, *J. Exp. Med.* **157**, 1660 (1983).
49. "Second Annual Report on Carcinogens" (NTP 81-43), p. 73 (1981).
50. L. J. Cole and P. C. Nowell, in "Biological Effects of Neutron and Proton Irradiations," Vol. 2, p. 129. Int. Atomic Energy Agency, Vienna, 1964.
51. T. A. Dragani, G. Manenti, G. Della Porta, and I. B. Weinstein, *Cancer Res.* **47**, 795 (1987).
52. G. D. Stoner, M. B. Simkin, A. J. Kniazeff, J. H. Weisburger, and G. B. Gori, *Cancer Res.* **33**, 3069 (1973); National Cancer Institute, Carcinogenesis Program, "Bioassay of 5-Azacytidine for Possible Carcinogenicity," NIH Publ. 78-842. Department of Health, Education and Welfare, Washington, D.C., 1978.
53. W.-L. Hsiao, S. Gattoni-Celli, and I. B. Weinstein, *MCBiol* **5**, 1800 (1984).
54. W.-L. Hsiao, S. Gattoni-Celli, and I. B. Weinstein, *J. Virol.* **57**, 1119 (1986).
55. L. R. Boone, G. S. Boone, C. L. Innes, W. K. Yang, and R. W. Tennant, *Carcinogenesis* **7**, 529 (1986).

VI. Molecular Analysis of Chromosomal Translocation and Gene Insertion

Molecular Genetics of Lymphoid Tumorigenesis 269
F. G. HALUSKA, Y. TSUJIMOTO, G. RUSSO, M. ISOBE, AND C. M. CROCE

Molecular Analysis of Chromosome Breakpoints 281
JOHN GROFFEN, ANDRÉ HERMANS, GERARD GROSVELD, AND NORA HEISTERKAMP

Homologous Recombination in Mammalian Somatic Cells 301
RAJU S. KUCHERLAPATI

Gene Transfer into Primates and Prospects for Gene Therapy in Humans 311
KENNETH CORNETTA, ROBERT WIEDER, AND W. FRENCH ANDERSON

Molecular Genetics of Lymphoid Tumorigenesis

F. G. HALUSKA
Y. TSUJIMOTO
G. RUSSO
M. ISOBE AND
C. M. CROCE[1]

The Wistar Institute, Philadelphia, Pennsylvania 19104

Human lymphoid tumors provide a fertile system for the analysis of tumorigenesis. Several factors contribute to the central place occupied by the lymphoid system in experimental oncology. First, the normal biology of the immune system is itself the subject of intensive investigation. Techniques for the identification of lymphoid cell lineages, for the establishment of lymophocyte cultures, and for the elucidation of molecular immunogenetics, have been well worked out. Consequently, the handling of lymphoid cell systems has become routine. Second, lymphoid cells lend themselves well to cytogenetic analyses. Thus, there is a wealth of literature regarding the normal and pathogenetic cytogenetics of lymphoid tissue. Third, and most importantly, lymphoid tumors exhibit a high frequency of consistent, nonrandom, tumor-specific cytogenetic abnormalities. The presence of these chromosome aberrations (most commonly chromosome translocations) has suggested fruitful approaches to the molecular understanding of these tumors. Thus, we have taken advantage of lymphoid-specific chromosome abnormalities to analyze on the molecular level the mechanisms of oncogenesis in B and T cells.

I. The Cytogenetics of Lymphoid Tumors

We have studied a number of chromosome translocations specifically associated with various lymphoid malignancies. The best understood B-cell malignancy in both cytogenetic and molecular terms is Burkitt's lymphoma (1). These extremely aggressive tumors carry one of three chromosome translocations. Approximately 80% of them

---

[1] Speaker.

exhibit t(8;14)(q24;q32) translocations. The remainder carry t(2;8)(p11;q24) or t(8;22)(q24;q11) translocations (2). The precise chromsome bands at which these translocations take place is of considerable significance because genes of central importance to B-cell ontogeny map to these chromosome regions. The immunoglobulin heavy-chain locus maps to chromosome 14q32. The immunoglobulin κ and λ light-chain genes map to chromosomes 2p11 and 22q11, respectively. Each of these chromosome regions is a site of chromosome translocation in Burkitt's lymphomas. The other involved chromosome in each of these chromosome aberrations is chromosome 8, at band q24. This is the site at which the human homolog of the avian myelocytomatosis virus transforming gene, c-*myc*, is located. Thus, cytogenetic evidence alone suggests that the translocations in Burkitt's lymphoma juxtapose the c-*myc* gene and immunoglobulin loci.

Another B-cell malignancy commonly exhibiting a nonrandom chromosome translocation is follicular lymphoma (3). Follicular lymphomas are probably the most common form of B-cell malignancy, and over 85% of them carry t(14;18)(q32;q21) translocations. Again, the site of the immunoglobulin heavy-chain locus is implicated in this translocation. At the site of chromosome breakage on chromosome 18 lies a gene called *bcl*-2, initially identified in our laboratory through the molecular analyses of these translocations (4), as will be discussed below (Section III).

The t(11;14)(q13;q32) translocation is also commonly observed in B-cell malignancies (3). It occurs predominantly in chronic lymphocytic leukemias, but also is found in diffuse small- and large-cell lymphomas and in a small proportion of multiple myelomas. The IgH gene on chromosome 14 and a locus designated *bcl*-1 are involved in this translocation (5).

T-cell malignancies also carry characteristic chromosome translocations, despite the fact that T-cell malignancies are usually less common than B-cell tumors and are not as intensely studied. The most common site of translocation in T cells is the T-cell-receptor α-chain locus (TCRα), which maps to chromosome 14q11 (6–9). Embedded within the TCRα locus is the gene for TCRδ (10). Thus, both of these genes are involved in a number of translocations. Some t(8;14)(q24;q11) translocations juxtapose c-*myc* and TCRα (11). Other t(8;14)(q24;q11) translocations, as well as the t(10;14)(q24;q11) (9,12) and t(11;14)(p13;q11) (7,13) translocations, involve the TCRδ gene. Clearly, an analogy exists between B- and T-cell malignancies: in both cases, rearranging immunoglobulin superfamily genes are involved in translocations.

II. The Molecular Genetics of Burkitt's Lymphoma

Burkitt's lymphoma is perhaps the most completely understood malignancy in molecular terms. The most common translocation in this tumor, t(8;14)(q24;q32), is also the best-described tumor-specific chromosome abnormality. The organization of the t(8;14) translocation is illustrated in Fig. 1. Several important features of the translocated chromosomes have emerged from many studies and are summarized in Fig. 1.

First, the chromosome breakpoints join the immunoglobulin heavy-chain locus on chromosome 14 to the c-*myc* gene on chromosome 8 in all cases. This suggests that this molecular architecture has a functional consequence. The c-*myc* gene is involved in the pathways leading to cellular proliferation. The immunoglobulin genes, in contrast, code for antibody proteins expressed in a strictly regulated tissue-specific and temporal fashion. Juxtaposition of the c-*myc* gene

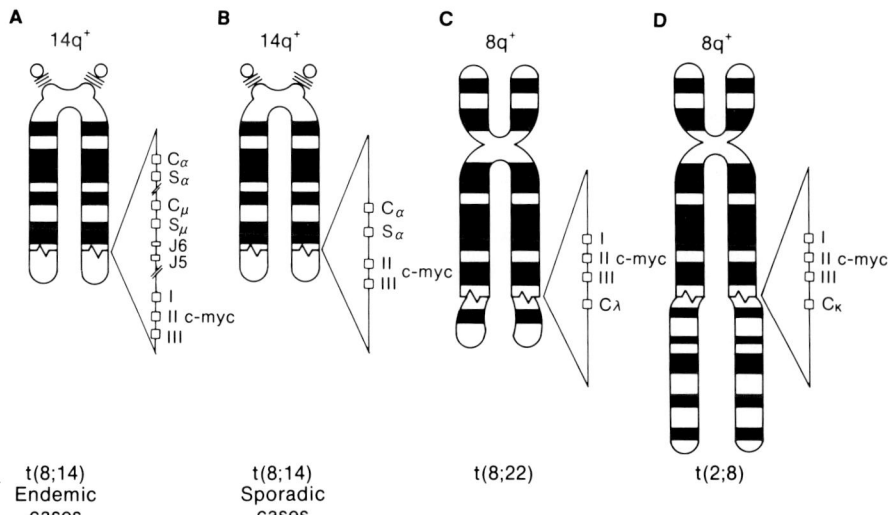

FIG. 1. Chromosomal translocations observed in Burkitt's lymphomas. The configurations of the immunoglobulin and juxtaposed c-*myc* genes resulting from the translocations, as revealed by molecular cloning studies, are illustrated. The 14q+ chromosome resulting from the t(8;14) translocation is shown in A and B for endemic and sporadic cases, respectively. Note the differences in the molecular organization of the breakpoints, as described in the text. C and D illustrate the variant translocations.

and the IgH control elements results in inappropriate expression of the c-*myc* gene and in neoplasia. The evidence for this hypothesis is extensive; we have recently reviewed it in great detail (*14,15*).

The second general feature of the translocated chromosome is that the breakpoints are heterogeneous. However, the breakpoints are not distributed randomly. On chromosome 14 (Fig. 2), they are all situated at sites subject to normal, physiologic genomic rearrangement. Recall that immunologic diversity is generated by the shuffling of immunoglobulin and TCR gene segments during lymphocyte development (*16*). This occurs in two ways. The first involves rearrangement of variable (V), diversity (D), and joining (J) segments to generate a diverse repertoire of antigenic recognition. The process is carried out by the V–D–J joining system of recombination enzymes. The second process of isotype switching allows for the diversification of the immune response in terms of effector function. This is presumably carried out by a second set of switching enzymes. Both the recombination and switching enzymes act at specific sites in the IgH locus; it is precisely at these sites that translocations occur. This localization of breakpoints has suggested that the translocations may be mediated through the same enzymatic systems that function during physiologic Ig recombination. Breakpoints are also dispersed widely on chromosome 8. They all occur on the 5' side of the c-*myc* coding unit. That is, all translocations leave the c-*myc* protein itself unchanged (the variant translocations all occur 3' of c-*myc*, without altering the c-*myc* gene). Thus, translocation results in deregulation of c-*myc* expression, but not in c-*myc* mutation (*17*).

The third salient feature of the organization of translocated chromosomes is that the locations of chromosome breakpoints on both

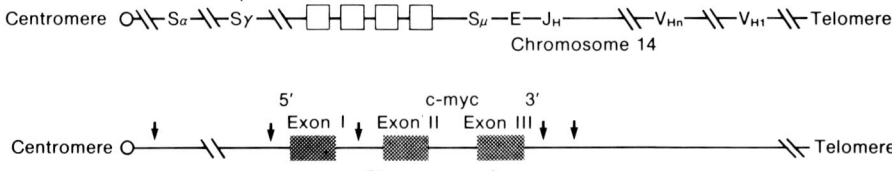

FIG. 2. Heterogeneity of translocation breakpoints in Burkitt's lymphomas. On chromosome 14 (top), chromosome breakage has been demonstrated in Sα, Sγ, Sμ, J$_H$, and D$_H$ regions, as indicated by arrows. On chromosome 8 (bottom), breakpoints are 5' of c-*myc* or within the first intron in cases carrying t(8;14) translocations. Thus, c-*myc* may assume either an intact or a truncated configuration. The variant translocations result in breakage 3' of an intact c-*myc*.

chromosomes 8 and 14 correlate with the type of Burkitt's lymphoma being studied (15). Endemic Burkitt's lymphomas carry translocations that join regions far 5' of c-*myc* and the immunoglobulin heavy-chain joining (J_H) or diversity (D_H) segments on chromosome 14 (Fig. 3). In contrast, the translocations found in sporadic Burkitt's tumors join sequences immediately 5' of c-*myc*, or within the first exon or intron of the gene, to isotype switching regions in the 3' portion of the IgH locus.

III. The Molecular Genetics of Follicular Lymphomas

A striking cytogenetic similarity between the Burkitt's lymphomas discussed above and several other lymphoid malignancies is the consistent involvement of chromosome 14q32 in chromosome translocations. Nearly 90% of follicular lymphomas exhibit t(14;18)(q32;q21) translocations. These translocations uniformly involve the J_H segments of the IgH locus. On chromosome 18, the translocation breakpoints are also consistently clustered. The breakpoint cluster is found in the 3' region of a transcriptional unit initially designated *bcl-2* (B-cell leukemia/lymphoma 2) (18). This gene comprises two widely separated exons that are transcribed into three transcripts, 3.5, 5.5, and 8.5 kb in length, respectively (19). These transcripts are, in turn, translated into two proteins, *bcl-2α* and *bcl-2β*, which are 239

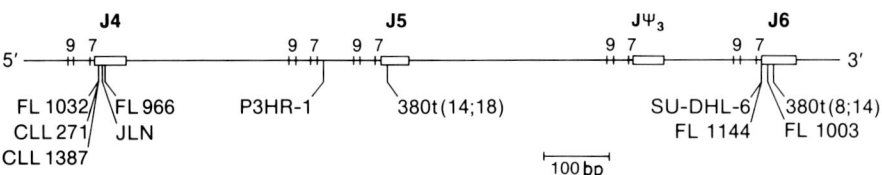

FIG. 3. Clustering of chromosome breakpoints 5' of J_H segments at the immunoglobuoin heavy-chain locus on chromosome 14. The close spacing of breakpoints at the nucleotide level strongly suggests that the "recombinase" is implicated in the mechanism of translocation in these various B-cell malignancies. Chromosome translocation occurs at or near the 5' end of each J_H segment, and downstream from the heptamer–nonamer signal sequences (indicated by 9 and 7). These are precisely the sites of physiological V_H–D–J_H joining. Breakpoints designated FL, JLN, and SU-DHL-6 are from follicular lymphoma cases; CLL is chronic lymphocytic leukemia; 380 is an acute pre-B-cell leukemia carrying t(8;14) and t(14;18) translocations; P3HR-1 is an African Burkitt's lymphoma (note that while the site of translocation in this tumor is upstream of J5, a heptamer–nonamer is situated at the break, and nucleotides 3' of the breakpoint demonstrate some features of a pseudo-J_H segment (15)).

and 205 amino acids in length, respectively. The proteins differ only at their carboxyl termini. The expression of these proteins is influenced by mitogenic stimuli, and they are significantly conserved across species boundaries, strongly suggesting an important role for this gene in the control of cell proliferation (20, 21).

As in Burkitt's lymphomas, translocation results in the deregulation of *bcl*-2 expression. Deregulation takes place as a consequence of enhanced *bcl*-2 transcription. It is likely that this occurs because of the abnormal proximity of the *bcl*-2 gene to enhancer elements within the IgH locus. Because the chromosome breakpoints all fall within the J_H semgnets, the IgH enhancer, which lies immediately 3' of J_H, is always positioned near the chromosome junction. Thus, the consistent transcriptional deregulation of *bcl*-2 in the setting of follicular lymphoma strongly argues for the oncogenic potential of *bcl*-2 in B cells.

IV. The Molecular Genetics of the t(11;14) Translocation

As noted above (Section I), a variety of tumors exhibit the t(11;14) translocation. Most commonly, this abnormality is observed in chronic lymphocytic leukemia (CLL), an indolent lymphoproliferative disorder of phenotypically mature B cells. Again, this translocation involves the IgH locus. Cloning and sequencing analyses of translocation breakpoints from several of these tumors have demonstrated the interruption of the J_H segments by these translocations (Fig. 4) (5). Thus, the J_H segments are involved in translocations in a spectrum of B-cell tumors. Breakpoints also cluster on chromosome 11 in CLL (22). We have termed the cluster region *bcl*-1 (B-cell leukemia/lymphoma 1). By analogy with previously described translocations, one would expect a gene with oncogenic potential to lie near the site of consistent chromosome translocation; however, the hypothesized *bcl*-1 gene has as yet eluded detection.

V. Summary of B-Cell Translocations

The analyses of B-cell-specific chromosome translocations described above (Sections I–IV) have facilitated the deduction of several general principles regarding tumorigenesis in these cells. First, the immunoglobulin loci, and in particular the heavy-chain locus, are prone to repeated involvement in chromosome translocations. These genes are subject to physiologic recombination during the normal ontogeny of every B cell, and the recombination process inherently carries a risk of recombinatorial error. Such errors may result in

LYMPHOID TUMORIGENESIS

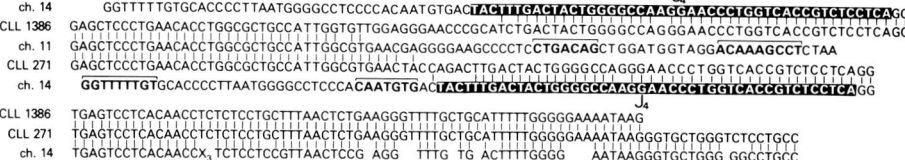

FIG. 4. DNA sequences of the joining sites between chromosomes 11 and 14 in CLL 271 and CLL 1386 and of corresponding normal chromosome 11. Identical nucleotide sequences are shown by vertial lines. The boxed region indicates the J4 coding segment of the immunoglobulin heavy-chain gene. The DNA sequences shown by brackets on chromosome 14 indicate the conserved heptamer–nonamer sequences.

translocations. Second, the Ig loci are expressed specifically and at very high levels in mature B cells. Thus, they have the capacity to adversely influence transcriptional control of genes translocated into their proximity. Clones of cells carrying translocations that juxtapose the Ig-activating elements and genes having an important function in the control of cell proliferation possess a significant proliferative advantage. Third, such translocationally activated genes function as oncogenes. The c-*myc* gene has oncogenic potential by many different criteria. The *bcl*-2 gene is expressed at high levels almost exclusively in lymphomas. And in some translocations, such as the t(11;14), identification of involved transcriptional units is still under way.

VI. The Molecular Biology of T-Cell Malignancies

The molecular biology of the T-cell-receptor (TCR) genes recently has been demonstrated to bear remarkable similarity to that of the Ig genes. Although the TCR proteins differ in both form and function from the Ig molecules, the genetic organization of the genes coding for the TCR subunit chains is quite like the architecture of Ig loci. V, D, and J segments rearrange to form a full variable exon during early T-cell development, and transcription of the variable and constant exons follows. The TCR loci, like the Ig loci, are thus subject to abnormal recombination. Furthermore, the tissue-specific TCR expression promotes activation of translocated oncogenes. The general principles of T-cell oncogenesis as a result of translocation differ little from those already elucidated for B cells.

The t(8;14)(q24;q11) translocation is the best understood of the T-cell-specific translocations (*11*). Molecular analyses of this translocation, found in a number of T-cell neoplasms, show that this translocation joins the c-*myc* gene with the TCRα locus on chromosome 14. Somatic cell genetic studies initially demonstrated interruption of the TCRα locus by the translocation. Subsequently, cloning and sequencing of the translocation breakpoint using probes from the region surrounding c-*myc* facilitated more detailed analysis. The region 3' of an intact c-*myc* gene is translocated to the 5' end of one of the Jα segments. Moreover, c-*myc* is deregulated in these translocations (8) in a manner similar to that observed in the variant Burkitt's translocations, which also involve breakage of chromosome 8 3' of the c-*myc* gene (17). These studies strengthen the analogy between chromosome translocations in T cells and in B cells by demonstrating that the TCR loci may function to translocate and activate oncogenes.

Other T-cell translocations have been analyzed utilizing these techniques. The TCRα locus is split between the Cα and Vα segments in both the t(10;14) and t(11;14) translocations (9, 23). However, this does not necessarily implicate the TCRα gene itself in the translocation. We and others have recently demonstrated that the TCRδ gene lies embedded within the extended TCR Jα region (12). That is, the Jδ and Cδ segments lie upstream of the Jδ segments on chromosome 14, yet downstream of the Vα segments. It is likely that the two translocations noted above both involve the TCR Jδ segments. We have hypothesized that genes important to T-cell oncogenesis, *tcl*-3 (9) and *tcl*-2 (7), map to the involved portions of chromosomes 10 and 11, respectively.

Another important T-cell-specific chromosome aberration involves chromosome 14q32. This is, of course, the location of the IgH gene. In fact, it has been demonstrated that an inv(14)(q11;q32) from the T-cell line SUPT1 joins a TCR Jα segment to an Ig V_H segment (24). At this point, it is not clear that this translocation has oncogenic consequences. This is suggested by the finding that many of the chromosome aberrations in T cells involving 14q32 [i.e., t(14;14)(q11;q32) translocations or inv(14)(q11;q32) inversions] actually join the TCRα locus to a separate locus centromeric to IgH (25). We have demonstrated this to be true in T-cell malignancies in ataxia telangiectasia (26); this has also been suggested by others. We hypothesized that the involved region of chromosome 14q32 carries an oncogene, *tcl*-1 (6).

Thus, T-cell malignancies exhibit chromosome translocations with many of the properties of B-cell translocations. Consistent chromosome bands are involved in the translocations, the Ig and TCR genes are repeatedly implicated in tumorigenesis, and known or putative oncogenes are activated by the chromosome rearrangements. We now turn to a discussion of the common mechanism by which these translocations arise in both B- and T-lymphoid tumors.

VII. The Mechanism of Chromosome Translocation in Lymphoid Tumors

The multitude of tumor-specific translocations in B and T cells and the fact that many of these translocations involve immunoglobulin superfamily genes together imply that these cells, and these genes, possess an intrinsic proclivity for translocation. This tendency for translocation is a consequence of the normal mechanisms of physiologic rearrangement employed by the cells expressing these genes.

An examination of the distribution of chromosome breakpoints in B-cell tumors immediately suggests that the same mechanisms that facilitate normal recombination might also catalyze chromosome translocation (Fig. 5). B-cell translocations involving chromosome 14q32 in follicular lymphomas carrying t(14;18), in CLLs carrying t(11;14), and in endemic Burkitt's lymphomas carrying t(8;14), all exhibit chromosome breakpoints in the J_H region. Many of these breakpoints map precisely to the 5' ends of normal J_H segments. This is the site of normal V–D–J recombination catalyzed by the B- and T-cell "recombinase" (which is one and the same). At least one endemic Burkitt's tumor has a breakpoint in the D_H segments, at another site of V–D–J joining (27). In T cells, breakpoints in the TCRα region involve the 5' ends of Jα segments. Those translocations that involve TCRδ exhibit recombination at the 5' ends of Jδ segments. And finally, translocations in the TCRβ locus involve the 5'

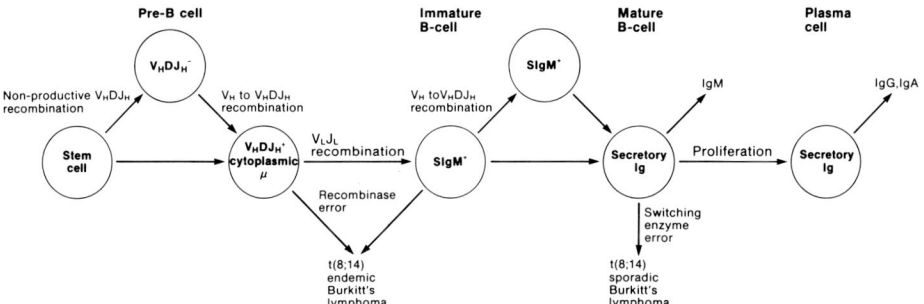

FIG. 5. Correlation between B-lymphocyte ontogeny and the point at which translocations leading to malignancy occur. The recombinase functions in pre-B cells or immature B cells. At these stages of differentiation, its physiologic function is to implement V_H–D–J_H joining, V_H to V_H–D–J_H recombination, or V_L–J_L joining. Errors at these steps may give rise to translocations, as observed in follicular lymphoma, chronic lymphocytic leukemia, acute lymphoblastic leukemia, and endemic Burkitt's lymphoma. In more mature cells, isotype switching takes place following antigenic stimulation. Thus, in mature B cells, translocations of the type seen in sporadic Burkitt's lymphomas may occur.

end of Jβ (12). The structural features of all of these translocation breakpoints, when taken together, very strongly suggest that these translocations arise as a consequence of errors made by the V–D–J recombinase.

Closer scrutiny of the breakpoints by sequence analyses further strengthens this supposition. Sequences related to those employed by the V–D–J recombinase in physiologic recombination have been found at sites of translocation on the chromosomes aberrantly joined to Ig or TCR regions. This observation was first made in the t(11;14) of CLL (22) and later extended to the t(14;18) of follicular lymphomas (28), endemic Burkitt's lymphomas (24), and T-cell tumors (11). Additionally, most of the translocation breakpoints have associated N regions, or stretches of nucleotides deriving from neither parent chromosome, at the joining sites. These sequences are thought to be added by terminal deoxynucleotidyl transferase during lymphocyte ontogeny. The molecular characteristics of the breakpoints support the view that the translocations arise through V–D–J joining mistakes.

The implication of the V–D–J joining enzymes in the genesis of chromosome translocations has ramifications for our understanding of the first steps in tumorigenesis itself. This is best exemplified by Burkitt's lymphomas. These B-cell tumors have two presentations. Endemic Burkitt's lymphomas are largely confined to equatorial Africa. They are almost uniformly associated with infection by the Epstein–Barr virus (EBV) and are comprised of tumor cells that have a relatively immature B-cell phenotype. In contrast, sporadic Burkitt's lymphomas occur throughout the world, are associated with EBV in a minority of cases, and exhibit a more mature B-cell phenotype.

The molecular features of t(8;14) chromosome translocations from endemic and sporadic Burkitt's lymphomas provide a possible explanation for these differences (15, 29). Endemic Burkitt's lymphomas carry t(8;14) translocations that join regions far 5' of c-*myc* to J_H segments and thus appear to arise from V–D–J joining mistakes (29). Since V–D–J joining occurs early in B-cell ontogeny, it is likely that the translocations also occur early during B-cell development. Further, this scenario provides a rationale for the observed association with EBV: it is possible that EBV leads to clonal expansion of early B-cell populations, resulting in an increased likelihood of recombination error. In contrast, sporadic Burkitt's lymphomas usually involve chromsome breakage in or near the c-*myc* gene and in isotype switching sequences. Thus, they probably arise later in B-cell ontogeny during isotype switching and exhibit relatively mature B-cell phenotype.

VIII. Summary

The body of this work illustrates the utility of the combined cytogenetic and molecular approach to lymphoid tumorigenesis. A number of tumor-specific translocations have proven amenable to dissection by molecular techniques. We have a firm grasp of the general principles that underlie lymphoid neoplasia; in particular, the activation of cellular oncogenes by translocation into genes of the immunoglobulin superfamily is a widespread phenomenon. However, numerous lymphopoietic malignancies are only poorly understood. These remain a challenge for the continued application of these methodologies.

Acknowledgments

This work was supported in part by National Institutes of Health (NIH) Grant CA 39860, Outstanding Investigator Grant (to C.M.C.). F.G.H. is a trainee of the Medical Scientist Trainee Program at the University of Pennsylvania School of Medicine, under NIH Grant T32 GM 07170.

References

1. J. L. Ziegler, *N. Engl. J. Med.* **305**, 735 (1981).
2. C. M. Croce and P. C. Nowell, *Adv. Immunol.* **38**, 245 (1986).
3. J. J. Yunis, *Science* **221**, 227 (1983).
4. Y. Tsujimoto, J. Cossman, E. Jaffe, and C. M. Croce, *Science* **228**, 1440 (1985).
5. Y. Tsujimoto, J. Yunis, L. Onorato-Showe, J. Erikson, P. C. Nowell, and C. M. Croce, *Science* **224**, 1403 (1984).
6. C. M. Croce, M. Isobe, A., Palumbo, J. Puck, J. Erikson, M. Davis, and G. Rovera, *Science* **227**, 1044 (1985).
7. J. Erikson, D. L. Williams, J. Finan, P. C. Nowell, and C. M. Croce, *Science* **229**, 784 (1985).
8. J. Erikson, L. Finger, L. Sun, A. Ar-Rushdi, K. Nishikura, J. Minowada, J. Finan, B. S. Emanuel, P. C. Nowell, and C. M. Croce, *Science* **232**, 844 (1986).
9. J. Kagan, J. Finan, J. Letofsky, E. C. Besa, P. C. Nowell, and C. M. Croce, *PNAS* **84**, 4543 (1987).
10. Y. Chien, M. Iwashima, D. Kaplan, J. F. Elliott, and M. M. Davis, *Nature (London)* **327**, 682 (1987).
11. L. R. Finger, R. C. Harvey, R. C. A. Moore, L. C. Showe, and C. M. Croce, *Science* **234**, 892 (1986).
12. M. Isobe, G. Russo, F. G. Haluska, and C. M. Croce, *PNAS* **85**, 3933 (1988).
13. T. Boehm, L. Buluwela, D. Williams, L. White, and T. Rabbitts, *EMBO J.* **7**, 2011 (1988).
14. F. G. Haluska, Y. Tsujimoto, and C. M. Croce, *Annu. Rev. Genet.* **21**, 321 (1987).
15. F. G. Haluska, Y. Tsujimoto, and C. M. Croce, *Trends Genet.* **3**, 11 (1987).
16. S. Tonegawa, *Nature* **302**, 575 (1983).
17. J. Erikson, A. Ar-Rushdi, H. L. Drwinga, P. C. Nowell, and C. M. Croce, *PNAS* **80**, 820 (1983).

18. Y. Tsujimoto, L. R. Finger, J. Yunis, P. C. Nowell, and C. M. Croce, *Science* **226**, 1097 (1984).
19. Y. Tsujimoto and C. M. Croce, *PNAS* **83**, 5214 (1986).
20. J. C. Reed, Y. Tsujimoto, J. D. Alpers, C. M. Croce, and P. C. Nowell, *Science* **236**, 1295 (1987).
21. M. Negrini, E. Silini, C. Kozak, Y. Tsujimoto, and C. M. Croce, *Cell* **49**, 455 (1987).
22. Y. Tsujimoto, E. Jaffe, J. Cossman, J. Groham, P. C. Nowell, and C. M. Croce, *Nature (London)* **315**, 340 (1985).
23. J. Erikson, J. Finan, Y. Tsujimoto, P. C. Nowell, and C. M. Croce, *PNAS* **81**, 4144 (1984).
24. R. Baer, K.-C. Chen, S. D. Smith, and T. H. Rabbitts, *Cell* **43**, 705 (1985).
25. L. Mengle-Gaw, H. F. Willard, C. I. E. Smith, L. Hammarstrom, P. Fisher, P. Sherrington, G. Lucas, P. W. Thompson, R. Baer, and T. H. Rabbitts, *EMBO J.* **6**, 2273 (1987).
26. G. Russo, M. Isobe, L. Pegoraro, J. Finan, P. C. Nowell, and C. M. Croce, *Cell* **53**, 137 (1988).
27. F. G. Haluska, Y. Tsujimoto, and C. M. Croce, *PNAS* **84**, 6835 (1987).
28. Y. Tsujimoto, J. Gorham, J. Cossman, E. Jaffe, and C. M. Croce, *Science* **229**, 1390 (1985).
29. F. G. Haluska, Y. Tsujimoto, S. Finver, and C. M. Croce, *Nature (London)* **324**, 158 (1986).

Molecular Analysis of Chromosome Breakpoints

JOHN GROFFEN[1]
ANDRÉ HERMANS*
GERARD GROSVELD* AND
NORA HEISTERKAMP

Section of Molecular Genetics,
Division of Medical Genetics,
Children's Hospital of Los
Angeles, Los Angeles, California
90027, and
* Department of Cell Biology and
Genetics, Erasmus University,
3000 DR Rotterdam, The
Netherlands

The Philadelphia (Ph') chromosome, an abnormal chromosome 22, is one of the best-known examples of a specific human chromosomal abnormality strongly associated with one form of human leukemia, chronic myelocytic leukemia (CML). The finding that a small region of chromosome 9 that includes the c-*abl* oncogene is translocated to chromosome 22 prompted studies to elucidate the molecular mechanism involved in this disease. All Ph'-positive CML patients examined to date have a breakpoint within a 5.8-kb region on chromosome 22 that is part of a gene for which we have proposed the name "breakpoint cluster region" gene (BCR). Using BCR cDNA sequences, we obtained data strongly suggesting the presence of a chimeric BCR/ABL mRNA in the leukemic cells of Ph'-positive CML patients. The recent isolation of cDNA clones containing BCR and ABL sequences confirms this finding. Moreover, our results indicated that the abnormal ABL protein, found in patients with CML, consists of a BCR/ABL fusion protein.

Chromosomal aberrations may be generated by specific events involving recombination-prone DNA sequences. Alternatively, such recombination events could occur almost at random. In either case, a very limited number of translocations will result in gene alterations leading to the disruption of normal growth and differentiation. In the

[1] Speaker.

Ph' translocation, we have found that breakpoints on chromosome 9 are spread over a region of over 100 kilobases (kb), while most breakpoints on chromosome 22 occur within a smaller region of around 5.8 kb. Nonetheless, no straightforward sequence homology can be found between breakpoint regions of different CML patients or coding regions of ABL and BCR genes.

In two patients, more detailed data were obtained through DNA sequencing of the breakpoints of the 9q+, 22q−, and corresponding normal chromosomes 9 and 22. In both patients, limited stretches of homology near the breakpoint could be identified. A DNA search revealed homology of this region to human *Alu* repetitive sequences. The fact that both of the 9;22 translocations occurred within *Alu* sequences is interesting. Four separate thalassemias have been caused by "illegal" recombination within *Alu* sequences; so has one case of familial hypercholesterolemia. All of these had breakpoints at different points in the *Alu* sequence. Because *Alu* sequences constitute 3–5% of the human genome, these occurrences could be by chance; however, it is possible that *Alu* sequences are hotspots for illegitimate recombination.

I. v-*abl* and the Mouse c-*abl* Gene

Abelson murine leukemia virus (A-MuLV) is a recombinant between Moloney murine leukemia virus (M-MuLV) and cellular sequences of mouse origin designated v-*abl* when they occur in the virus. Abelson and Rabstein isolated A-MuLV from a thymectomized mouse inoculated with Moloney murine leukemia virus (M-MuLV) (*1*). One mouse developed an acute B-cell leukemia. The virus isolated from this mouse differed from M-MuLV in that it was replication-defective and could transform murine NIH/3T3 fibroblasts and lymphoid cells *in vitro* (*2, 3*). Analysis of viral RNAs and proteins demonstrated that A-MuLV produced a fusion protein encoded by M-MuLV *gag* polyprotein sequences and sequences of nonviral origin (*4*). The non-M-MuLV component was derived from a normal cellular gene (mouse c-*abl*). Molecular cloning and DNA sequence studies have confirmed that A-MuLV arose from recombination between M-MuLV and the normal mouse c-*abl* gene (*5–7*).

The v-*abl* sequences encode a protein with tyrosine-specific protein kinase activity (*8, 9*). The intrinsic tyrosine kinase activity of the protein is intimately connected with the ability of A-MuLV to transform cells because mutants with reduced kinase efficiency have lower transforming efficiency; moreover, mutants that lack tyrosine kinase are transformation-deficient (8). Tyrosine kinase has also been associated with the transforming proteins of other retroviruses and with the cellular receptors for several growth factors, e.g., epidermal growth factor (EGF), platelet-derived growth factor (PDGF) and insulin (10–12). Nucleotide sequence comparison of the genes encoding these viral and cellular tyrosine kinases show that they are highly homologous in their catalytic domains (6, 13–15).

The cellular *abl* gene of the mouse represents the progenitor of the transforming v-*abl* sequences in the genome of Abelson murine leukemia virus. The mouse c-*abl* locus spans at least 40 kb of mouse genomic DNA (7). The c-*abl* gene is expressed as two equally abundant, distinct mRNA species of 6.5 and 5.3 kb in most murine cell types. The two mRNAs are generated by alternative splicing events at the 5' end of the mouse c-*abl* gene (16). Based on the cloning and characterization of four different mouse c-*abl* cDNAs, it appears that at least four c-*abl* proteins with alternative N-termini exist (16). No amino-acid homology between these extreme N-termini and those of c-*src* and c-*yes* has been found, suggesting that these specific domains might contribute unique, important characteristics to the proteins.

II. Human ABL and CML

A. Characterization of Human ABL

The human genome contains sequences highly homologous to the mouse v-*abl* oncogene. Based on Southern hybridization experiments of v-*abl* probes with human genomic DNA, the human ABL oncogene was estimated to span a region of at least 30 kb (17).

To facilitate further characterization of v-*abl* homologous human DNA sequences, a cosmid library of human carcinoma DNA was utilized (18). Hybridization of this library with a v-*abl* probe resulted in the isolation of multiple cosmid clones, which were characterized by restriction enzyme mapping (17). These experiments confirmed that human ABL is a large gene; within the gene, v-*abl* homologous sequences are distributed discontinuously over a region of 32 kb and are dispersed over nine exons (17, 19).

The nucleic-acid sequences of the regions encoding the tyrosine phosphorylation acceptor sites of v-*abl* and human ABL (exon designated Tyr; Fig. 1) have been determined (6, 13). Extensive homology between this region of ABL and the acceptor domains of the v-*src*, v-*yes*, and v-*fps/fes* family of viral oncogenes was found, as well as more distant relatedness to regions within the catalytic chain of the mammalian cAMP-dependent protein kinase. These findings and the

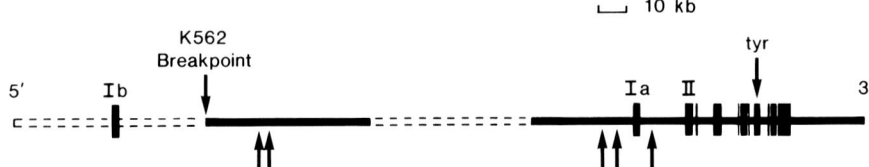

FIG. 1. The genomic DNA of the human ABL locus is represented by the horizontal line, with dashed and solid lines indicating the regions uncloned and cloned to date. The positions of the variable exons Ib and Ia and the common exon II are indicated, as well as the location of the exon containing the phosphotyrosine acceptor site (tyr). Arrows beneath the schematic diagram indicate the position of the chromosomal breakpoints on chromosome 9 in DNAs from five CML patients; the position of the breakpoint in the CML cell line K562 is indicated by an arrow above the schematic diagram.

results of other comparative studies suggest that all the homologs of retroviral oncogenes with tyrosine-specific protein kinase activity were probably derived from a common progenitor and may represent members of a diverse family of cellular protein kinases.

As found in A-MuLV, v-*abl* sequences lack introns and may be regarded as incomplete cDNA copies of the murine c-*abl* oncogene. Whereas v-*abl* is approximately 3 kb long, v-*abl* homologous mRNAs of 6 and 7 kb have been detected in human cells; indeed, the isolation of cDNAs corresponding to these mRNAs established the presence of additional human ABL exons, located 5' to the v-*abl* homologous exons (*19, 20*). One exon is located immediately 5' of the most 5' v-*abl* homologous exon (II in Fig. 1). Two more *abl* exons are present 5' of exon II and are designated exon Ia and Ib (Fig. 1).

Exon Ia is the most 5' exon present within the 6.0-kb RNA species, whereas exon Ib represents the 5' terminus of the 7.0-kb *abl* mRNA (Fig. 1); both mRNAs contain exon II. Both exons Ia and Ib contain coding sequences for different peptides (*19*). This indicates that the two ABL mRNAs initiate at different promoters and give rise to proteins that differ in their N-termini. A similar observation has been made for the *abl* mouse mRNAs of 5.3 and 6.5 kb. The coding regions of human exons Ia and Ib are closely homologous to their murine counterparts; this is a strong indication that the two ABL proteins have distinct functions.

Alternative splicing has been found in a number of DNA viruses (*21*) and in cellular genes. In the latter, alternative splicing could affect the level of gene expression by the selection of different promoters without affecting the coding sequence (*22, 23*), or, alternatively, could generate different proteins (24–26). For ABL, two distinct transcriptional promoters are present; moreover, since the coding regions begin within the two alternative first exons, exon selection will determine which of the two protein products will be made. To date, there is no evidence that the two proteins are expressed differentially, as both ABL mRNAs have been detected in approximately equal quantities in all tissues analyzed (27; G. Grosveld, personal communication). The low expression of ABL in many tissues is reminiscent of the expression of some genes considered to have a "housekeeping" function such as hypoxanthine phosphoribosyl-transferase (EC 2.4.2.8). ABL, like these genes (28), lacks TATA and CAAT boxes, and its two promoter regions are (G+C)-rich and contain multiple GGGCGG repeats. This sequence was identified as the core within the consensus binding site of the Sp1 factor, which was found to regulate transcription *in vitro* (29).

As mentioned, the ABL locus is large; the v-*abl* homologous exons alone were found to be dispersed over a region of 32 kb. Exon Ia lies 18 kb 5' of the most 5' v-*abl* homologous exon (19). Although considerable stretches of DNA immediately 5' of this exon have been cloned (see Fig. 1), none of these cloned DNAs contains exon Ib (19 and unpublished data). Based on the data of DNA molecularly cloned to date, exon Ib should lie more than 100 kb 5' of exon II (30). From recent experiments using pulsed-field gel electrophoresis, exon Ib is situated approximately 200 kb (see Fig. 1) 5' of the first common exon, exon II (31). These data indicate that the ABL locus has some unusual features. First, splicing must occur over very long distances; second, when transcription initiates at exon Ib, exon Ia is spliced out and exon Ib is fused to exon II. Experiments *in vivo* (32) and *in vitro* (33) argue strongly against a progressive scanning mechanism for splicing. In this mechanism, the splicing enzymes attach at one position and move along the RNA chain. Instead, normal splicing and alternative splicing are thought to be determined by secondary structures within intervening sequences that juxtapose particular exons and exclude others. The current model proposes that regions of the RNA that could be spliced in more than one way could assume different conformations, possibly stabilized by interactions with *trans*-acting factors such as RNAs or proteins (34). Whether this model can explain the different ABL transcripts has yet to be established.

B. Chromosomal Breakpoints in Human ABL

Using somatic cell hybrids and *in situ* hybridization, human ABL has been located in chromosome 9 band q34 (35–37). This finding is of interest, because of the involvement of the long arm of chromosome 22 (22q11–qter) in a specific translocation with the long arm of chromosome 9 (band 9q34), the Philadelphia translocation (Ph'), occurring in CML. The abnormal chromosomes are designated 9q+ and 22q−; of these, the 22q− (Ph') chromosome is oberved in 96% of all cases of CML (Fig. 2).

We investigated the chromosomal location of the human ABL gene in cases of CML where the Ph' translocation is present. Southern blot analyses with ABL and v-*abl* probes were performed on *Eco*RI-digested DNAs from somatic cell hybrids segregating the 9q+ and 22q− chromosomes (*36*).

These experiments establish that in CML, ABL is translocated to the 22q− chromosome and that the Ph' translocation is reciprocal, a general assumption that is demonstrated unequivocally by these results. Moreover, the data raise the possibility of involvement of ABL in the generation of CML.

In addition, the translocation of ABL in variant forms of CML was investigated. Employing *in situ* hybridization techniques (*37*), no translocation of human ABL to chromosome 22 was found in two Ph'-negative CML patients; however, in two patients with complex Ph' translocations, a t(9;11;22), the c-*abl* oncogene was translocated to the Ph' chromosome. This indicates that ABL moves consistently to chromosome 22 in patients with complex Ph' translocations. Moreover, recent *in situ* hybridization experiments demonstrated that in yet another variant form of CML, in which chromosome 22 was previously thought not to be involved, ABL is translocated to chromosome 22 (*38, 39*).

These results indicate strongly that the ABL oncogene may be involved in Ph'-positive CML. If ABL were actively involved in the development of this disease, one would expect that the breakpoint on chromosome 9 would occur either within or in relatively close proximity to ABL. To investigate this possibility, DNA of high molecular weight was isolated from biopsy samples of three CML patients. Each DNA was digested with *Bam* HI and subjected to Southern blot analysis. Extensive analysis with probes isolated from the v-*abl* homologous region of ABL revealed no abnormalities or rearrangements in the DNAs of these patients. A probe specific for the region 5' of exon II (Fig. 1) detected a normal fragment in normal human DNA and in the DNAs of two CML patients; however, an abnormal restriction fragment was identified in the DNA of one of the patients (*40*). The most likely explanation for this finding is that one allelic copy of DNA sequences immediately 5' of the v-*abl* homol-

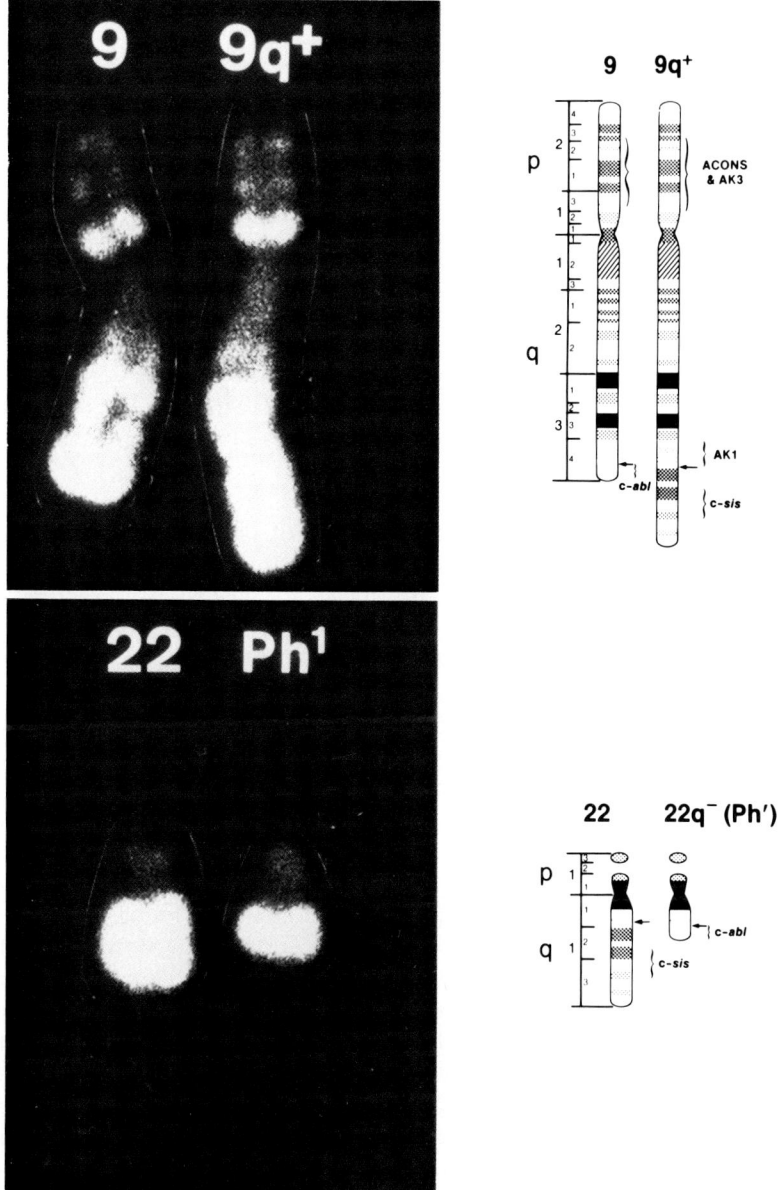

FIG. 2. The panel on the left shows the chromosomes 9 and 22 that are involved in the Philadelphia chromosome (Ph', or Ph[1]) translocation, and the resulting abnormal chromosomes 9q+ and 22q−. A schematic drawing of these chromosomes is shown on the right, with the approximate locations of the ABL and SIS oncogenes. Arrows indicate the positions of the chromosomal breakpoints.

ogous sequences in this patient is normal, whereas the second is rearranged. Molecular cloning of the rearranged sequences confirmed this hypothesis. DNA sequences in the 5' region of this fragment completely matched with the ABL locus on chromosome 9. In contrast, sequences more to the 3' exhibited no homology to chromosome 9 sequences.

To determine the origin of the latter sequences, a probe corresponding to this region was isolated and hybridized to DNAs isolated from rodent/human somatic-cell hybrids. Under stringent washing conditions, this probe hybridized to one fragment in normal human DNA, but not to sequences in mouse DNA or in rodent/human somatic-cell hybrids containing a human chromosome 9 (40). However, the probe detected a restriction enzyme fragment in a somatic-cell hybrid containing chromosome 22 sequences as its only human component. Thus, the cloned rearranged fragment of the one CML patient represents a chimeric chromosome 9/22 DNA fragment.

Later studies confirmed the presence of Ph' breakpoints in other CML DNAs within the 5' region of the c-*abl* gene. Other breakpoint fragments have been cloned, and the location of their corresponding breakpoints within ABL have been determined (see Fig. 1) (41–43); none of the different CML DNAs examined to date have a breakpoint at an identical location. Breakpoints appear to be distributed more or less at random within the 5' region of ABL.

C. The BCR Gene on Chromosome 22

The molecular cloning of a chimeric restriction enzyme fragment containing the t(9;22) breakpoint from one patient enabled us to use the chromosome 22 sequences as a molecular probe. Hybridization of the chromosome 22-specific part of the cloned chimeric DNA fragment to DNA isolated from a second CML patient revealed the presence of abnormal restriction fragments (41). This finding suggested that in both CML patients a breakpoint had occurred within the limited stretch of chromosome 22 DNA that could be examined with this probe. To investigate the possibility of clustering of CML breakpoints on chromosome 22, approximately 45 kb of chromosome 22 corresponding to the chromosome 22-specific probe was cloned from a normal human cosmid library. No homology was apparent between the restriction map of this region and that of human SIS (44), an oncogene situated on chromosome 22 but translocated to chromosome 9 in the Ph' translocation (45). Subsequent *in situ* hybridization studies indicated that SIS is not located in the immediate proximity of the Ph' breakpoint (46). The immunoglobulin λ light-chain constant region (Cλ) and the Ph' chromosomal breakpoint are in chromosome 22 band q11 (47–49); this suggested that analogous to the t(8;14) of c-*myc* in Burkitt's lymphoma, ABL could be translocated into Cλ in CML. However, a probe isolated from the immunoglobulin λ light-chain constant region showed no cross-homology with the above-described chromosome 22 sequences. Moreover, no hybridization to a murine-λ variable region probe (50) was observed. As the breakpoint in the t(9;22) on chromosome 22 was situated within a common region in two CML DNAs, other CML DNAs were examined for a breakpoint in this region.

Initially, 17 independent CML DNAs were subjected to Southern blot analysis, using the chromosome 22-specific probe; all Ph'-positive CML DNAs contained additional DNA fragments hybridizing to this probe. In contrast, no additional fragments were detected in over 20 random non-CML DNAs (41). Moreover, extra DNA fragments were not found in DNA isolated from cultured fibroblasts of a Ph'-positive CML patients, although DNA isolated from the leukemic cells of these patients clearly contained extra DNA fragments. Finally, in DNA isolated from leukemic cells of a Ph'-negative CML patient and a 2-year-old child with juvenile Ph'-negative CML, no visible rearrangements were found, confirming the results of our previous experiments, (46) in which no translocations involving ABL or SIS were found in Ph'-negative CML. Based on these and additional data, all Ph'-positive CML patients appear to have a breakpoint in a very small region (~5 kb) of chromosome 22 (41). The name "major breakpoint cluster region" (Mbcr) has been proposed for this region (51).

Two to five percent of CML patients lack a clearly identifiable Ph' chromosome. The lack of a cytogenetically visible Ph' chromosome is not an unambiguous indication for the lack of involvement of Mbcr and ABL: in a Ph'-negative CML patient with a t(9;12)(q34;q21) and two apparently normal chromosomes 22, Mbcr and ABL were found to be jointly translocated to chromosome 12; this patient had an unusually long clinical remission as compared to other Ph'-negative patients (38).

Similar data have been obtained in a study of five Ph'-negative CML patients (39); two of these patients were found to have a breakpoint within Mbcr, and it is likely that the Ph' translocation has occurred in these patients, although it could not be detected cytogenetically.

It is obvious that the Ph' translocation has a profound influence on the cellular ABL oncogene, resulting in disruption of the linkage between the gene and sequences 5' to it. In the following studies, the question was addressed as to how the sequences on chromosome 22 are affected by this chromosomal break. One possibility was that these sequences could be a gene or part of a gene disrupted as a consequence of the Ph' translocation.

To examine whether Mbcr contains protein-encoding regions, probes from the Mbcr were tested for their ability to hybridize to cDNA sequences. One of the probes tested hybridized to sequences in a human fibroblast cDNA library and was subsequently used for the isolation of several cDNA clones. The largest cDNA, containing an insert of 2.2 kb, was characterized in detail by restriction-enzyme mapping and sequence analysis (42). Northern blot analysis revealed that this cDNA was incomplete at its 5' end; mRNAs of 4.5 and 6.5 kb were detected in a human fibroblast cell line and in various tissues (G. Grosveld et al., unpublished results).

The orientation of this breakpoint-cluster-region gene (BCR) (51) on chromosome 22 was determined by preparing 5' and 3' probes from the cDNA, followed by hybridization to cosmids containing human chromosome 22 sequences; the 5' end of the BCR gene is toward the centromere of chromosome 22 and remains on the Ph' chromosome, while the 3' end is translocated to chromosome 9 in the t(9;22) (42).

Computer programs were used to examine whether any homology could be detected between the BCR gene and oncogenes, growth factors, or other previously isolated proteins; however, no significant homology was found, indicating that, at present, the BCR protein sequence yields no clues as to its cellular function and its possible role in CML. Moreover, recent analysis of the complete 5' end of the 4.5-kb mRNA did not result in the detection of significant homology with any other peptide sequences (52).

The origin of the normal 6.5-kb BCR transcript remains obscure; probes isolated from different coding regions of the BCR gene hybridize to both transcripts. Hence, the 6.5-kb mRNA appears to contain most, if not all, of the sequences present in the 4.5-kb mRNA. The additional sequences in the 6.5-kb transcript might be located at its 5' end, internally, or at its 3' end; for example, the 6.5-kb BCR transcript may be generated by a more distal polyadenylation signal or an alternative splice in the 3' region.

The 5' untranslated region of the 4.5-kb BCR mRNA has a high GC content and is likely to display considerable secondary structure. Such structures may be involved in post transcriptional regulation, since increasing the secondary structure of a 5' noncoding region has been shown to reduce the translational efficiency of eukaryotic mRNA, both *in vitro* and *in vivo*.

D. Breakpoints within Mbcr

The isolation of BCR cDNA sequences allowed a further characterization of Mbcr on chromosome 22. The exact position of five BCR exons in the region affected by the Ph' translocation were determined (see Fig. 3; the exons are numbered 1–5). These exons are relatively small and vary in size from 76 to 105 base-pairs (*42*).

Once the number and the position of the exons within Mbcr were known, the breakpoints in CML could be redefined as to whether they occur in exon or intron regions. The breakpoint regions of six chimeric 9q+ and/or 22q− restriction-enzyme fragments were sequenced. The exact location of the breakpoints was determined by sequencing of the corresponding normal chromosome 22 regions. of six breakpoints analyzed in detail, none were found within an exon, indicating that in the Ph' translocation, breakpoints occur within introns of the BCR gene. Although the exact point of translocation on chromosome 22 differs for each CML DNA, there seems to be little variation in the amount of BCR exons remaining on the Ph' chromosome. Four DNAs retain exon 2 and all sequences 5' to it, whereas the DNAs of two CML patients included exon 3 of the major cluster region.

These results indicate that the Ph' translocation results in the fusion of two genes in a head-to-tail fashion; the BCR gene at the 5' end has lost its 3' sequences, which are replaced by ABL sequences from chromosome 9. To examine the effect of the translocation on the expression of these genes, poly(A) RNA was isolated from the CML cell line K562 (*53*) and control HeLa cells. A probe isolated from the most 5' v-*abl* homologous exon of human ABL detected mRNAs of 8.5, 7.0, and 6.0 kb in K562. The 7.0- and 6.0-kb mRNAs were also

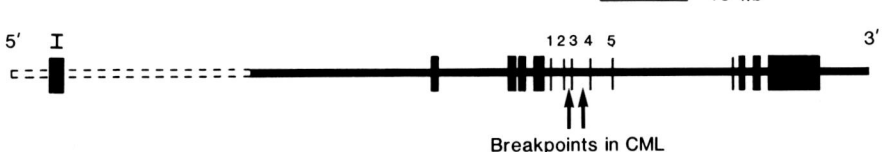

FIG. 3. The solid horizontal line indicates the cloned genomic sequences of the human BCR locus; the bracket line at the 5' end represents genomic sequences (not cloned to date) that separate exon I from the main body of the BCR exons. The breakpoints in CML DNAs are indicated with arrows and are located in the introns between exons numbered 1–5 of the major breakpoints cluster region.

detected in RNA isolated from HeLa cells; these represent transcripts from the normal, unrearranged ABL allele on chromosome 9. The 8.5-kb ABL mRNA, however, is found only in K562 and not in non-CML cells (54). Indeed, abnormally sized ABL-homologous mRNA seems to be characteristic for Ph'-positive CML cells; an mRNA of ~8.5 kb was also detected in RNA isolated from the leukemic cells of CML patients (55–57). These findings, in combination with the knowledge that the ABL gene is affected by the Ph' translocation, resulted in the following hypothesis: as a result of the Ph' translocation and the consequent head-to-tail fusion of BCR and ABL genomic sequences, a chimeric gene might be generated that could be transcribed into a chimeric BCR/ABL mRNA. To test this hypothesis, two probes were isolated from the BCR cDNA, probe A corresponding to the 5' region of the cDNA ending in exon 2 of the Mbcr, and probe B starting in Mbcr exon 5 and encompassing 3' BCR cDNA sequences.

If the 8.5-kb *abl* hybridizing mRNA is chimeric, it should hybridize to probe A, which contains BCR exons 5' to the breakpoint; however, the mRNA should not hybridize to probe B, as 3' exons of BCR were translocated to chromosome 9. This hypothesis proved to be correct, as probe A hybridized with the 8.5-kb mRNA, whereas no hybridization with this mRNA could be detected with probe B in mRNA isolated from the cell line K562 (38). These results unambiguously show that the abnormal 8.5-kb c-*abl* mRNA as found in the CML cell line K562 is chimeric and represents 5' BCR sequences fused to ABL sequences. Confirming evidence for the existence of a BCR/ABL chimeric mRNA as a direct consequence of the Ph' translocation was obtained for five independent CML patients (58).

The structure of the chimeric *bcr/abl* mRNA of K562 could be examined in more detail by the molecular cloning of a chimeric cDNA. As predicted from the genomic DNA organization, it encompasses all BCR exons to the 5' including exon 3 (Fig. 3); the BCR sequence is then joined to that of ABL exon II (Fig. 1) in such a fashion that it is in frame (20). Although one ABL exon 5' to exon II (exon Ia) is present on the Ph' chromosome in the genomic DNA of K562, it has apparently been deleted by splicing from the mature chimeric mRNA; the CML cell line EM2 has a similarly structured chimeric mRNA, with Mbcr exon 3 joined to ABL exon II (20). Although most Ph'-positive CML cells seem to contain a similarly sized abnormal transcript of 8–8.5 kb [all of 25 patients (59), all of five patients (58)], two cases with an additional transcript of 9.0 kb have been reported (59). This indicates that in some cases additional BCR and/or ABL exons may be included in the transcript.

E. Are Specific DNA Sequences Required for the Ph' Translocation?

Chromosomal aberrations may be generated by specific events involving recombination prone DNA sequences. In at least one case of familial hypercholesterolemia, an abnormal LDL receptor resulted from an unequal crossing-over between homologous *Alu*-repetitive DNA sequences (60). Moreover, at least four different cases of thalassemia have been caused by illegitimate recombination within *Alu* sequences (61). In some types of human cancer involving translocations between an immunoglobulin gene and a second gene, the translocation seems to be a mistake of the immunoglobulin VDJ joining mechanism (62), which is controlled by specific heptamer–nonamer DNA sequences.

We have sequenced the region around the breakpoints of the 9q+, 22q−, and corresponding normal chromosomes 9 and 22. In some patients, we could find *Alu*-repetitive sequences near or as a part of the breakpoint (see Fig. 4). In contrast, no *Alu*-repetitive sequences were found associated with the breakpoint of at least one patient (42). These observations suggest that *Alu*-repetitive sequences may facilitate the Ph' chromosomal translocation. However, it is not likely that *Alu* sequences are a prerequisite for this type of chromosomal translocations.

F. The Chimeric BCR/ABL mRNA Is Translated

Using v-*abl* antisera, the normal cellular gene product of human ABL has been identified as a phosphoprotein with a mass of 145,000 daltons (P145). In the cell line K562, v-*abl* antisera precipitate two abnormally large phosphoproteins of 210,000 and 190,000 daltons (P210 and P190) in addition to P145 (63).

These results indicate that the chimeric BCR/ABL mRNA present in Ph'-positive CML cell lines and patient material is translated into a chimeric protein; the difference in mass between P210 and the normal ABL P145 is caused in part by the addition of a BCR moiety to the amino terminus of P145. Using antisera specific for the segment of the BCR protein included in the BCR/ABL protein, the existence of a BCR/ABL P210 fusion protein was demonstrated definitively (64, 65).

As the properties of human BCR/ABL and v-*abl* P160 seem similar, the specific contribution of the BCR or *gag* moieties of these fusion proteins (apart from furnishing promoter sequences and a novel 5' end) to the *in vivo* protein kinase activity is unclear. Interestingly, the normal BCR gene product is a phosphoprotein (P160) with an associated serine or threonine protein kinase activity, making it

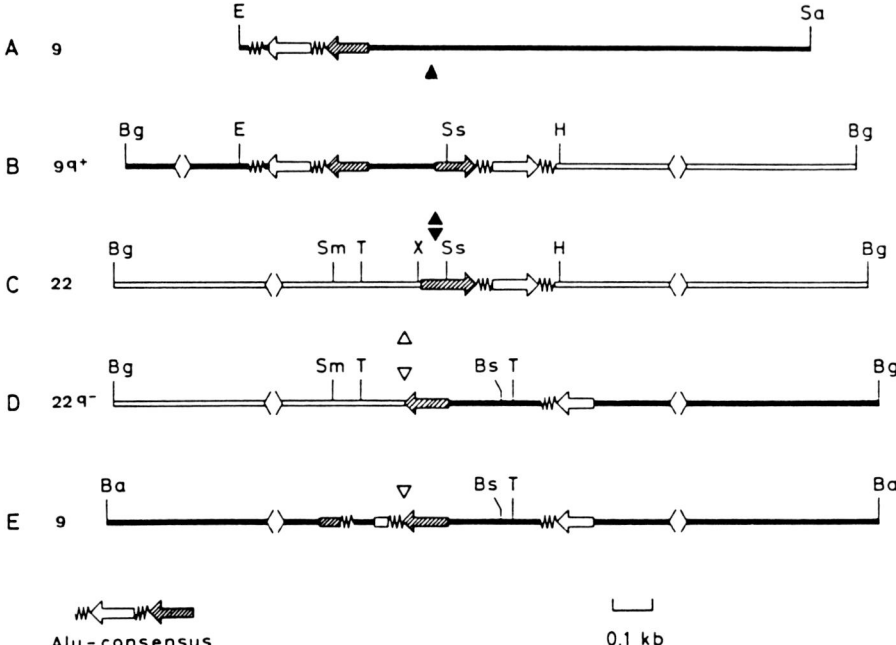

FIG. 4. Restriction enzyme maps of the subcloned breakpoint (B, D) and corresponding normal chromosome 9 (A, E) and 22 (C) fragments. Solid bars indicate chromosome 9 sequences and open bars indicate chromosome 22 sequences. Only the restriction enzyme maps and *Alu*-repetitive sequences around the breakpoints are shown (—<>— indicates known DNA regions omitted from the figure). The position and orientation of *Alu*-repetitive sequences are indicated in the maps using the *Alu*-consensus symbol shown at the bottom. The 300-bp *Alu*-consensus consists of two nearly homologous halves (arrows), each followed by an A-rich tract (zigzags). Restriction enzymes: Ba, *Bam*HI; Bg, *Bgl*II; Bs, *Bst*II; E, *Eco*RI; H, *Hin*dIII; Sa, *Sal*I; Sm, *Sma*I; Ss, *Sst*1; T, *Taq*1; X, *Xho*I.

entirely dissimilar to *gag* (64). As the sequence of the BCR cDNA bears no homology to the region conserved among the members of the tyrosine-specific protein kinase family, it is not likely that the BCR gene product has intrinsic kinase activity.

G. The Ph' Translocation in Acute Lymphoblastic Leukemia (ALL)

The Ph' chromosome also exists in other malignant hematopoietic disorders, including different subtypes of acute leukemias in which no preceding chronic phase has been observed (66, 67). The Ph' chromosome is found in the leukemic cells of 2% to 3% of the patients exhibiting acute myeloid leukemia (AML) and a similar incidence (2% to 6%) is reported for childhood ALL (68). In adult ALL, the Ph' chromosome is the most frequent chromosomal abnormality, with an incidence of 17% to 25% (69). The clinical distinction between blast crisis of CML and *de novo* Ph'-positive acute leukemia is not always clear (66). Nevertheless, these two disorders have fundamental differences, the salient feature of CML being the rapid proliferation of mature myeloid cells, whereas that of ALL is a greatly increased number of immature lymphoid cells. Furthermore, the presence of Ph'-negative cells in the bone marrow during the acute phase and the elimination of Ph'-positive cells from the bone marrow during remission are typical features of cases presenting as Ph'-positive acute leukemias with no known prior CML. These unique clinical manifestations of Ph'-positive CML compared with ALL suggest discrete events in their pathogenesis.

In contrast with CML, not all Ph'-positive (Ph+) ALL DNAs appear to have a rearrangement in the major cluster region (70, 71). These results indicate that Ph+ ALL is heterogeneous with respect to the involvement of this region of the BCR gene. Nonetheless, the translocation of ABL to chromosome 22 has been shown in some Ph+ ALLs, indicating that, in analogy with CML, ABL is activated. Additional evidence for the involvement of c-*abl* was obtained in studies using ABL DNA probes and antisera (72, 73). In contrast to CML, where a BCR/ABL fusion of P210 is characteristic, a phosphoprotein with a mass of 190,000 daltons (P190) was detected in three Ph+ ALL patients with no rearrangement of the BCR. Recent results indicate that in these Ph+ ALL patients, the Ph' translocation breakpoint on chromosome 22 occurs more to the 5' side of the BCR but is still within the BCR gene (74). For at least three Ph+ ALL patients, it could be demonstrated that the chimeric BCR/ABL mRNA consists of BCR exon I fused to ABL exon II. Why the P190 BCR/ABL fusion protein is specifically found in some cases of Ph+ ALL and not in Ph+ CML is presently unclear. Nonetheless, these data suggest that, as in CML, the immediate result of the Ph' translocation in Ph+ ALL is a BCR/ABL fusion protein.

III. Summary and Future Directions

We have described an increasing insight into the molecular events occurring as a consequence of the first translocation identified as specifically associated with one type of leukemia: the Philadelphia chromosome translocation. A hybrid gene is generated by the translocation, consisting of 5' regulatory, promoter, and exon sequences of the BCR gene on chromosome 22 fused to 3' exons and polyadenylation/termination sequences of the ABL oncogene originating from chromosome 9 (see Fig. 5). Therefore, it is likely that such a hybrid gene will be regulated by an as-yet-unidentified transcriptional regulatory signals directed at the normal BCR gene.

The attachment of a BCR moiety to the ABL protein has resulted in increased tyrosine kinase. In analogy to this increased activity and the transforming potential of v-*abl*, it is likely that this enzymatic activity of the BCR/ABL protein is involved in cellular transformation.

Circumstantial evidence indicates that, at minimum, the production of an active P210 must confer some selective advantage on cells expressing it; our results indicate no obvious reason (such as "illegitimate" homologous recombination) for the t(9;22) to occur. Lacking contrary evidence, one could assume that many rearrangements may take place in cells that are genetically unstable. Although random recombinations between ABL and the BCR gene would then occur in all introns and exons (with the presence of *Alu* repeats and the larger size of introns favoring recombination in introns), only certain rare configurations, allowing the production of a functional protein, would lead to an abnormal clonal expansion of Ph'-positive cells. The presence of large introns 5' within the ABL gene would increase the chance of generating functional chimeric BCR/ABL genes.

Research has reached a point in which emphasis is shifting way from molecular cloning; it seems well established that translocation and a consequent chimeric gene product play a crucial role in this type of leukemia. Future studies should address questions such as how the altered ABL protein is involved in leukemia. It seems obvious that the BCR/ABL protein will differ from the v-*abl* gene product; no significant homology between BCR and A-MuLV *gag* sequences have been detected. Moreover, the BCR/ABL fusion protein has not been associated with types of cancer other than leukemia, in contrast to the v-*abl* gene product. Indeed, the evidence suggests that BCR/ABL lacks the ability to transform NIH/3T3 cells (G. Grosveld *et al.*, unpublished results), although these cells can be transformed efficiently by v-*abl*. The ABL part of the BCR/ABL protein is highly homologous to v-*abl*. It therefore seems highly probable that this region of human ABL has the potential to transform NIH/3T3 cells; the specific involvement of the BCR/ABL protein in leukemia must, at least in part, be contributed through the amino-terminal BCR gene sequences.

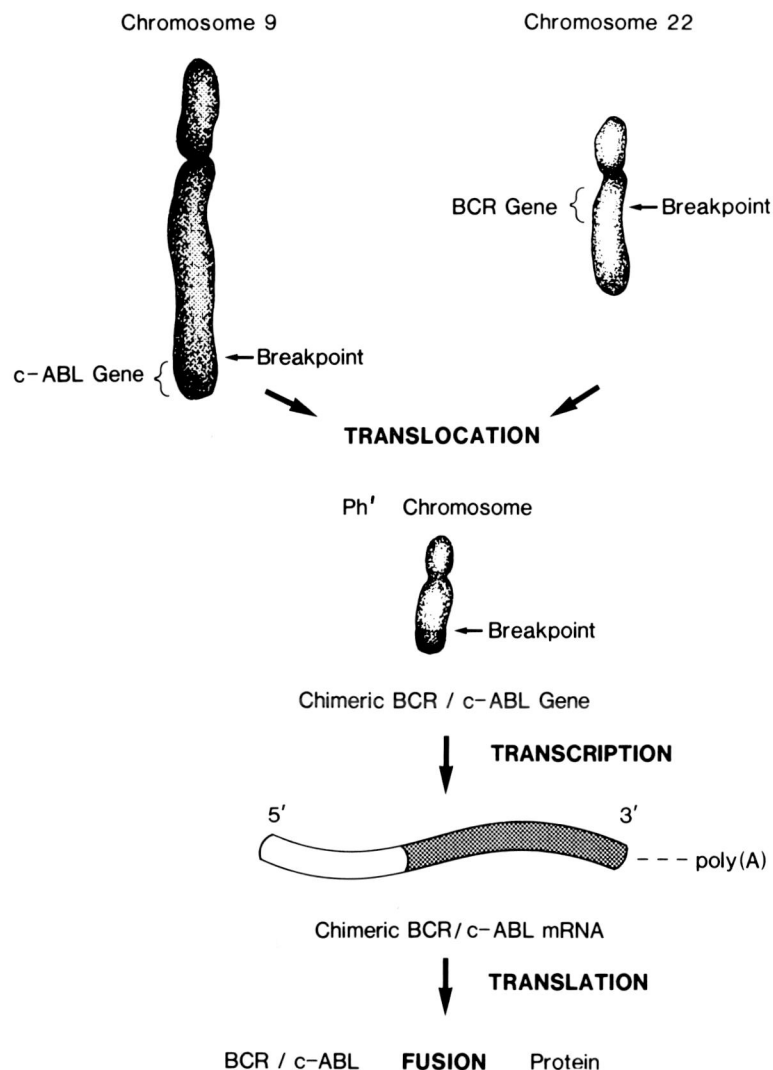

FIG. 5. Schematic representation of the molecular genetic effects of the Ph' translocation in CML.

Experiments directed toward resolving the question of how the BCR/ABL fusion protein is involved in leukemia are not easily designed. Presently, we do not know the normal cellular function of either the ABL or the BCR gene products. In addition, little, if anything, is known about their interactions with other cellular substrates. Therefore, research aimed at investigating the normal function of ABL and BCR may be crucial in efforts to understand their abnormal functioning in leukemia and to increase our understanding of the disease. Perhaps the development of BCR/ABL but not v-*abl* transgenic mice will yield an animal model for the study of CML.

Ackowledgment

Part of this work was supported by The Margaret E. Early Medical Research Trust. We wish to thank the secretarial staff of the Division of Medical Genetics, Children's Hospital of Los Angeles, for typing this manuscript.

References

1. H. T. Abelson and L. S. Rabstein, *Cancer Res.* **30**, 2213 (1970).
2. C. D. Scher and R. Siegler, *Nature (London)* **253**, 729 (1975).
3. N. Rosenberg, D. Baltimore, and C. D. Scher, *PNAS* **72**, 1932 (1975).
4. F. H. Reynolds, T. L. S. Sacks, D. N. Deobaghar, and J. R. Stephenson, *PNAS* **75**, 3974 (1978).
5. S. P. Goff, E. Gilboa, O. N. Witte, and D. Baltimore, *Cell* **22**, 777 (1980).
6. E. Reddy, M. Smith, and A. Srinivasan, *PNAS* **80**, 7372 (1983).
7. Y. J. Wang, F. Ledley, S. Goff, R. Lee, Y. Groner, and D. Baltimore, *Cell* **36**, 349 (1984).
8. O. N. Witte, A Dasgupta, and D. Baltimore, *Nature (London)* **283**, 826 (1980).
9. O. N. Witte, S. P. Goff, N. Rosenberg, and D. Baltimore, *PNAS* **77**, 4993 (1980).
10. H. Ushiro and S. Cohen, *JBC* **255**, 8363 (1980).
11. B. Ek, B. Westermark, A. Wasteson, and C. D. Heldin, *Nature (London)* **295**, 419 (1982).
12. M. Kasuga, Y. Zick, D. Blithe, M. Crettaz, and C. R. Kahn, *Nature (London)* **298**, 667 (1982).
13. J. Groffen, N. Heisterkamp, F. H. Reynolds, Jr., and J. R. Stephenson, *Nature (London)* **304**, 364 (1983).
14. J. Downward, Y. Yarden, E. Mayes, G. Scrace, N. Totty, P. Stockwell, A. Ulrich, J. Schlesinger, and M. D. Waterfield, *Nature (London)* **307**, 521 (1984).
15. A. Ullrich, J. R. Bell, E. Y. Chen, R. Herrera, L. M. Petruzzelli, T. J. Dull, A. Gray, L. Coussens, Y. C. Liao, M. Tsubokawa, A. Mason, P. H. Seeburg, C. Grunfeld, O. M. Rosen, and J. Ramachandran, *Nature (London)* **313**, 756 (1985).
16. Y. Ben-Neriah, A. Bernards, M. Paskind, G. Q. Daley, and D. Baltimore, *Cell* **44**, 577 (1986).
17. N. Heisterkamp, J. Groffen, and J. R. Stephenson, *J. Mol. Appl. Genet.* **2**, 57 (1983).
18. J. Groffen, N. Heisterkamp, F. Grosveld, W. J. M. Van de Ven, and J. R. Stephenson, *Science* **216**, 1136 (1982).
19. E. Shtivelman, B. Lifshitz, R. P. Gale, B. A. Roe, and E. Canaani, *Cell* **47**, 277 (1986).
20. E. Shtivelman, B. Lifshitz, R. P. Gale, and E. Canaani, *Nature (London)* **315**, 550 (1985).
21. E. Ziff, *Int. Rev. Cytol.* **93**, 327 (1985).
22. C. Benyajati, N. Spoerel, H. Haymerie, and M. Ashburner, *Cell* **33**, 125 (1983).
23. R. A. Young, O. Hagenbuchle, and U. Schibler, *Cell* **23**, 451 (1981).

24. P. Early, J. Rogers, M. Davis, K. Calame, M. Bond, R. Wall, and L. Hood, *Cell* **20**, 313 (1980).
25. C. E. Rozek and N. Davidson, *Cell* **32**, 23 (1983).
26. G. S. Basi, M. Boardman, and R. V. Storti, *MCBiol* **4**, 2828 (1984).
27. E. H. Westin, F. Wong-Staal, E. P. Gelman, R. Dalla-Favera, T. Papas, J. A. Lautenberger, A. Eva, E. P. Reddy, S. R. Tronick, S. A. Aaronson, and R. C. Gallo, *PNAS* **79**, 2490 (1982).
28. D. W. Melton, C. McEwan, A. B., McKie, and A. M. Reid, *Cell* **44**, 319 (1986).
29. W. S. Dynan, S. Sazer, R. Tjian, and R. Schimke, *Nature (London)* **319**, 246 (1986).
30. G. Grosveld, T. Verwoerd, T. van Agthoven, A. de Klein, K. L. Ramachandran, N. Heisterkamp, K. Stam, and J. Groffen, *MCBiol* **6**, 607 (1986).
31. C. A. Westbrook, C. M. Rubin, M. M. Le Beau, L. S. Kaminer, S. D. Smith, F. D. Rowley, and M. O. Diza, *PNAS* **84**, 251 (1987).
32. T. Kuhne, B. Wieringa, J. Weiser, and C. Weissmann, *EMBO J.* **2**, 727 (1983).
33. D. Solnick, *Cell* **42**, 157 (1985).
34. R. E. Breitbart, H. T. Nguyen, R. M. Medford, A. T. Destree, V. Mahdavi, and B. Nadal-Ginard, *Cell* **41**, 67 (1985).
35. N. Heisterkamp, J. Groffen, J. R. Stephenson, N. K. Spurr, P. N. Godfellow, E. Solomon, B. Carrit, and W. F. Bodmer, *Nature (London)* **299**, 747 (1982).
36. A. de Klein, A. Geurts van Kessel, G. Grosveld, C. R. Bartram, A. Hagemeijer, D. Bootsma, N. K. Spurr, N. Heisterkamp, J. Groffen, and J. R. Stephenson, *Nature (London)* **300**, 765 (1982).
37. C. R. Bartram, A. de Klein, A. Hagemeijer, T. van Agthoven, A. G. van Kessel, D. Bootsma, G. Grosveld, M. A. Ferguson-Smith, T. Davies, M. Stone, N. Heisterkamp, J. R. Stephenson, and J. Groffen, *Nature (London)* **306**, 277 (1983).
38. C. R. Bartram, E. Kleihauer, A. de Klein, G. Grosveld, J. J. Teyssier, N. Heisterkamp, and J. Groffen, *EMBO J* **4**, 683 (1985).
39. C. M. Morris. A. E. Reeve, P. H. Fitzgerald, P. E. Hollings, M. E. J. Beard, and D. C. Heaton, *Nature (London)* **330**, 281 (1986).
40. N. Heisterkamp, J. R. Stephenson, J. Groffen, P. F. Hansen, A. de Klein, C. R. Bartram, and G. Grosveld, *Nature (London)* **306**, 239 (1983).
41. J. Groffen, J. R. Stephenson, N. Heisterkamp, A. de Klein, C. R. Bartram, and G. Grosveld, *Cell* **36**, 93 (1984).
42. N. Heisterkamp, K. Stam, J. Groffen, A. de Klein, and G. Grosveld, *Nature (London)* **314**, 758 (1985).
43. A. de Klein, T. van Agthoven, C. Groffen, N. Heisterkamp, J. Groffen, and G. Grosveld, *NARes* **14**, 7071 (1986).
44. R. Dalla Favera, E. P. Gelman, R. C. Gallo, and F. Wong-Staal, *Nature (London)* **292**, 31 (1981).
45. J. Groffen, N. Heisterkamp, J. R. Stephenson, A. van Kessel, A. de Klein, G. Grosveld, and D. Bootsma, *J. Exp. Med.* **158**, 9 (1983).
46. C. R. Bartram, A. de Klein, A. Hagemeijer, G. Grosveld, N. Heisterkamp, and J. Groffen, *Blood* **306**, 223 (1984).
47. J. D. Rowley, *Nature (London)* **243**, 290 (1973).
48. O. W. McBride, P. A. Hieter, G. F. Hollis, D. Swan, M. C. Otey, and P. Leder, *J. Exp. Med.* **155**, 1480 (1982).
49. J. J. Yunis, *Science* **221**, 227 (1983).
50. J. Miller, A. Bothwell, and U. Storb, *PNAS* **78**, 3829 (1982).
51. UCLA Colloquium on CML, Annapolis, Maryland, December 1–4, 1987.
52. I. K. Hariharan and J. M. Adams, *EMBO J.* **6**, 115 (1987).

53. C. B. Lozzio and B. B. Lozzio, *Blood* **45**, 321 (1975).
54. S. J. Collins and M. T. Groudine, *PNAS* **80**, 4813 (1983).
55. E. Cannani, R. P. Gale, and D. Steiner-Saltz, *Lancet* **1**, 593 (1984).
56. S. J. Collins, I. Kubonishi, I. Miyoshi, and M. T. Groudine, *Science* **225**, 72 (1984).
57. R. P. Gale and E. Cannani, *PNAS* **81**, 5648 (1984).
58. K. Stam, N. Heisterkamp, G. Grosveld, A. de Klein, R. S. Verma, M. Coleman, H. Dosik, and J. Groffen, *N. Engl. J. Med.* **313**, 1429 (1985).
59. R. P. Gale and E. Cannani, *Br. J. Haematol.* **60**, 395 (1985).
60. M. Lehrman, W. Schneider, T. Sudhof, M. Brown, J. Goldstein, and D. Russell, *Science* **227**, 140 (1985).
61. E. Vanin, P. Henthorn, D. Kioussis, F. Grosveld, and O. Smithies, *Cell* **35**, 701 (1983).
62. J. Tsujimoto, J. Gorham, J. Cossman, E. Jaffe, and C. M. Croce, *Science* **229**, 1390 (1985).
63. J. B. Konopka, S. M. Watanabe, and O. N. Witte, *Cell* **37**, 1035 (1984).
64. K. Stam, N. Heisterkamp, F. H. Reynolds, Jr., and J. Groffen, *MCBiol* **7**, 1955 (1987).
65. Y. Ben-Neriah, G. Q. Daley, A. M. Mes-Masson, O. N. Witte, and D. Baltimore, *Science* **223**, 212 (1986).
66. M. E. J. Beard, J. Durrant, D. Catovsky, E. Wiltshaw, J. L. Amess, R. L. Brearley, B. Kirk, P. F. M. Wrigley, G. Janossy, M. F. Greaves, and D. A. G. Galton, *Br. J. Haematol.* **34**, 167 (1976).
67. A. Oshimura and A. A. Sandberg, *Cancer* **40**, 1149 (1977).
68. J. M. Chessells, G. Janossy, S. D. Lawlery, L. M. Secker, and L. M. Walker, *Br. J. Haematol.* **41**, 25 (1979).
69. R. E. Champlin and D. W. Golde, *Blood* **65**, 1039 (1985).
70. A. de Klein, A. Hagemeijer, C. R. Bartram, R. Houwen, L. Hoefsloot, F. Carbonell, L. Chan, M. Barnett, M. Greaves, E. Kleihauer, N. Heisterkamp, J. Groffen, and G. Grosveld, *Blood* **68**, 1369 (1986).
71. J. Erikson, C. A. Griffin, A. ar-Rushdi, M. Valtieri, J. Hoxie, J. Finan, B. S. Emanuel, G. Rovera, P. C. Nowell, and C. M. Croce, *PNAS* **83**, 1807 (1986).
72. L. C. Chan, K. K. Karhi, S. I. Rayter, N. Heisterkamp, S. Eridani, R. Powles, S. D. Lawler, J. Groffen, J. G. Foulkes, M. F. Greaves, and L. M. Wiedemann, *Nature (London)* **325**, 635 (1987).
73. R. Kurzrock, M. Shtalrid, P. Romero, W. S. Kloetzer, M. Talpas, J. M. Trujillo, M. Blick, M. Beran, and J. U. Gutterman, *Nature (London)* **335**, 631 (1987).
74. A. Hermans, N. Heisterkamp, M. von Lindern, S. van Baal, D. Meijer, D. van der Plas, L. M. Wiedemann, J. Groffen, D. Bootsma, and G.Grosveld, *Cell* **51**, 33 (1987).

Homologous Recombination in Mammalian Somatic Cells

Raju S. Kucherlapati

Department of Genetics, University
of Illinois College of Medicine,
Chicago, Illinois 60612

Genetic recombination plays a very important role in several different cellular processes. An understanding of homologous recombination in mammalian somatic cells is expected to play an important role in developing a new genetic tool for the study of animal cells. Using viral or plasmid DNA molecules, it has been possible to demonstrate that somatic mammalian cells have the ability to catalyze recombination between molecules introduced into cells. Knowledge gained from such studies has permitted the design of experiments to detect recombination between specific chromosomal sequences and their counterparts introduced by DNA transfection methods. These experiments have permitted the correction of defective chromsomal genes, inactivation of chromosomal genes, and targeted integration of DNA sequences. The use of homologous recombination strategies along with the newer methods to culture pluripotent mouse embryonic stem cells is expected to play an important role in understanding gene structure-function relationships.

The ability to introduce defined segments of DNA into yeast cells and have them undergo homologous recombination with their cellular counterparts (1) provided means by which cellular genes could be inactivated ("gene knock-out") or corrected in this organism. Such gene manipulation has proved to be a very valuable tool in the genetic analysis of yeast. The development of efficient gene transfer systems, if associated with homologous recombination in mammalian cells, would prove extremely valuable in analysis of gene structure–function relationships and gene therapy, and in developing animal models for certain human diseases.

I. Gene Transfer Systems

Efficient gene transfer by means of a calcium phosphate coprecipitation method has been reported for adenoviral DNA (2). Based on early observations, efficient methods of DNA-mediated gene transfer were developed (*3, 4*). After the successful introduction of a variety of genes by this method, several new methods for gene introduction have been developed. These include use of viral vectors (for reviews, see 5 and 6), microinjection into somatic cells (*7, 8*) and into fertilized mouse embryos (*9*) DEAE–dextran-mediated transfer (*10*) and electroporation (*11*). Each of these methods is useful in transferring genetic information into mammalian cells at varying efficiencies. It was also observed that the genetic information introduced by most of these methods became incorporated into the cellular genome (*12*). The sites at which the DNA sequences integrate, however, seem to be random. Though the introduced genes are expressed, reliable regulation of them is not oberved. There is substantial evidence to indicate that this lack of qualitatively and quantitatively proper regulation of gene expression is the result of the random integration of the introduced genetic information. We felt that if the genes could be introduced into the homologous sites in the chromosome, they would be expressed and regulated properly.

Several years ago, we embarked on a program to see if somatic mammalian cells have the ability to catalyze homologous recombination, and if so, to understand the processes and mechanisms by which such events are catalyzed. We wished to utilize the knowledge gained from such experiments to develop methods that would permit targeted integration of DNA sequences at specific sites in the mammalian genome by homologous recombination. In this paper, I summarize the current status of these efforts, using our experimental systems to illustrate the features of recombination.

II. Homologous Recombination of Extrachromosomal Molecules

The fact that somatic mammalian cells have all of the enzymes required to catalyze homologous recombination was first illustrated using viral substrates. Oligomeric simian virus 40 (SV40) DNA sequences that cannot be packaged into viral envelopes were introduced into permissive monkey cells. Intramolecular homologous recombination resulted in the formation of unit-sized molecules that were packaged into viral envelopes and released to form plaques (*13, 14*). Similar work involving other viruses has been reported (*15, 16*). However, these experiments held out the possibility that the homologous recombination events are catalyzed by viral gene products and not by the cellular machinery. Definitive evidence that recombination can be mediated by cellular enzymes was obtained by introducing two plasmids, each of which carried a mutant or incomplete selectable gene (*17, 18*). Recombination between the two plasmids at an appropriate location within the region of homology would result in the reconstruction of an intact gene that would enable the cells to grow in the appropriate selective medium. Examination of the cells that acquired the selectable phenotype revealed that homologous recombination had indeed occurred.

The system we developed to study recombination was based on a eukaryotic–prokaryotic shuttle vector, pSV2*neo* (*19*). This plasmid contains a neomycin phosphotransferase gene (*neo*) that can be expressed in mammalian cells by virtue of the SV40 transcriptional signals associated with the gene in the plasmid. The presence of the *neo* gene product renders the cells resistant to an aminoglycoside analog, G418. Since mammalian cells do not carry *neo* or its analog, the G418 resistance acts as a dominant selectable phenotype. To study recombination, deletion substrates of pSV2*neo* were constructed. One substrate, "pSV2*neo*-deletion-left" (DL), was generated by removing a 248-bp *Nar*I fragment from the 5' end of the gene. A second substrate, "pSV2*neo*-deletion-right" (DR), was obtained by deleting a 283-bp *Nae*I fragment from the 3' end of the *neo*. Each of these deletions rendered *neo* inactive in bacterial as well as in mammalian cells. To study recombination, these plasmids were introduced into human and mouse cells individually or as a mixture. Neither of the deletion plasmids yielded any resistant (G418^R) colonies, but when the DL and DR plasmids were introduced as a mixture by the calcium-phosphate coprecipitation method, G418^R colonies were obtained (*20*). The rate at which the G418^R colonies were obtained with the mixture of deletion plasmids was 0.1–0.05 that obtained when wild-type pSV2*neo* plasmid was used for transfection. Similar results were obtained by other investigators using other substrates (*21–24*). Examination of the DNA from the G418^R cells as well as plasmids rescued from these cells revealed that homologous recombination was indeed responsible for the generation of intact *neo* (*20*).

Since transfection, selection of G418^R colonies, and examination of their DNA require a substantial amount of time, we developed an alternative assay that provided a rapid and efficient method to study recombination and to examine the products of recombination. The strategy involved introduction of the pSV2*neo* deletion plasmids into monkey COS cells (*25*), where they can replicate autonomously. Forty-eight hours after transfection, low-molecular-weight DNA, which includes the autonomously replicating plasmids, was isolated and used to transform a strain of *E. coli* carrying a mutation in a key gene involved in recombination (*recA*). Individual bacterial colonies that grow in the presence of neomycin or kanamycin harbor plasmids that are the result of individual recombination events. As such, analysis of the plasmids would reveal the structure of the recombined plasmid, from which it is possible to deduce the nature of the recombination event (26).

Studies of recombination conducted in our laboratory (26) as well as those from other laboratories (27) revealed several features of recombination, which can be summarized as follows. (*a*) Homologous recombination between extrachromosomal plasmids occurs at a high efficiency. (*b*) Recombination events do not require integration of the plasmids. (*c*) Recombination events can be detected as early as 4 hours after introduction of DNA. (*d*) Recombinant molecules are generated by reciprocal as well as nonreciprocal events. Studies of autonomously as well as nonautonomously replicating plasmids provided convincing evidence that somatic mammalian cells can indeed catalyze homologous recombination.

III. Cell-Free Systems for Study of Homologous Recombination

Further studies of mechanisms of recombination were pursued by introduction of a variety of recombination substrates into cells grown in culture, or incubation of the cells with nuclear extracts. In experiments involving nuclear extracts, nuclear proteins free of nucleic acid were incubated with bacterial plasmids carrying mutant genes (28) or bacteriophages carrying mutant genes (29). Following the incubations, the DNA was isolated and the presence of recombinant molecules was detected by gel electrophoresis or by introduction of the molecules into an appropriate bacterial strain in which the recombinant molecules could be easily detected. Successful experiments of this nature provided additional evidence that all of the enzymes required for homologous recombination are present in a number of mammalian cells tested.

Based upon observations in yeast, several investigators examined the effect of double-stranded DNA breaks on homologous recombination in mammalian cells. To study this feature, we digested pSV2*neo* DL or DR with a number of restriction endonucleases, each of which cuts the plasmid once in *neo* or outside *neo*. The effect of such breaks on recombination were tested by either introducing a mixture of the linear and its circular partner into cells and observing the yield of

G418^R colonies, or by co-incubating the plasmids with nuclear extracts and counting the number of neo^R colonies obtained. We observed that double-strand breaks within the region of homology, especially at the site of the deletion in the DL or DR molecules, increased recombination 10- to 100-fold. Breaks close to the site of deletion also increased recombination, but if the breaks were outside neo, there was no effect on the frequency of reconstruction of intact neo (20, 26).

Examination of the fate of polymorphic markers flanking neo in the two substrates revealed that (1) the recombination process resulted in the correction of neo in the cut molecule, (2) nonreciprocal recombination (gene conversion) events played an important role in the generation of wild-type neos, (3) the gene conversion event sometimes included closely linked markers (coconversion), (4) the double-strand breaks acted as initiation sites for recombination, and (5) the cut molecule acted as a recipient of information in the recombination process (28, 30). Similar conclusions were drawn from studies conducted by others (23). These observations, that double-strand DNA breaks within the region of homology increase the frequency of homologous recombination, played a central role in the design of experiments to target genes to specific sites in the mammalian genome.

IV. Modification of Chromosomal Genes by Homologous Recombination

The information that was gathered from studies of inter- and intraplasmid recombination in mammalian cells provided confidence that it may be possible to modify cellular genes by homologous recombination. Experiments designed to modify cellular genes by homologous recombination can be divided into two categories. In one class of experiments, a plasmid containing a mutant selectable gene such as neo or thymidine kinase (tk) was introduced into mammalian cells. The integrated defective selectable gene was then used as a target for gene modification by transfecting those cells in a second round with another plasmid carrying the same selectable gene but carrying a mutation at a different site.

For experiments conducted in our laboratory, a derivative of pSV2gpt, SV2neo (19), was used. This plasmid carries two bacterial dominant selectable genes, neo and the xanthine–guanine phosphoribosyltransferase gene, gpt. We introduced a deletion into neo identical to that present in the pSV2neo DL plasmid described earlier. The resulting plasmid, pSV2*gpt* SV2*neo* DL, was introduced into hamster and human cells. Cells expressing *gpt* were isolated and examined for the presence of *neo* sequences. Several cell lines carrying two to four copies of *neo* in the integrated state were chosen for further experimentation. Each of the cell lines was transfected with pSV2*neo*, pSV2*neo* DL, and pSV2*neo* DR and selected in G418. All cells yielded G418^R colonies when transfected with plasmid pSV2*neo* containing a normal *neo*. None of the cell lines gave G418^R colonies when transfected with pSV2*neo* DL, which carries a deletion identical to that present in the chromosomal plasmid. Most of the cell lines yielded G418^R colonies when transfected with pSV2*neo* DR. It can be assumed that when pSV2*neo* DR was used for transfection, the G418^R colonies could be obtained only through homologous recombination between the chromosomal and plasmid DNA sequences. Thus, the frequency with which G418^R colonies were obtained with this substrate can be considered to be the frequency of homologous recombination. pSV2*neo* can integrate anywhere in the genome (a nonhomologous recombination event) to yield G418^R cells. The ratio of the number of G418^R colonies obtained with pSV2*neo* DR and pSV2*neo* would represent the ratio of homologous to nonhomologous recombination events. In our experiments, this ratio ranged from a high of 1:75 to a low of 1:500 (*31*). Similar successes of modification of an integrated plasmid gene were reported by several investigators (*32–34*).

To examine the mechanism of generation of intact *neo* by homologous recombination, we utilized molecular and genetic methods to analyze the products of recombination (*31*). Genetic analysis of the recombination events was facilitated by the use of polymorphic markers flanking *neo*. The integrated *neo* DL gene was flanked by a *Hin*dIII site at the 5′ end and a *Bam*HI site at the 3′ end. The pSV2*neo* DR used for recombination contained a *Sma*I site at its 5′ end and a *Xho*I site at the 3′ end. A nonreciprocal event or a double-crossover event on either side of the DL lesion would result in a wild-type *neo* flanked by *Hin*dIII and *Bam*HI sites (Class I). A single reciprocal recombination event would yield a *neo* flanked by *Sma*I and *Bam*HI sites (Class II). If a correction of the gene borne on the DR plasmid is responsible for the G418^R phenotype, *neo* resulting from that event would be flanked by *Sma*I and *Xho*I sites.

Examination of the plasmids carrying a wild-type *neo* from 11 independently derived G418^R colonies from a single primary transfectant revealed that four are of Class I, three of Class II, three of Class III, and one had an origin which could not unambiguously be identified. These results indicated that all three possible mechanisms of recombination played a role in generating wild-type *neo*. It is of interest that Class I events are the most desirable when gene therapy is attempted, and Class II events are required for insertional inactivation of genes by homologous recombination. Methods to achieve Class I or Class II events exclusively would prove most useful tools for geneticists. Different strategies to achieve these goals are being explored.

Since the above-described experiments involve the modification of a mutant *neo* introduced into mammalian chromosomes, it is reasonable to ask if the knowledge gained from such systems is equally applicable to true chromosomal genes. To address this issue, we have attempted to modify the human β-globin locus. To detect rare recombination events between the chromosomal gene and a β-globin gene located on a plasmid, we have used a combination of molecular biological strategies.

The input plasmid carried an SV2*neo* gene. This gene facilitated the isolation of colonies representing cells that have stably integrated the input DNA. To distinguish between homologous and nonhomologous events, the input plasmid was designed to have only a part of the globin gene; specifically, the second intervening sequence and sequences to the 3' end of it were deleted. In addition, we have introduced a bacterial tRNA suppressor gene (*supF*) at the 5' end of the globin gene. Recombination between the input plasmid and the chromosomal sequence results in a unique restriction fragment that is not present in the chromosome or in the input plasmid and that could be generated only through a homologous recombination event. This fragment, in addition to its characteristic size, would carry a complete globin gene along with *supF* at the 5' end. Such a DNA fragment is easily identified by virtue of its ability to suppress amber mutations in some essential genes of a bacteriophage vector (Charon 3A) and by hybridizing to the globin IVS-2 DNA probe. Those fragments that carry *supF* as well as a complete β-globin gene would represent homologous recombination events and those that contain only *SupF* will represent nonhomologous events.

When we introduced the above-described plasmid into human cells or into a mouse–human hybrid cell line carrying a single copy of an active human adult β-globin gene, we were able to detect DNA fragments that would have resulted from homologous recombination events. We estimated that the ratio of homologous to nonhomologous events in this case is about 1 : 1,000. Using a sib-selection procedure, we isolated a cell line whose DNA contained the modification of the globin gene predicted from the homologous recombination event (35). These results provided convincing evidence that cellular genes can be modified by homologous recombination at efficiencies comparable to the artificial targets such as *neo*

V. Animal Models for Human Disease

As described above, experiments from several different laboratories have clearly established that somatic mammalian cells have the ability to carry out recombination between extrachromosomal plasmids as well as between chromosomal sequences and introduced plasmid DNA sequences. It is now possible to utilize these methodologies in conjunction with mouse embryonic stem cell growth methodologies to address a number of interesting and important issues about the role of specific genes in development and disease. Embryonic stem cells are derived from the inner cell mass of mouse blastocysts and can be maintained in long-term cell culture. These cells can be manipulated and injected into the blastocoel of appropriate recipient blastocysts. Implantation of these embryos would yield chimeric mice in which the introduced ES cells are capable of contributing to the development of all organs, including the germ line (36). Chimeric mice that harbor the modified cells in their germ line can be bred so that the modified gene can be obtained in a homozygous state.

Kuehn *et al.* (37) obtained a mouse in which the hypoxanthine phosphorsibosyltransferase (*hprt*) gene was inactivated by insertion of a retrovirus. Recent reports that *hprt* in ES cells can be modified by homologous recombination (38, 39) pave the way for developing animals in which specific genes are modified. Such modifications, should in turn permit us to address issues about the role of specific genes in development and differentiation, as well as in disease processes.

VI. Perspectives

Homologous recombination is expected to become a valuable tool in the genetic analysis of mammalian systems. It is now clear that mammalian genes can be modified by homologous recombination. Strategies to increase the frequencies with which homologous recombination occurs and methods to detect them are the subjects of intensive investigation in several different laboratories. An immediate use of this methodology is that it permits inactivation ("knock-out") of genes whose functions can then be assessed. Application of this methodology to pluripotent mouse embryonic stem cells should permit us to assess the developmental or disease significance of specific genes. As the methods for gene modification become perfected, we expect that it will be possible to correct a variety of lesions in chromosomal genes, introduce specific mutations into chromosomal genes, and finally, generate specific deletions in mammalian chromosomes. It is possible that efficient correction of gene mutations in primary cells may eventually permit a safe and effective way of treating a number of human genetic diseases.

Acknowledgments

Work reported in this paper is supported by funds from the National Institutes of Health and the March of Dimes Birth Defects Foundation. A number of individuals from my laboratory contributed to the work described here. They include D. Ayares, L. Chekuri, S. Ehrlich, S. Rauth, and K.Y. Song. My collaborators in some of the work described are P. Moore and O. Smithies. The manuscript was prepared by V. Cummins.

References

1. A. Hinnen, J. B. Hicks, and G. R. Fink, PNAS 75, 1929 (1978).
2. F. L. Graham and A. J. Van der Eb, Virology 52, 456 (1973).
3. M. Wigler, S. Silverstein, L. S. Lee, A. Pellicer, Y. C. Chang, and R. Axel, Cell 11, 223 (1977).
4. N. Maitland and J. K. McDougall, Cell 11, 233 (1977).
5. V. R. Baichwal and B. Sugden, in "Gene Transfer" (R. Kucherlapati, ed.), p. 117. Plenum, New York, 1986.
6. H. M. Temin, in "Gene Transfer" (R. Kucherlapati, ed.), p. 149. Plenum, New York, 1986.
7. W. F. Anderson, L. Killos, L. Sanders-Haigh, P. J. Kretschmer, and E. G. Diakumakos, PNAS 77, 5399 (1980).
8. M. R. Capecchi, Cell 22, 479 (1980).
9. J. W. Gordon, G. A. Scangos, D. J. Plotkin, J. A. Barbosa, and F. H. Ruddle, PNAS 77, 7380 (1980).
10. J. H. McCutchan and J. S. Pagano, JNCI 41, 351 (1968).
11. H. Potter, L. Weir, and P. Leder, PNAS 81, 7161 (1984).
12. A. Pellicer, M. Wigler, R. Axel, and S. Silverstein, Cell 14, 133 (1978).
13. C. T. Wake and J. H. Wilson, PNAS 76, 2876 (1979).
14. C. T. Wake and J. H. Wilson, Cell 21, 141 (1980).
15. P. Upcroft, B. Carter, and C. Kidson, NARes 8, 2725 (1980).
16. F. C. Volkert and C. S. H. Young, Virology 125, 175 (1983).
17. J. Small and G. Scangos, Science 219, 174 (1983).
18. B. R. de Saint-Vincent and G. M. Wahl, PNAS 80, 2002 (1983).
19. P. J. Southern and P. Berg, J. Mol. Appl. Gent. 1, 327 (1982).
20. R. S. Kucherlapati, E. M. Eves, K. Y. Song, B. S. Morse, and O. Smithies, PNAS 81, 3153 (1984).
21. L. Miller and H. M. Temin, Science 220, 606 (1984).

22. G. Shapira, J. L. Stachelek, A. Letsou, L. Soodak, and R. M. Liskay, *PNAS* **80**, 4827 (1983).
23. D. A. Brenner, A. C. Smigocki, and R. D. Camerini-Otero, *MCBiol* **5**, 684 (1985).
24. S. Subramani and P. Berg, *MCBiol* **3**, 1040 (1983).
25. Y. Gluzman, *Cell* **23**, 175 (1981).
26. D. Ayares, L. Chekuri, K. Y. Song, and R. Kucherlapati, *PNAS* **83**, 5199 (1986).
27. S. Subramani and J. Rubnitz, *MCBiol* **5**, 659 (1985).
28. R. S. Kucherlapati, J. Spencer, and P. D. Moore, *MCBiol* **5**, 714 (1985).
29. V. Darby and F. Blattner, *Science* **226**, 1213 (1984).
30. K. Y. Song, L. Chekuri, S. Rauth, S. Ehrlich, and R. S. Kucherlapati, *MCBiol.* **5**, 3331 (1985).
31. K. Y. Song, F. Schwartz, N. Maeda, O. Smithies, and R. Kucherlapati, *PNAS* **84**, 6820 (1987).
32. A. J. H. Smith and P. Berg, *CSHSQB* **49**, 171 (1984).
33. F. L. Lin, K. Sperle, and N. Sternberg, *PNAS* **82**, 1391 (1985).
34. K. R. Thomas, K. R. Folger, and M. R. Capecchi, *Cell* **44**, 419 (1986).
35. O. Smithies, R. G. Gregg, S. S. Boggs, M. A. Koralewski, and R. Kucherlapati, *Nature (London)* **317**, 230 (1985).
36. M. J. Evans and M. H. Kaufman, *Nature (London)* **292**, 154 (1981).
37. M. R. Kuehn, A. Bradley, E. J. Robertson, and M. J. Evans, *Nature (London)* **326**, 295 (1987).
38. K. R. Thomas and M. R. Capecchi, *Cell* **51**, 503 (1987).
39. T. Doetschman, R. G. Gregg, N. Maeda, M. L. Hooper, D. W. Melton, S. Thompson, and O. Smithies, *Nature (London)* **330**, 576 (1987).

Gene Transfer into Primates and Prospects for Gene Therapy in Humans

> KENNETH CORNETTA[1]
> ROBERT WIEDER AND
> W. FRENCH ANDERSON
>
> Laboratory of Molecular
> Hematology, National Heart,
> Lung and Blood Institute,
> National Institutes of Health,
> Bethesda, Maryland 20892

The ability to transfer and express genetic material in mammalian cells presents a new approach to the treatment of many genetic diseases. Retroviral-mediated gene transfer has proven to be an efficient system of gene transfer *in vitro* (1–3). While *in vivo* expression has been obtained in the mouse (4–8), vectors expressing in murine systems do not necessarily express in primates (9). Similarly, complications seen in mice may not occur in other species. To address these issues in primates, we have studied the expression and safety of a murine-based retroviral vector containing the gene for human adenosine deaminase (ADA), whose absence results in a clinical syndrome well suited to an initial attempt at gene therapy (10). To date, we have attained transient *in vivo* expression of the human ADA gene in monkey hematopoietic cells infected by means of a bone marrow transplant protocol. In addition to our efforts to attain *in vivo* expression, we also present our preliminary results on the risks of retroviral-mediated gene transfer, specifically the lack of clinical consequences to primates from exposure to a replication-competent murine retrovirus.

I. Gene Therapy for ADA Deficiency

ADA converts adenosine to inosine, and deoxyadenosine to deoxyinosine. The absence of functional ADA results in toxic levels of deoxyadenosine, which leads to a virtual absence of mature T cells.

[1] Speaker.

Affected patients have a severe combined immunodeficiency and usually die from infections by the age of 2 years. ADA deficiency is a disease well suited for attempts at gene therapy, because it is a single-gene defect that has a wide range of expression, with patients expressing 0.05 to 50 times the normal ADA levels, maintaining essentially a normal phenotype. ADA deficiency is cured by bone marrow transplantation, indicating that the target organ in the disease is one that is accessible and well suited to *ex vivo* manipulations. In addition to the technical advantages, the ability to treat bone marrow *ex vivo* protects other organs, most notably the germ line, from exposure to retroviral insertion. Finally, the lethal nature of the disease is a consideration, as the risk of complications from retroviral gene transfer, including malignancy, is not known.

II. Retroviral Vectors

For a gene transfer system to be clinically useful, it must transfer genetic material at high efficiency into a large number of cells. Retroviral vectors meet this requirement, as the infection efficiency can be as high as 100% and there is the capability of infecting the number of cells required for clinical applications (i.e., 10^7–10^9 cells for a bone marrow transplant protocol).

Retroviruses suited for gene therapy can be divided into three functional regions: (1) the long-terminal-repeats (LTRs), which are required for integration and which contain a strong promoter and enhancer function; (2) the ψ, or packaging, sequence required for RNA packaging into virions; and (3) the *gag, pol,* and *env* regions, which code for proteins that are responsible for virion assembly and ultrastructure. The N2 (*11*) and SAX (*12*) vectors are derived from the Moloney murine leukemia virus (MoMuLV) (see Fig. 1). The vector N2 retains the MoMuLV LTRs and packaging sequences but the *pol, env,* and most of the *gag* region are deleted and replaced by the bacterial gene responsible for neomycin resistance (neo^R). The vector SAX was constructed by subcloning a fragment containing the SV40 early promoter and the human ADA cDNA (*13*) into the *Xho*I site of N2 (Fig. 1).

Both SAX and N2 are replication-defective, as the genes required for virion formation are deleted. In order to package vector RNA, a DNA transcript of the vector is transfected into cells expressing the *gag, pol,* and *env* sequences. Wild-type virus (helper virus) can serve as a source of these gene sequences, but the virions produced will

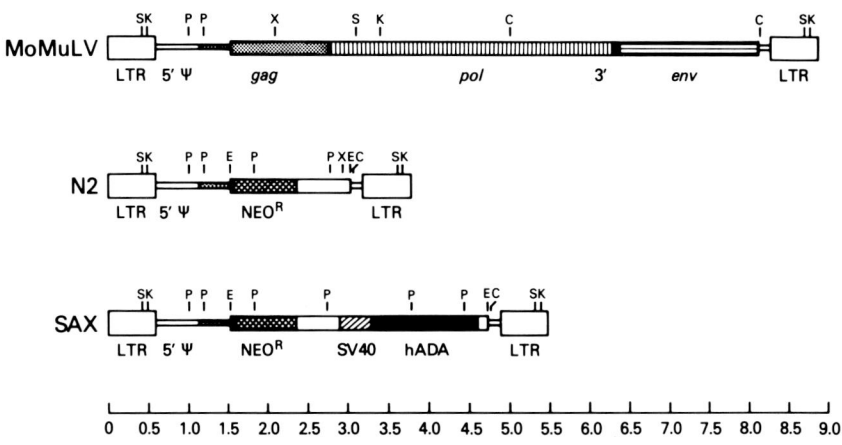

FIG. 1. The Moloney murine leukemia virus (MoMuLV) and the retroviral vectors N2 and SAX. LTR, Long terminal repeat; 5′, retroviral 5′ splice donor site; ψ, viral packaging signal; NEO^R, neomycin resistance gene from Tn5 transposon (crosshatching represents coding sequence); SV40, KpnI–HindIII fragment of the SV40 early promoter; hADA, human ADA cDNA. Restriction sites: S, SacI; K, KpnI; P, PstI; X, XhoI; C, ClaI; E, EcoRI.

contain both replication-defective vector RNA and replication-competent helper-virus RNA. To avoid exposing patients to helper virus, specially designed "packaging" cell lines have been constructed that contain retroviral genomes lacking the ψ, or packaging, sequence. These defective viral genomes express the gag, pol, and env gene products but are unable to package their own RNA.

SAX plasmid was transfected into the ecotropic packaging cell line ψ-2 (14) and then was transinfected into the amphotropic packaging cell line PA12 (15). SAX-producer cell lines generate virions conferring G418 resistance to 3T3 cells at a titer of 5×10^6 colony-forming units/ml. Infection of target cells can be accomplished by cocultivation with viral producers, or by incubating the target cells with the supernatant solution (which contains the virions) collected from producer cells.

Murine and primate cell lines infected with the SAX vector express both the human ADA and neo^R genes *in vitro*. Human ADA-negative T cells infected with the SAX vector produce ADA at levels similar to normal T cells, and become resistant to the toxic effects of deoxyadenosine (12).

III. Bone Marrow Transplant Protocol

Cynomolgus and rhesus monkeys were used for autologous bone marrow transplantation (Fig. 2). Animals were anesthetized and 40 to 60 ml of bone marrow was removed from the long bones. Mononuclear cells were obtained by Ficoll/Hypaque separation (cynomolgus marrow received prior treatment with 3% gelatin sedimentation). Mononuclear bone marrow cells were incubated with vector-producing cells or vector-containing supernatant solution at 37°C in

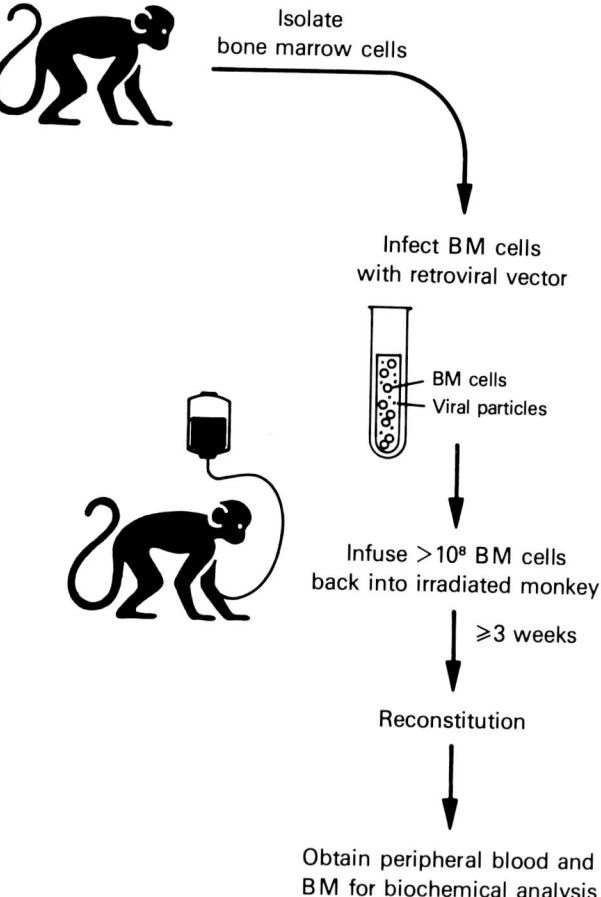

FIG. 2. Primate autologous bone marrow transplant/gene therapy protocol.

the presence of Polybrene (8 µg/ml). While the marrow was being infected with the retroviral vector, the animal received total body irradiation at doses shown to produce permanent aplasia in unreconstituted animals (1000 rads). Immediately after irradiation, the treated marrow was infused intravenously. Animals were supported with blood products, antibiotics, and hyperalimentation.

Once reconstitution occurred, samples of peripheral blood mononuclear cells were assayed for human ADA activity using column chromatography to separate the human isozyme from endogenous monkey ADA (16). Column fractions were incubated with [^{14}C]adenosine and the conversion to inosine was measured by thin-layer chromatography; neo^R expression was determined by a neomycin phosphotransferase assay (17). Additional data were obtained by performing Southern blot analysis, T-cell cloning in G418, and *in situ* hybridization.

Table I summarizes our findings in 14 animals undergoing autologous bone marrow transplantation and infection with the retroviral vector SAX (9). (Experiments were performed at the National Institutes of Health in collaboration with Dr. Arthur Nienhuis and at the Memorial Sloan–Kettering Cancer Center with Dr. Richard O'Reilly.) Of the animals undergoing transplantation, four died before analysis could be performed. Of the 10 surviving animals, six demonstrated evidence of human ADA activity in peripheral blood mononuclear cells, and four of these had the phosphotransferase activity. Only one animal tested positive by Southern blot analysis. The amount of human ADA activity was small compared to the monkey's endogenous ADA activity, with the maximal human ADA activity observed accounting for only 0.5% of the total ADA activity in our most successful animal (Robert; Fig. 3). *In situ* hybridization demonstrated activity in 0.8% of peripheral blood mononuclear cells, suggesting that cells expressing ADA produce close-to-endogenous levels of the enzyme.

The gene expression observed was transient, with peak activity occurring between 2 and 3 months posttransplantation. No activity was detected after day 160. Still unresolved is whether the disappearance of activity is a result of gene inactivation or an immune response, or whether the initial infection occurred only in committed progenitor cells without infecting a renewable totipotential stem cell. Since the initial infections occurred at levels below the level of detection of Southern blot analysis (limit of detection: 1 in 20 cells), the answer has not been attainable. Polymerase chain-reaction analysis should be helpful in answering this question in future experiments.

TABLE I
Primate Bone Marrow Transplant/Gene Transfer: SAX Vector

							ADA Analysis	
No.	Name[a]	Method[b]	Reconstitution	DNA	NPT[c]	hADA	%A-I[d]	% Endogenous[e]
1	Bill (C)	C	No	Pos	Pos	Pos	1	<0.01
2	Mork (R)	C	Yes	—	—	—		
3	Mindy (R)	C	No	—	—	—		
4	Kate (C)	C	No	—	—	—		
5	Ethel (R)	S	Yes	Neg	Neg	Pos	3	<0.01
6	Robert (C)	S	Yes	Neg	Pos	Pos	66	0.5
7	Kyle (C)	S	Yes	Neg	Pos	Pos	17	0.2
8	Venus (R)	C	No	Neg	Pos	Pos	0.5	<0.01
9	George (C)	S	Yes	—	Neg	Neg		
10	Ken (C)	S	Yes	—	Neg	Pos	2	<0.01
11	Oppie (C)	S	No	—	—	—		
12	Barney (C)	S	Yes	—	Neg	Neg		
13	Arthur (R)	S	Yes	Neg	Neg	Neg		
14	Lancelot (R)	S	Yes	Neg	Neg	Neg		

[a] C, Cynomolgus; R, rhesus.
[b] C, Cocultivation; S, supernatant solution.
[c] NPT, Neomycin phosphotransferase activity.
[d] Percentage of conversion of [^{14}C]adenosine to [^{14}C]inosine after isolation of human ADA by column chromatography.
[e] Human ADA activity was a percentage of endogenous monkey ADA activity.

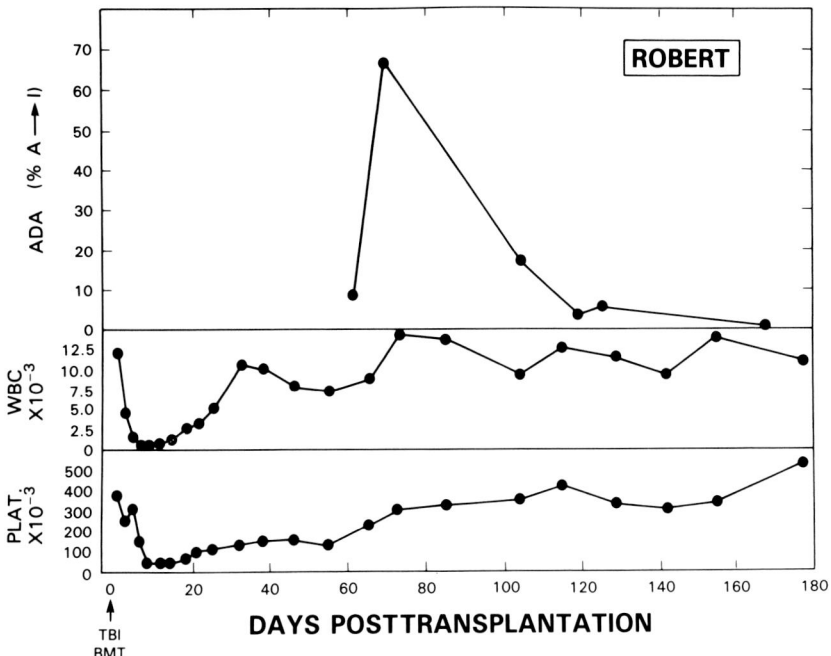

FIG. 3. Clinical course of Robert (monkey 6) after *ex vivo* infection of bone marrow cells during transplant protocol. ADA, Percentage of conversion of [^{14}C]adenosine to [^{14}C]inosine after isolation of human ADA by column chromatography; WBC, peripheral white blood cell count; PLAT., platelet count; TBI, total body irradiation; BMT, bone marrow transplantation (see 9 for details).

IV. Safety

Retroviral-mediated gene transfer is associated with a number of potential complications. Attempts to understand these risks should be addressed before clinical trials are begun. Two possible complications are (*a*) the development of malignancy by insertional mutagenesis, and (*b*) retroviremia with replication-competent murine helper virus.

A. Insertional Mutagenesis

MoMuLV induces T-cell lymphoma when injected into some species of newborn mice. Cell transformation occurs by integration of the retrovirus near protooncogenes, often T-cell-specific protein kinases, leading to abnormal regulation of the protooncogenes (*18–20*).

The aberrant regulation is usually mediated through the enhancer function of the MoMuLV LTR (18). While MoMuLV exhibits a specific tropism for mouse T cells, it is unclear what, if any, primate cell will be susceptible to malignant transformation. These viruses appear to have a low transformation frequency, even in mice. Tumors are usually clonal and develop predominantly in mice exposed during the newborn period, a time when persistent viremia will develop. Taken together, these factors suggest that the malignant potential of retroviral vectors in primates should be small. While we hope this will be the case, it will probably require clinical trials, and many years of follow-up, before the true risk of malignancy is realized.

B. Exposure to Replication-Competent Helper Virus

Retroviral vectors are produced in packaging cell lines designed to prevent helper-virus contamination, but recombination can occur with generation of replication-competent virus (helper virus). This has been the case for the SAX vector generated in PA12 packaging cells used to treat the primates in our bone marrow transplant protocol.

As one means to evaluate the consequences of helper-virus exposure, we retrospectively evaluated animals exposed to helper virus soon after they were lethally irradiated. The five monkeys available for analysis are numbers 6, 7, 9, 10, and 13 listed in Table I and include three animals (numbers 6, 7, and 10) positive for vector expression earlier in their course. All five had bone marrow cells exposed to contaminated SAX vector (helper virus titer approximately 1.3×10^6 focus-forming units/ml when tested by S+/L− assay on pG4 cells). Serum obtained up to 712 days posttransplantation was negative for helper virus by the S+/L− assay. It was also not possible to demonstrate neutralizing antibodies in their serum. Evaluation of peripheral blood and bone marrow cells for amphotropic envelope sequences was negative by Southern blot analysis. These studies suggest that there is no ongoing retroviremia.

If helper virus were to contaminate supernatant preparations used to treat patients, retrovirus could be infused intravenously when the treated bone marrow cells are infused. To observe the clinical consequences of viral infusion, we injected four animals with viral supernatant containing both replication-competent amphotropic retrovirus (4070A) and the SAX retroviral vector. Animals received $4.6-11 \times 10^7$ focus-forming units and $2.5-19 \times 10^7$ colony-forming units intravenously. The protocol was modified for the fourth animal, which received immunosuppression prior to infusion in an attempt to exacerbate any adverse effects of the virus. Retrovirus was detected in the circulation up to 12 minutes postinfusion, but serial serum samples evaluated by S+/L− assay remained negative, with a mean follow-up of 5.2 months. Attempts to rescue helper virus from serum by exposing 3T3 cells to monkey serum have also been negative. The rapid disappearance of helper virus is associated, at least in part, with the ability of primate complement to inactive murine retroviruses (Table II) and agrees with a similar finding for human complement (21).

Clinically, the four virus-infused animals have had no change in temperature, blood chemistries, complete blood counts, or weight. The only abnormality detected was the appearance of bilateral lymphadenopathy in animals 2 and 3. The lymph nodes appeared 7 and 8 days postinfusion and were associated with no other symptoms. The adenopathy resolved spontaneously 7 to 10 days later. Excision of the abnormal lymph node tissue was performed, and the tissue was analyzed for the presence of vector and viral DNA sequences, human ADA activity, and neomycin phosphotransferase activity. All analyses were negative. Lymph node lymphocytes were cultured and serial samples were taken for S+/L− analysis, but no replicating helper virus was detected. The inability to detect virus in the lymph nodes suggests that the adenopathy may not have been in response to replicating virus. Antigens, either viral or contained in the viral supernatant (which included 10% fetal calf serum), may have been responsible. Peripheral blood lymphocytes were isolated from the immunocompromised animal, but no evidence of replicating virus was detected.

TABLE II
SERUM INACTIVATION OF AMPHOTROPIC HELPER VIRUS[a]

Sample	S+/L− Titer on PG4 Cells: Virus Concentration	
	Undiluted	1:100
Amphotropic virus alone	2×10^5	1.8×10^3
Virus + monkey serum	1.0×10^3	0
Virus + monkey serum preheated 56°C	1.0×10^5	1.2×10^3

[a] Supernatant was collected from helper virus-contaminated SAX producer cells and incubated for 30 minutes at 37°C with media alone, monkey serum collected on ice, or monkey serum preheated to 56°C for 20 minutes. After incubation, serial dilutions of the mixtures were plated on dextran-treated pG4 cells and the number of focus-forming units per milliliter was determined on day 4.

Our preliminary work suggests that primates clear murine retroviruses very rapidly from their circulation; to date, we have seen no clinical consequences of such an exposure even when the animal has been immunosuppressed. Complement appears to be an efficient mechanism for inactivating murine retroviruses in primates. Unfortunately, it may also mask ongoing viral production. We are in the process of determining the time course of antibody production in response to the amphotropic envelope protein. A persistently high antibody titer would suggest a continuing antigen exposure. Organ biopsies are planned to evaluate the possibility of localized viral production. We also hope to use the polymerase chain-reaction technology to improve our ability to detect retroviral envelope sequences.

While it is important to study the clinical consequences of murine retroviral exposure in primates, it is critical that steps be taken to minimize the risk of such an exposure. New packaging cell lines are now available that should greatly decrease the likelihood of recombination (22–24); these will be very important in minimizing the risks associated with human gene therapy.

V. Summary

Retroviral vectors infect primate bone marrow cells and express *in vivo* the transferred genes (the human ADA gene and the bacterial gene for neomycin resistance). The SAX vector appears to express human ADA at normal levels, but the infection efficiency is low (<1%) so that the gene product is only detectable in the peripheral blood at low levels. Vector expression disappears after 5 months (except for occasional T cells), presumably due to a failure to infect a renewal stem cell.

While the level of ADA expression obtained in primates would not appear to be sufficient to correct outright the disease caused by ADA deficiency, it is possible that T-cell progenitors in the marrow will have a selective advantage. T cells expressing an ADA vector would then be able to expand and potentially restore immune function. Unfortunately, this hypothesis will go untested until an animal model for ADA deficiency is found or a human clinical trial is performed.

At present, consideration of gene therapy as a treatment for ADA deficiency would only be appropriate if all conventional forms of treatment were unsuccessful. If such a scenario should present itself, the critical question becomes one of safety, to both the patient and those in contact with the patient. We have begun to address the safety issues associated with gene therapy. Five animals exposed to replication-competent retrovirus during bone marrow transplantation show no evidence of helper virus, with a mean follow-up of 18.3 months. Four animals injected with replication-competent helper virus cleared the virus rapidly and, after the initial clearance, have shown no evidence of retroviremia, with a mean follow-up of 5.2 months.

Our preliminary findings suggest that murine retroviruses do not cause a productive infection *in vivo*. These results, combined with the availability of better producer cell lines free of helper virus, are encouraging, and suggest that the risk of clinical disease from murine retrovirus introduced by a gene therapy protocol should be small. Unfortunately, high infection efficiency and long-term vector expression still must be obtained before retroviral-mediated gene transfer can be considered as first-line therapy for ADA deficiency.

References

1. A. D. Miller, D. J. Jolly, T. Friedmann, and I. M. Verma, *PNAS* **80**, 4709 (1983).
2. A. Joyner, G. Keller, R. A. Phillips, and A. Bernstein, *Nature (London)* **305**, 556 (1983).
3. D. A. Williams, I. R. Lemischka, D. G. Nathan, and R. C. Mulligan, *Nature (London)* **310**, 476 (1984).
4. A. D. Miller, R. J. Eckner, D. J. Jolly, T. Friedmann, and I. M. Verma, *Science* **225**, 630 (1984).
5. J. E. Dick, M. C. Magli, D. Huszar, R. A. Phillips, and A. Bernstein, *Cell* **42**, 71 (1985).
6. G. Keller, C. Paige, E. Gilboa, and E. F. Wagner, *Nature (London)* **318**, 149 (1985).
7. M. A. Eglitis, P. Kantoff, E. Gilboa, and W. F. Anderson, *Science* **230**, 1395 (1985).
8. B. Lim, D. A. Williams, and S. H. Orkin, *MCBiol.* **7**, 3459 (1987).
9. P. W. Kantoff, A. P. Gillio, J. R. McLachlin, C. Bordignon, M. A. Eglitis, N. A. Kernan, R. C. Moen, D. B. Kohn, S. Yu, E. Karson, S. Karlsson, J. A. Zwiebel, E. Gilboa, R. M. Blaese, A. Nienhuis, R. J. O'Reilly, and W. F. Anderson, *J. Exp. Med.* **166**, 219 (1987).
10. W. F. Anderson, *Science* **226**, 401 (1984).
11. D. Armentano, S. Yu, P.W. Kantoff, T. von Ruden, W. F. Anderson, and E. Gilboa, *J. Virol.* **61**, 1647 (1987).
12. P. W. Kantoff, D. B. Kohn, H. Mitsuya, D. Armentano, M. Sieberg, J. A. Zwiebel, M. A. Eglitis, J. R. McLachlin, D. A. Wiginton, J. J. Hutton, S. D. Horowitz, E. Gilboa, R. M. Blaese, and W. F. Anderson, *PNAS* **83**, 6563 (1986).
13. G. S. Adrian, D. A. Wiginton, and J. J. Hutton, *MCBiol.* **4**, 1712 (1984).
14. R. Mann, R. C. Mulligan, and D. Baltimore, *Cell* **33**, 153 (1983).
15. A. D. Miller, D. R. Trauber, and C. Buttimore, *Somatic Cell Mol. Genet.* **12**, 175 (1986).
16. J. R. McLachlin, S. C. Bernstein, and W. F. Anderson, *Anal. Biochem.* **163**, 143 (1987).
17. B. Reiss, R. Sprengel, H. Will, and H. Schaller, *Gene* **30**, 211 (1984).

18. G. Selten, H. T. Cuypers, B. Boelens, E. Robanus-Maandag, J. Verbeek, J. Domen, C. van Beveren, and A. Berns, *Cell* **46**, 603 (1986).
19. A. F. Voronova, J. E. Buss, T. Patshinsky, T. Hunter, and B. M. Sefton, *MCBiol* **4**, 2705 (1984).
20. J. D. Maith, R. Peet, E. G. Krebs, and R. M. Perlmutter, *Cell* **43**, 393 (1985).
21. B. Banapour, J. Sernatinger, and J. A. Levy, *Virology* **152**, 268 (1986).
22. A. D. Miller and C. Buttimore, *MCBiol.* **6**, 2895 (1986).
23. D. Markowitz, S. Goff, and A. Bank, *J. Virol.* **62**, 1120 (1988).
24. O. Danos and R. Mulligan. *J. Cell. Biochem., Suppl.* **12B**, 172 (1988).

Addendum: Abstracts

Asymmetrical Exchanges and Chromosomal Rearrangements in *Drosophila*   325
B. H. JUDD, E. A. MONTGOMERY, S.-M. HUANG, AND C. H. LANGLEY

The L1 Family of Repetitive Sequences in Mammals   327
M. H. EDGELL, D. D. LOEB, R. SHEHEE, M. B. COMER, N. C. CASAVANT, AND C. A. HUTCHINSON, III

Repetitive Sequences in the Human Genome   329
R. K. MOYZIS

Asymmetrical Exchanges and Chromosomal Rearrangements in *Drosophila*

> B. H. Judd
> E. A. Montgomery
> S.-M. Huang and
> C. H. Langley
>
> Laboratory of Molecular Genetics,
> NIEHS, Research Triangle Park,
> North Carolina 27709

We are examining the role of transposons in the generation of chromosomal rearrangements in *Drosophila melanogaster*. Earlier work shows that transposons inserted at different sites in homologous chromsomes can regularly pair and undergo exchange, producing reciprocal duplication/deficiency products. Ongoing experiments address this phenomenon in more genetic and molecular detail. Of particular interest is the influence of the interchromsomal effect on the asymmetrical exchange process. Parallel experiments were conducted, beginning with two X chromosomes marked to facilitate recovery of asymmetrical exchange products in the region of the *white* locus. One series had standard-configuration autosomes. The other was heterozygous for autosomal inversions, a condition known to increase meiotic exchange in X chromosomes. Five classes of deficiency chromosomes and an inversion class have been recoverd from the autosomal inversion series; two deficiency classes were found in the standard arrangement series. About half of the rearrangement classes appear to be generated by intrachromosomal exchange; the other half can be explained by asymmetrical exchange between homologs. The nature and origin of the products will be discussed in relation to the role of transposons in the rearrangement of chromosome structure. The regulation of transposon copy number will also be considered.

REFERENCES

P. S. Davis, M. W. Shen, and B. H. Judd, *PNAS* **84**, 174 (1987).
M. L. Goldberg, J.-Y. Sheen, W. J. Gehring, and M. M. Green, *PNAS* **80**, 5017 (1983).

The L1 Family of Repetitive Sequences in Mammals

M. H. Edgell
D. D. Loeb
R. Shehee
M. B. Comer
N. C. Casavant and
C. A. Hutchinson, III

*Department of Microbiology,
University of North Carolina,
Chapel Hill, North Carolina
27599*

All mammals tested carry a family of large dispersed repetitive sequences in their genomes called L1. This element is a transposon that encodes proteins important for L1. The L1 family is different from other characterized transposons in a number of ways. L1 has no long terminal repeats. Most members of the family carry 5' polar truncations of various lengths. These differences clearly suggest that L1 must replicate in a novel fashion. Arguments can be made both for and against a selfish mode of replication. The element transposes at a very high rate in the mouse and must generate a substantial genetic load for the species. This raises the possibility that L1 may play a noticeable role in the extinction of species.

References

M. H. Edgell, S. C. Hardies, D. D. Loeb, W. R. Shehee, R. W. Padgett, F. H. Barton, M. B. Comer, N. C. Casavant, F. D. Fank, and C. A. Hutchinson III, *Dev. Cont. Glob. Gene Exp.* 107 (1987).
M. F. Singer and J. Skowronski, *TIBS* **10**, 119 (1985).

Repetitive Sequences in the Human Genome

R. K. Moyzis

Genetics Group, Los Alamos
National Laboratory, Los Alamos,
New Mexico 87545

An ultimate goal of human genetics is the generation of a complete physical map/sequence of the human genome. Twenty-five percent of human DNA, however, consists of repetitive DNA sequences. A general outline of the chromosomal organization of these repetitive sequences will be discussed. Our working hypothesis is that certain classes of human repetitive DNA sequences "encode" the information necessary for defining genomic structure.

Using a combination of biochemical, computational, and recombinant DNA approaches, the organization of interspersed, centromeric, and telomeric repetitive DNA in the human genome has been investigated. The distribution of interspersed repeats can be adequately described by models that assume a random spacing, with an average distance of 3 kb. This observed distribution for the "integration" of interspersed repetitive DNA is the expected result for sequences that transpose randomly throughout the genome. However, local regions of "preference" or "exclusion" for the integration of repetitive DNA are suggested by the data.

Centromeric repetitive sequences, on the other hand, can be highly chromosome-specific. Our laboratory has isolated three recombinant DNA clones of human repetitive DNA sequences that hybridize specifically to the heterochromatic positions 1qh, 9qh, and 16qh, respectively. These locations were determined by fluorescent *in situ* hybridization, and confirmed by DNA hybridizations to human chromosomes sorted by flow cytometry. *In situ* hybridizations to intact interphase nuclei showed a well-defined, localized organization for all three DNA sequences. Clamped homogeneous electrical field (CHEF) electrophoresis enabled us to resolve discrete high-molecular-weight DNA bands complementary to these clones. Collectively, these bands comprise at least 5400 kb of chromosome 9 DNA (~3.8% of chromosome 9) and greater than 4300 kb of chromsome 16 DNA (~4.5% of chromsome 16). Using flow-sorted chromosome 16 as starting material for CHEF electrophoresis, bands similar to those obtained from cellular DNA were produced. The presence of these large repetitive DNA clusters suggests that constructing a complete physical map of the human genome will present unique challenges. The ability to perform macrorestriction analyses on DNA from specific flow-sorted chromosomes, however, will aid this construction, especially where chromosomally dispersed repeats or pseudogenes complicate the analysis. Furthermore, the ability to localize defined human chromosome domains in intact interphase nuclei should allow the ultimate generation of a three-dimensional human genome map to complement the linear map.

Finally, a highly conserved repetitive DNA sequence has been isolated from a human repetitive DNA sequence library. Quantitative hybridizations to chromosomes sorted by flow cytometry indicates that comparable amounts of this sequence are present on each human chromosome. Fluorescent *in situ* hybridization experiments indicate that the major clusters of this sequence occur at the telomeres of all mammalian chromosomes. The evolutionary conservation of this sequence, the conservation of its chromosomal location, and its similarity to telomeres isolated from lower eukaryotes, indicate that this sequence is a functional human telomere.

Acknowledgment

Research was supported by the Department of Energy under Contract W-7405-ENG-36.

Index

A

Abelson murine leukemia virus, chromosome breakpoints and, 282-284
ABL, chromosome breakpoints and, 281, 282, 285-288, 295-298
  characterization, 283-285
  chronic myelocytic leukemia, 289, 292
  mRNA, 293, 294
Acute lymphoblastic leukemia, chromosome breakpoints and, 294, 295
Acute myeloid leukemia, chromosome breakpoints and, 295
Adenosine deaminase deficiency, gene transfer into primates and, 311, 312, 320, 321
  bone marrow transplant, 315
  retroviral vectors, 313
  safety, 318
Adenovirus, homologous recombination and, 301
Albumin, 131, 132, 136
  linkage, 140, 141
  regulatory elements
    distal elements, 139, 140
    linkage, 140, 141
    promoter proximal elements, 138, 139
  tissue specificity, 132, 133
Alcohol dehydrogenase, albumin and α-fetoprotein and, 133
Alleles
  chromosome breakpoints and, 286, 292
  Drosophila and
    gene cloning, 100, 106, 107
    suppressible insertion-induced mutations, 87, 90-97

methylation mosaic model and, 145
  data, 150
  gamete-of-origin, 147
  specificity, 151-153
murine leukemia virus and, 229
retroviral insertion in dilute locus and
  dilute suppressor, 216
  genetic analysis, 207-209, 211
  molecular analysis, 211, 215
variable number of tandem repeat sequences and, 189, 198
X-chromosome inactivation and, 120, 121, 127
Amino acids
  chromosome breakpoints and, 283
  Drosophila and
    foldback elements, 6, 7
    P element transposition, 52
  lymphoid tumorigenesis and, 274
Ampicillin, *Drosophila* gene cloning and, 106
Amplification, intracisternal A-particle genes and, 176-178
Antibodies
  Drosophila foldback elements and, 8
  gene transfer into primates and, 318, 320
  intracisternal A-particle genes and, 181, 182
  lymphoid tumorigenesis and, 271
  transgenic mice, *c-fos* expression and, 240
Antigens
  endogenous retroviral genes and, 254, 260
  gene transfer into primates and, 319, 320
  lymphoid tumorigenesis and, 272
AP-1, transgenic mice, *c-fos* expression and, 240, 241, 243

331

A

ATP, intracisternal A-particle genes and, 182, 184, 185
Azacytidine, endogenous retroviral genes and, 262
5-Azacytidine, X-chromosome inactivation and, 122

B

B cells
  chromosome breakpoints and, 282
  lymphoid tumorigenesis and, 269
    chromosome translocation, 277
    cytogenetics, 269, 270
    follicular lymphomas, 274
    t(11;14) translocation, 274-276
Bacteria
  endogenous retroviral genes and, 257
  gene transfer into primates and, 312
  homologous recombination and, 303-305, 307
  intracisternal A-particle genes and, 176, 179, 181
Bone, transgenic mice, c-fos expression and, 236, 238, 239, 242-244
Bone marrow transplant, gene transfer into primates and, 311, 321
  adenosine deaminase deficiency, 312
  protocol, 314-317
  safety, 318
Brain
  albumin and α-fetoprotein and, 132, 139
  retroviral insertion in dilute locus and, 213, 215
  transgenic mice, c-fos expression and, 242
Breakpoint cluster region, chromosome breakpoints and, 281, 282, 296-298
  ABL, 293, 294
  acute lymphoblastic leukemia, 295
  chronic myelocytic leukemia, 288-292
Breakpoints, see Chromosome breakpoints
Burkitt's lymphoma
  chromosome breakpoints and, 289
  lymphoid tumorigenesis and, 271-273
    chromosome translocation, 277, 278
    cytogenetics, 269, 270
    t(11;14) translocation, 276

C

c-fos expression in transgenic mice, see Transgenic mice, c-fos expression and
Cadmium chloride, transgenic mice, c-fos expression and, 236
Calcium phosphate, homologous recombination and, 301, 303
cAMP, chromosome breakpoints and, 283
Cancer, variable number of tandem repeat sequences and, 201
Carbon tetrachloride, endogenous retroviral genes and, 260, 261, 263
Carcinogenesis, endogenous retroviral genes and, 247, 248, 259, 260, 262, 263
Catalysis, Drosophila gene cloning and, 102
cDNA
  chromosome breakpoints and, 281, 283
    ABL, 284, 294
    chronic myelocytic leukemia, 290-292
  Drosophila P element transposition and, 55
  endogenous retroviral genes and, 259
  gene transfer into primates and, 312
  intracisternal A-particle genes and, 174
  retroviral insertion in dilute locus and, 213, 215, 218
  transgenic mice, insertional mutation and, 167
Centromeres
  chromosome breakpoints and, 290
  lymphoid tumorigenesis and, 276
  repetitive sequences in human genome and, 329
Chloramphenicol acetyltransferase, endogenous retroviral genes and, 257
Chondrosarcomas, transgenic mice, c-fos expression and, 236, 239
Chromatin
  albumin and α-fetoprotein and, 134
  X-chromosome inactivation and, 123
Chromosome breakpoints, 281, 282, 295-298
  ABL, 285-288
  characterization, 283-285

INDEX
333

mRNA, 293, 294
acute lymphoblastic leukemia, 294, 295
chronic myelocytic leukemia
  BCR gene, 288-291
  Mbcr, 291, 292
  DNA sequences, 293
  v-*abl*, 282, 283
Chromosomes
  Drosophila and
    foldback elements, 3, 4, 7
    gene cloning, 99, 100, 102, 104-108
    hobo element, 39, 42, 43
    molecular lesions, 111, 113-115
    P element transposition, 48, 50
    P-transposable elements, 71-75, 79-84
    rearrangements, 325
    suppressible insertion-induced mutations in, 90
  Drosophila, transposable elements and, 25, 34, 35
    population dynamics, 30-32
    population statics, 26-28
  endogenous retroviral genes and, 247-249, 262, 263
    cloning, 249, 250, 252-254
    environmental insults, 259
    tissue-specific expression, 254, 257
  homologous recombination and, 301, 309
    gene modification, 305-308
    gene transfer systems, 302
  inactivation, *see* X-chromosome inactivation
  lymphoid tumorigenesis and, 269
    Burkitt's lymphoma, 271-273
    cytogenetics, 269, 270
    follicular lymphomas, 273, 274
    t(11;14) translocation, 274, 276
    translocation, 277, 278
  methylation mosaic model and, 147-149, 152-154
  molecularly tagged P elements and, 60, 62, 64
  primate L1 elements and, 9
  repetitive sequences in human genome and, 329, 330
  retroviral insertion in, *see* Retroviral insertion in dilute locus
  transgenic mice and, 13
    insertional mutation, 166, 168

  variable number of tandem repeat sequences and, 187, 199, 201, 202
    DNA fingerprinting, 188-194
    neoplasia, 193-200
Chronic lymphocytic leukemia, lymphoid tumorigenesis and, 270, 274, 277, 278
Chronic myelocytic leukemia, chromosome breakpoints and, 281, 282, 295, 297, 298
  ABL, 286, 288, 293
  BCR gene, 288-291
  Mbcr, 291, 292
Clamped homogeneous electrical field electrophoresis, repetitive sequences in human genome and, 329, 330
Cloning
  albumin and α-fetoprotein and, 134
  chromosome breakpoints and, 282, 283, 296
    ABL, 283, 288
    chronic myelocytic leukemia, 288, 290, 292
  Drosophila and
    foldback elements, 3
    gene cloning, *see* *Drosophila*, gene cloning and
    P element transposition, 49, 50, 54, 55
    transposable elements, 32
  endogenous retroviral genes and, 249, 263
    environmental insults, 262
    MRL elements, 252-254
    MuLV, 249-252
    tissue-specific expression, 257
  gene transfer into primates and, 318
  intracisternal A-particle genes and, 174, 179, 181
  lymphoid tumorigenesis and, 274-276, 278
  methylation mosaic model and, 155
  murine leukemia virus and, 221
  repetitive sequences in human genome and, 329
  retroviral insertion in dilute locus and, 211, 218, 219
  transgenic mice and, 20-22
    insertional mutation, 159-161, 167
    variable number of tandem repeat sequences and, 187, 194, 198, 202

Collagenase, transgenic mice, *c-fos* expression and, 242
Copia, Drosophila and
 foldback elements, 3
 suppressible insertion-induced mutations, 91–93
 transposable elements, 32
Cycloheximide, transgenic mice, *c-fos* expression and, 236, 239
Cytoplasm
 Drosophila gene cloning and, 99
 intracisternal A-particle genes and, 176, 179, 182
 murine leukemia virus and, 233

D

Decapentaplegic gene, *Drosophila, hobo* element and, 38–42
Deletion
 homologous recombination and, 303, 305, 309
 insertional mutation in transgenic mice and, 160
 intracisternal A-particle genes and, 173–175
 variable number of tandem repeat sequences and, 199
Deletion mapping, retroviral insertion in dilute locus and, 217, 218
Demethylation, methylation mosaic model and, 149, 155
Dilute locus, retroviral insertion in, *see* Retroviral insertion in dilute locus
Dilute suppressor, retroviral insertion in dilute locus and, 216, 217
DNA
 albumin and α-fetoprotein and, 134, 136, 138, 140, 141
 chromosome breakpoints and, 281–283
  ABL, 283–286, 288
  acute lymphoblastic leukemia, 295
  chronic myelocytic leukemia, 288, 289, 291, 292
  sequences, 293
 Drosophila and
  foldback elements, 5–7
  gene cloning, 99–101, 106

hobo element, 42, 43
 molecular lesions, 113, 114
 P element transposition, 48–50, 52–54
P-transposable elements, 72, 74, 81
 suppressible insertion-induced mutations, 88
 transposable elements, 25, 32
endogenous retroviral genes and, 247, 262, 263
 cloning, 249, 252–254
 environmental insults, 260, 262
 tissue-specific expression, 254, 255, 257
gene transfer into primates and, 312, 319
homologous recombination and, 301
 cell-free systems, 304, 305
 chromosomal gene modification, 307
 extrachromosomal molecules, 302–304
 gene transfer systems, 302
 human disease, 308
intracisternal A-particle genes and, 173, 176, 185
 myeloma cells, 184
 open reading frame, 179
 proviral elements, 178, 179
 RNA expression, 176
methylation mosaic model and
 consequences, 152, 153
 data, 148–151
 gamete-of-origin, 147, 148
 imprint, 155
molecularly tagged P elements and, 62, 66
murine leukemia virus and
 ecotropic proviruses, 222, 224, 226, 228–230
 future prospects, 233, 234
primate L1 elements and, 8, 10–12
repetitive sequences in human genome and, 329, 330
retroviral insertion in dilute locus and, 211–213, 217, 218
transgenic mice and, 13, 14, 20, 21
 c-fos expression, 236, 240, 242
transgenic mice, insertional mutation and, 159–161, 168
 analysis, 167

INDEX

identification, 161, 163
variable number of tandem repeat
  sequences and, 188-194
neoplasia, 195-200
DNase I, albumin and α-fetoprotein
  and, 134, 135, 139, 140
Drosophila
  albumin and α-fetoprotein and,
    133, 136
  asymmetrical exchanges in, 325
  chromosomal rearrangements in, 325
  molecularly tagged P elements and,
    59-61, 65-67
  retroviral insertion in dilute locus
    and, 216
  X-chromosome inactivation and, 119,
    121, 127
Drosophila, foldback elements and, 3-8
  primate L1 elements, 9, 12
  transgenic mice, 21
Drosophila, gene cloning and
  insertion library, 107, 108
  insertion lines, 105, 106
  insertion site DNA, 106, 107
  mutagenesis, 101-105
Drosophila, hobo element of, 37
  decapentaplegic gene, 39-42
  germ-line transformation, 42, 43
  mobilization, 38, 39
  molecular organization, 37, 38
  P mobilization systems, 43, 44
  transformation vector, 44, 45
Drosophila, P element transposition
  and, 47, 48
  genetic control, 51-54
  hybrid dysgenesis, 48, 49
  molecular analysis, 49-51
  tissue specificity, 54-56
Drosophila, P-transposable elements
  and, 71, 72
  methods, 72-74
  natural populations, 84
  P cytotype, 83
  proliferation, 81, 82
  results
    P-element sites, 76-81
    P-element stocks, 74, 75
Drosophila, suppressible insertion-
  induced mutations in, 87
  allele specificity
    fly, 94-97

transposon, 90-93
transposon parasites, 88-90
Drosophila, transposable elements in,
  25, 26, 34, 35
population dynamics
  infinite population, 28-31
  models, 31-34
population statics
  element frequencies, 27, 28
  element numbers, 26, 27
Drosophila melanogaster
  albumin and α-fetoprotein and, 133
  insertional mutation in transgenic
    mice and, 159
  molecular lesions and, 111-116
  murine leukemia virus and, 222, 231
  X-chromosome inactivation and, 119

E

Ecotropic viruses, murine leukemia
  virus and, 221, 222
  gene susceptibility determinants,
    231, 232
  inbred strains of mice, 222-224
  insertional mutagenesis, 231
  new germ-line, 224-230
  spontaneous germ-line virus infection,
    232
Electroporation, homologous
  recombination and, 302
Embryonic stem cells, homologous
  recombination and, 308, 309
Emv-3, retroviral insertion in dilute
  locus and, 212, 213, 217
Emv virus, murine leukemia virus and,
  225, 226, 229, 230, 232, 233
Endogenous retroviral genes, 247-249,
  262-264
  cloning, 249
  MboI repeat, 252-254
  MuLV, 249-252
  environmental insults, 259, 260
    azacytidine, 262
    carbon tetrachloride, 260, 261
  tissue-specific expression, 254, 255
    MRL elements, 257, 259
    MuLV, 255, 256
    negative regulatory element, 257,
      258

Endometrium, endogenous retroviral
  genes and, 256
Endonuclease
  endogenous retroviral genes and, 254
  intracisternal A-particle genes
    and, 175
  myeloma cells, 182-185
  open reading frame, 179, 181
  murine leukemia virus and, 222
Endoplasmic reticulum, intracisternal
  A-particle genes and, 173
Enzymes, *Drosophila* foldback elements
  and, 7, 8
Epithelial cells, transgenic mice, *c-fos*
  expression and, 240, 242
Epstein-Barr virus, lymphoid
  tumorigenesis and, 278
Escherichia coli
  Drosophila and
    gene cloning, 101, 106
    transposable elements, 28
  homologous recombination and, 304
  transgenic mice, 21
Estrogen, albumin and α-fetoprotein
  and, 135
Ethyl methanesulfonate, molecularly
  tagged P elements and, 65
Euchromatin, *Drosophila* and
  gene cloning, 107
  P-transposable elements, 75
  transposable elements, 27
Eukaryotes
  chromosome breakpoints and, 291
  Drosophila and
    foldback elements, 3
    molecular lesions, 115
    P element transposition, 56
    transposable elements, 28
  endogenous retroviral genes and, 248
  homologous recombination and, 303
  methylation mosaic model and, 145
  molecularly tagged P elements and,
    59, 65, 66
  repetitive sequences in human
    genome and, 330
  retroviral insertion in dilute locus
    and, 216
Exons
  chromosome breakpoints and, 296
  ABL, 283-286
  acute lymphoblastic leukemia, 295
  chronic myelocytic leukemia,
    291, 292
  lymphoid tumorigenesis and,
    273, 275
  retroviral insertion in dilute locus
    and, 213, 215
Extrachromosomal molecules,
  homologous recombination and,
  302-304

F

α-Fetoprotein, 131, 132, 138
  developmental regulation, 141
  linkage, 140, 141
  regulatory elements
    distal regulatory elements, 135-138
    identification, 134, 135
    promoter proximal elements, 138
  tissue specificity, 132, 133
Fibroblasts
  chromosome breakpoints and, 282,
    289, 290
  methylation mosaic model and, 152
Friend leukemia virus, intracisternal A-
  particle genes and, 184

G

Galago, primate L1 elements and, 11,
  12
Gamete-of-origin-dependent DNA
  methylation, methylation mosaic
  model and, 147, 148
Gamete-specific methylation,
  methylation mosaic model and,
  150, 151, 155
GD sterility, *Drosophila*, P-transposable
  elements and, 73, 74, 76, 79, 84
Gene dosage, X-chromosome
  inactivation and, 119-128
Gene susceptibility determinants,
  murine leukemia virus and,
  231, 232
Gene therapy
  gene transfer into primates and, 311,
    312, 320, 321
  homologous recombination and, 307
Gene transfer into primates, 311, 320, 321

adenosine deaminase deficiency, 311, 312
  bone marrow transplant, 314-317
  retroviral vectors, 312, 313
  safety, 317
  insertional mutagenesis, 317, 318
  replication-competent virus, 318-320
Gene transfer systems, homologous recombination and, 301, 302
Genetic drift, *Drosophila*, transposable elements and, 30, 31
Genetic mapping, *Drosophila* gene cloning and, 99
Genomic characterization, retroviral insertion in dilute locus and, 214-216
Genotype, methylation mosaic model and, 154, 155
Germ-line viruses, murine leukemia virus and, 224, 225
  future prospects, 234
  gene susceptibility determinants, 231, 232
  spontaneous, 232, 233
  Srev loci, 226-230
  viral RNA expression, 230
Globin genes
  albumin and α-fetoprotein and, 132, 135, 140
  homologous recombination and, 307, 308
  insertional mutation in transgenic mice and, 162, 163, 165, 168
Glucocorticoids, albumin and α-fetoprotein and, 135
Glucose-6-phosphate dehydrogenase, X-chromosome inactivation and, 123, 127
Gonadal dysgenesis, *Drosophila*, hobo element and, 41
Gypsy alleles, *Drosophila*, suppressible insertion-induced mutations and, 92, 96, 97

H

Hairy wing alleles, *Drosophila*, suppressible insertion-induced mutations and, 92

HB elements, *Drosophila* foldback elements and, 6
Heat shock, *Drosophila*, transposable elements and, 28
Heat-shock genes, transgenic mice and, 21
Heat shock protein, *Drosophila* P element transposition and, 54
Hemoglobin, insertional mutation in transgenic mice and, 162, 165
Hepatitis B virus, albumin and α-fetoprotein and, 137
Hepatoma cells, albumin and α-fetoprotein and, 137, 138
Heterochromatin, *Drosophila*, P-transposable elements and, 72, 74, 81, 82
Hobo elements, *Drosophila* and foldback elements, 6
  molecular lesions, 112
  P element transposition, 49
Homologous recombination in mammalian somatic cells, 301
  cell-free systems, 304, 305
  chromosomal gene modification, 305-308
  extrachromosomal molecules, 302-304
  gene transfer systems, 301, 302
  human disease, 308
  perspectives, 308, 309
Homology
  albumin and α-fetoprotein and, 137
  X-chromosome inactivation and, 119, 123, 124
Hormones
  albumin and α-fetoprotein and, 135, 137, 139, 140
  X-chromosome inactivation and, 124
Horse-radish peroxidase, insertional mutation in transgenic mice and, 162
Huntington's chorea, methylation mosaic model and, 148
Hybrid dysgenesis, *Drosophila* and gene cloning, 99, 107
  molecular lesions, 111-116
  P element transposition, 47-49, 51
  P-transposable elements, 72, 73, 84
Hybridization
  chromosome breakpoints and, 296
  ABL, 283, 285, 286, 288

chronic myelocytic leukemia,
    288-290, 292
Drosophila and
    foldback elements, 3
    gene cloning, 99, 100, 105, 107
    hobo element, 43
    P-transposable elements, 71-77,
        81, 82
    transposable elements, 27, 31, 32
endogenous retroviral genes and, 253,
    255, 256, 259, 260
gene transfer into primates and, 315
homologous recombination and, 307
intracisternal A-particle genes and,
    176, 177, 179
methylation mosaic model and, 148
murine leukemia virus and, 221, 222
    ecotropic proviruses, 229, 230
    gene susceptibility determinants,
        231, 232
    insertional mutagenesis, 231
    new germ-line, 224, 225
    spontaneous germ-line virus
        infection, 232
    Srev loci, 226-230
    viral RNA expression, 230
primate L1 elements and, 11
repetitive sequences in human
    genome and, 329, 330
retroviral insertion in dilute locus
    and, 208, 215, 217
transgenic mice and, 18, 21
    insertional mutation, 162, 163,
        165-168
variable number of tandem repeat
    sequences and, 188, 189, 195, 202
X-chromosome inactivation and, 120
Hybridoma, intracisternal A-particle
    genes and, 174
Hypercholesterolemia, chromosome
    breakpoints and, 282, 293
Hypermutability, *Drosophila*,
    P-transposable elements and, 73,
    79, 81, 82
Hypomethylation, methylation mosaic
    model and, 147
Hypoxanthine phosphoribosyltransferase
    gene, homologous recombination
    and, 308

I

Immunofluorescence
    endogenous retroviral genes and, 260
    intracisternal A-particle genes
        and, 182
Immunoglobulins
    albumin and α-fetoprotein and, 133,
        136, 137
    chromosome breakpoints and, 288,
        289, 293
    insertional mutation in transgenic
        mice and, 163
    lymphoid tumorigenesis and, 279
        Burkitt's lymphoma, 271-273
        chromosome translocation, 277, 278
        cytogenetics, 270
        follicular lymphomas, 273, 274
        t(11;14) translocation, 274-276
Imprinting
    methylation mosaic model and, *see*
        Methylation mosaic model
    X-chromosome inactivation and, 121
Inducer-reactor hybrid dysgenesis,
    Drosophila and, 111, 112, 115, 116
Insertase gene, endogenous retroviral
    genes and, 263
Insertion sequences, endogenous
    retroviral genes and, 247, 248, 253
Insertional mutagenesis
    Drosophila and
        foldback elements, 6
        gene cloning, *see Drosophila*, gene
            cloning and
        hobo element, 44
    gene transfer into primates and,
        317, 318
    murine leukemia virus and, 221, 225,
        231
    transgenic mice, 13-22
Insertional mutation, *Drosophila*,
    transposable elements and, 25,
    30-32
Interferon, albumin and α-fetoprotein
    and, 137
Interleukin-3, intracisternal A-particle
    genes and, 174
Invasiveness, *Drosophila*, suppressible
    insertion-induced mutations
    and, 89

J

Jumpstart element, *Drosophila* gene cloning and, 101, 102, 105, 106

K

Karyotype, variable number of tandem repeat sequences and, 194–196, 198
Kidney
  albumin and α-fetoprotein and, 132
  endogenous retroviral genes and, 255, 259, 262

L

L1 elements
  Drosophila foldback elements and, 4
  primate, 8–12
L1 family of repetitive sequences, 327
Lethal mutation in transgenic mice, 160
Leukemia
  chromosome breakpoints and, 281, 289, 295–298
  endogenous retroviral genes and, 247, 262
    cloning, 250, 252
    tissue-specific expression, 254
  lymphoid tumorigenesis and, 270, 274, 277, 278
Leukocytes, methylation mosaic model and, 152
Linkage group I, retroviral insertion in dilute locus and, 208
Lipopolysaccharides, endogenous retroviral genes and, 259
Liver
  albumin and α-fetoprotein and, 131–133, 137, 139, 141
  endogenous retroviral genes and, 255, 259–262
  transgenic mice and
    c-fos expression, 236, 240
    insertional mutation, 167
  X-chromosome inactivation and, 120
Long terminal repeats
  Drosophila, suppressible insertion-induced mutations and, 91
  endogenous retroviral genes and, *see* Endogenous retroviral genes
  gene transfer into primates and, 312, 318
  intracisternal A-particle genes and, 173, 178
  retroviral insertion in dilute locus and, 211, 212
  transgenic mice, *c-fos* expression and
    H2, 238–240
    metallothionein, 236, 238
    proliferation, 243
    specificity, 241, 242
Low-density lipoprotein, chromosome breakpoints and, 293
Lymph nodes, transgenic mice, *c-fos* expression and, 240
Lymphoid cells
  chromosome breakpoints and, 295
  endogenous retroviral genes and, 259
Lymphoid organs, endogenous retroviral genes and, 254
Lymphoid tumorigenesis, 269, 279
  Burkitt's lymphoma, 271–273
  chromosome translocation, 277, 278
  cytogenetics, 269, 270
  follicular lymphomas, 273, 274
  t(11;14) translocation, 274
    B cell, 274, 275
    T cell, 275, 276
Lymphoma
  gene transfer into primates and, 317
  murine leukemia virus and, 232

M

Macrophages, transgenic mice, *c-fos* expression and, 235
Magnesium, intracisternal A-particle genes and, 182
Major breakpoint cluster region, chromosome breakpoints and, 289–292
Mammalian genome imprinting, methylation mosaic model and, *see* Methylation mosaic model
Manganese, intracisternal A-particle genes and, 182

*Mbo*I repeat LTR (MRL) elements,
  endogenous retroviral genes
  and, 260, 263
  cloning, 252-254
  expression, 257, 259
Meiosis
  Drosophila and, 325
  transposable elements, 32
  methylation mosaic model and, 154
  X-chromosome inactivation and, 122, 124-126
Melanocytes, retroviral insertion in
  dilute locus and, 219
  dilute suppressor, 217
  genetic analysis, 207, 208
  molecular analysis, 213, 215
Melanoma, retroviral insertion in dilute
  locus and, 213
Melanosomes, retroviral insertion in
  dilute locus and, 207, 208
Metallothionein, transgenic mice,
  c-fos expression and, 236, 237, 240, 242
Methylation mosaic model, 145, 146
  consequences
    allele specificity, 151-153
    genetic traits, 153, 154
  data, 148-151
  gamete-of-origin, 147, 148
  genotype, 154, 155
  imprint, 155, 156
Microinjection
  homologous recombination and, 302
  murine leukemia virus and, 233
Mink-cytopathic-focus virus, endogenous
  retroviral genes and, 249
Mitogen
  lymphoid tumorigenesis and, 274
  transgenic mice, *c-fos* expression
    and, 242, 243
Mitosis
  Drosophila, suppressible insertion-
    induced mutations and, 95
  methylation mosaic model and, 153
Molecular lesions in *Drosophila
  melanogaster*, 111-116
Moloney murine leukemia virus
  chromosome breakpoints and, 282
  gene transfer into primates and, 312, 317, 318
  intracisternal A-particle genes and, 185

Monoclonal antibodies, *Drosophila* P
  element transposition and, 53
Mouse chromosome 9, retroviral
  insertion in, *see* Retroviral
  insertion in dilute locus
mRNA
  albumin and α-fetoprotein and, 131, 132, 136
  chromosome breakpoints and, 281, 283
    ABL, 284, 285, 293, 294
    acute lymphoblastic leukemia, 295
    chronic myelocytic leukemia, 290-292
  Drosophila and
    P element transposition, 48, 51, 52, 54-56
    suppressible insertion-induced
      mutations, 94, 96
  primate L1 elements and, 8
  retroviral insertion in dilute locus
    and, 214, 215, 218
  transgenic mice, *c-fos* expression and, 235, 236, 239
Murine leukemia virus, 221, 222
  chromosome breakpoints and, 296
  ecotropic proviruses, 222-224
    new germ-line, 224, 225
    Srev loci, 226-230
    viral RNA expression, 230
  endogenous retroviral genes and, 262-264
    cloning, 249-252
    environmental insults, 259, 260, 262
    tissue-specific expression, 254-257
  future prospects, 233, 234
  gene susceptibility determinants, 231, 232
  insertional mutagenesis, 231
  intracisternal A-particle genes
    and, 179
  retroviral insertion in dilute locus
    and, 211
  spontaneous germ-line virus infection, 232, 233
Mutagenesis, *see also* Insertional
  mutagenesis
  Drosophila gene cloning and, *see*
    Drosophila, gene cloning and
  endogenous retroviral genes and, 248, 259, 260

Mutation
  chromosome breakpoints and, 282
  Drosophila and
    foldback elements, 5, 6
    hobo element, 39-44
    molecular lesions, 111-115
    P element transposition, 48-51, 55
    P-transposable elements, 72-74, 76, 80, 82, 84
    transposable elements, 25, 29, 30, 34, 35
  endogenous retroviral genes and, 247
  homologous recombination and, 303-305, 308, 309
  insertional in transgenic mice, *see* Transgenic mice
  lymphoid tumorigenesis and, 272
  methylation mosaic model and, 153, 155
  molecularly tagged P elements and, 65, 66
  primate L1 elements and, 10
  retroviral insertion in dilute locus and
    deletion mapping, 217, 218
    dilute suppressor, 216, 217
    functional units, 219
    genetic analysis, 207, 208
    induced, 209-211
    molecular analysis, 215
    reverse, 211
    spontaneous, 208, 209
  suppressible insertion-induced in *Drosophila*, *see Drosophila*
  transgenic mice and, 13-22
  c-fos expression, 244
  variable number of tandem repeat sequences and, 201, 202
Myeloma
  intracisternal A-particle genes and, 174, 175, 185
    endonuclease, 182-185
    RNA expression, 176, 177
  lymphoid tumorigenesis and, 270
Myoglobin, variable number of tandem repeat sequences and, 198
Myotonic dystrophy, methylation mosaic model and, 148

N

Natural selection, *Drosophila*, transposable elements and, 29, 30

Negative regulatory element, endogenous retroviral genes and, 257
Neomycin
  gene transfer into primates and, 312, 313, 315, 319
  homologous recombination and, 303-308
Neoplasia
  endogenous retroviral genes and, 262, 263
  lymphoid tumorigenesis and, 272, 276, 279
  variable number of tandem repeat sequences and, 187, 202
    cellular models, 193-195
    DNA fingerprinting, 188, 195-200
Nerve growth factor, retroviral insertion in dilute locus and, 215
Neurological disorders, retroviral insertion in dilute locus and, 210, 215, 216
Neuromuscular disorders, retroviral insertion in dilute locus and, 209, 210
Neutrophils, transgenic mice, *c-fos* expression and, 235
Nucleosomes, albumin and α-fetoprotein and, 135
Nucleotides
  albumin and α-fetoprotein and, 137
  chromosome breakpoints and, 283
  Drosophila P element transposition and, 51
  endogenous retroviral genes and, 250, 252, 253, 257
  intracisternal A-particle genes and, 176, 181
  lymphoid tumorigenesis and, 278
  retroviral insertion in dilute locus and, 213, 215

O

Oncogenes
  chromosome breakpoints and, 281
    ABL, 283, 284, 286
    chronic myelocytic leukemia, 288, 290
  endogenous retroviral genes and, 247, 248, 250

lymphoid tumorigenesis and, 269, 274, 275, 279
variable number of tandem repeat sequences and, 202
Open reading frames (ORF)
Drosophila and
foldback elements, 6, 7
molecular lesions, 115
P element transposition, 50-52, 55, 56
intracisternal A-particle genes and, 179, 181
primate L1 elements and, 8
retroviral insertion in dilute locus and, 213
Ornithine transcarbamoylase (OTC), X-chromosome inactivation and, 120
Osteosarcomas, methylation mosaic model and, 152, 153
Ovary
endogenous retroviral genes and, 257, 259
murine leukemia virus and, 226, 228-230, 232, 233

P

P cytotype, Drosophila and
gene cloning, 100
P-transposable elements, 71-73, 76, 78, 79, 83, 84
P-element insertional mutagenesis, Drosophila and, see Drosophila, gene cloning and
P elements
Drosophila and
foldback elements, 6
hobo element, 37, 39, 41-43, 45
molecular lesions, 112
transposable elements, 30
transposition, see Drosophila, P element transposition and
molecularly tagged, 59, 60, 66, 67
genetic systems, 60-65
mutagen-induced mobilization, 65, 66
primate L1 elements and, 9
Packaging, gene transfer into primates and, 313

Pancreas
albumin and α-fetoprotein and, 132, 133
transgenic mice, c-fos expression and, 236, 240
Parasites, Drosophila and
allele specificity, 90, 91, 94, 95, 97
transposon, 88-90
Peptides
chromosome breakpoints and, 290
intracisternal A-particle genes and, 173, 182
Phenotype
Drosophila and
gene cloning, 102, 105-107
hobo element, 40
P element transposition, 51
suppressible insertion-induced mutations, 87, 92, 93
endogenous retroviral genes and, 262
gene transfer into primates and, 312
homologous recombination and, 303, 306
lymphoid tumorigenesis and, 274, 278
methylation mosaic model and, 147, 148, 153, 155
molecularly tagged P elements and, 59
murine leukemia virus and, 231, 233
transgenic mice and, 13-15, 19, 20, 22
c-fos expression, 238
insertional mutation, 160, 161, 165-168
variable number of tandem repeat sequences and, 194, 197, 198, 202
Philadelphia chromosome, chromosome breakpoints and, 281, 282, 295, 296
ABL, 286, 293
acute lymphoblastic leukemia, 294, 295
chronic myelocytic leukemia, 288-290, 292
DNA sequences, 293
Plasmid
Drosophila gene cloning and, 101, 102, 106
gene transfer into primates and, 313
homologous recombination and, 301, 303-308
intracisternal A-particle genes and, 176, 182

Polyadenylation, endogenous retroviral genes and, 248, 253
Polymorphism, homologous recombination and, 305, 306
Polypeptides, *Drosophila* and
 molecular lesions, 112
 P element transposition, 51, 52, 55
 suppressible insertion-induced mutations, 96
Primate L1 elements, 8-12
Proliferation, transgenic mice, *c-fos* expression and, 235, 242-244
Protein
 albumin and α-fetoprotein and, 131, 133, 137, 139
 chromosome breakpoints and, 281-283, 296, 297
  ABL, 284, 285, 293
  acute lymphoblastic leukemia, 295
  chronic myelocytic leukemia, 290
 Drosophila and
  foldback elements, 5-8
  gene cloning, 102
  hobo element, 38, 39
  molecular lesions, 112
  P element transposition, 48, 52-54, 56
  suppressible insertion-induced mutations, 95, 96
 endogenous retroviral genes and, 249, 257, 263
 homologous recombination and, 304
 intracisternal A-particle genes and, 176, 179, 182-185
 lymphoid tumorigenesis and, 271-273, 275
 methylation mosaic model and, 148
 primate L1 elements and, 8, 9
 retroviral insertion in dilute locus and, 215-217
 transgenic mice and, 14
  c-fos expression, 235, 236, 239, 240, 242, 244
Protein kinase
 chromosome breakpoints and, 282-284, 293, 294
 gene transfer into primates and, 317
Provirus
 endogenous retroviral genes and, 248, 262, 263
  cloning, 249-252, 254
  environmental insults, 259, 260, 262
  tissue-specific expression, 254, 255, 257
 intracisternal A-particle genes and, 173, 176, 185
  open reading frame, 179
  unintegrated, 178-180
 murine leukemia virus and
  ecotropic proviruses, 224-230
  future prospects, 233
  gene susceptibility determinants, 232, 233
  inbred strains of mice, 222-224
  insertional mutagenesis, 231
  retroviral insertion in dilute locus and, 211-214

R

Radiation, retroviral insertion in dilute locus and, 217, 218
Recombinase, lymphoid tumorigenesis and, 277, 278
Repetitive sequences
 in human genome, 329, 330
 L1 family of, 327
Replication
 chromosome breakpoints and, 282
 Drosophila and
  gene cloning, 101
  suppressible insertion-induced mutations, 88, 89
  transposable elements, 25
 endogenous retroviral genes and, 248
 gene transfer into primates and, 312
 homologous recombination and, 303, 304
 L1 family of repetitive sequences and, 327
 murine leukemia virus and, 229, 233
 primate L1 elements and, 10
 X-chromosome inactivation and, 120
Replication-competent virus, gene transfer into primates and, 311, 313, 317-321
Repressors, *Drosophila*, P-transposable elements and, 83

Restriction endonuclease, *Drosophila* gene cloning and, 106
Restriction-fragment-length polymorphism
  endogenous retroviral genes and, 252
  methylation mosaic model and, 152, 153
  variable number of tandem repeat sequences and, 188
Retinol-binding protein, albumin and α-fetoprotein and, 137
Retrotransposons
  Drosophila and
    molecular lesions, 112, 115
    suppressible insertion-induced mutations, 91
  endogenous retroviral genes and, 248, 249
  retroviral insertion in dilute locus and, 216
Retroviral genes, endogenous, *see* Endogenous retroviral genes
Retroviral insertion in dilute locus
  dilute suppressor, 216, 217
  functional units
    deletion mapping, 217, 218
    future directions, 218, 219
  genetic analysis
    first allele, 207, 208
    induced mutation, 209–211
    reverse mutation, 211
    spontaneous mutation, 208, 209
  molecular analysis
    Emv-3, 212, 213
    expression, 214–216
    reversion of original allele, 211, 212
Retroviremia, gene transfer into primates and, 317, 318, 321
Retrovirus
  chromosome breakpoints and, 283, 284
  Drosophila and
    foldback elements, 3, 5
    hobo element, 44
    molecular lesions, 112
    transposable elements, 28
  gene transfer into primates and, 311, 320
    adenosine deaminase deficiency, 312
    bone marrow transplant, 315
    safety, 317–320
    vectors, 312, 313
    homologous recombination and, 308
  intracisternal A-particle genes and, 173, 179, 185
  molecularly tagged P elements and, 59, 65, 66
  murine leukemia virus and, *see* Murine leukemia virus
  transgenic mice and
    c-fos expression, 236, 242, 244
    insertional mutation, 160, 168
Reverse transcriptase, intracisternal A-particle genes and, 185
Reverse transcription
  Drosophila melanogaster, molecular lesions and, 112, 114, 115
  endogenous retroviral genes and, 248, 250, 253, 263
  primate L1 elements and, 8, 9
RNA
  chromosome breakpoints and, 282
    ABL, 284, 285
    chronic myelocytic leukemia, 291, 292
  Drosophila and
    molecular lesions, 114, 115
    P element transposition, 51, 55, 56
    suppressible insertion-induced mutations, 91–93
  endogenous retroviral genes and, 247, 263
    environmental insults, 259, 260
    tissue-specific expression, 255, 257, 259
  gene transfer into primates and, 312, 313
  intracisternal A-particle genes and, 173, 174, 185
    expression, 176–178
    open reading frame, 179
  molecularly tagged P elements and, 66
  murine leukemia virus and, 229, 230, 232
  retroviral insertion in dilute locus and, 213, 214
  transgenic mice and
    c-fos expression, 236, 241, 242
    insertional mutation, 167
  X-chromosome inactivation and, 124
RNA polymerase
  Drosophila foldback elements and, 7

INDEX

endogenous retroviral genes and, 248
primate L1 elements and, 9, 10, 12
RNase, *Drosophila* P element
 transposition and, 55
RNase H, *Drosophila melanogaster*,
 molecular lesions and, 112
rRNA, albumin and α-fetoprotein
 and, 135

S

Short-ear complex, retroviral insertion
 in dilute locus and, 210
Small nuclear ribonucleoprotein
 particles, *Drosophila* P
 element transposition and, 56
Snell's waltzer, retroviral insertion in
 dilute locus and, 210, 219
Somatic dysgensis, *Drosophila* P
 element transposition and, 52
Spermatogenesis, X-chromosome
 inactivation and, 124-126
Spleen
  albumin and α-fetoprotein and, 139
  endogenous retroviral genes and, 255,
   256, 259
  murine leukemia virus and, 222, 226
  transgenic mice, *c-fos* expression and,
   236, 240, 241
Srev loci, murine leukemia virus and,
 226-231
Sterility, X-chromosome inactivation
 and, 125
Steroid sulfatase, X-chromosome
 inactivation and, 123, 124, 127, 128
Suppressible insertion-induced
 mutations in *Drosophila, see
 Drosophila*

T

T cell receptor, lymphoid tumorigenesis
 and
  Burkitt's lymphoma, 272
  chromosome translocation, 277, 278
  cytogenetics, 270
  t(11;14) translocation, 275, 276
T cells
  gene transfer into primates and, 311,
   313, 317, 320

lymphoid tumorigenesis and, 269,
 275-278
murine leukemia virus and, 232
Telomeres, repetitive sequences in
 human genome and, 330
Testis
  endogenous retroviral genes and, 257,
   259
  transgenic mice, *c-fos* expression and,
   236
Thalassemia, chromosome breakpoints
 and, 282, 293
Thymidine kinase, homologous
 recombination and, 305
Thymus
  endogenous retroviral genes and, 255,
   256
  transgenic mice, *c-fos* expression and,
   236, 240-243
Tissue specificity
  albumin and α-fetoprotein and,
   132-134, 138
  Drosophila P element transposition
   and, 47, 54-56
  endogenous retroviral genes and, 254,
   255
   MRL elements, 257, 259
   MuLV, 255, 256
   negative regulatory element, 257,
    258
  lymphoid tumorigenesis and, 271, 275
Titration, *Drosophila* and
  P element transposition, 52
  P-transposable elements, 82
Topoisomerase, intracisternal A-particle
 genes and, 182, 184
TPA, transgenic mice, *c-fos* expression
 and, 242, 243
TPA-responsive element, transgenic
 mice, *c-fos* expression and, 242
Transcription
  albumin and α-fetoprotein and, 131,
   132, 134-136, 138-141
  chromosome breakpoints and, 285,
   290, 292
  Drosophila and
   hobo element, 38-40
   P element transposition, 48, 51,
    54, 55
   suppressible insertion-induced
    mutations, 87, 91-97

endogenous retroviral genes and, 248
  environmental insults, 259, 260
  tissue-specific expression, 255-257, 259
  intracisternal A-particle genes and, 174, 176-178, 185
  lymphoid tumorigenesis and, 273-275
  primate L1 elements and, 9-12
  retroviral insertion in dilute locus and
    dilute suppressor, 217
    functional units, 218
    molecular analysis, 213-215
  transgenic mice and
    c-fos expression, 235, 236, 240, 242
    insertional mutation, 167
  X-chromosome inactivation and, 119
Transfection
  gene transfer into primates and, 313
  homologous recombination and, 301, 305, 306
Transferrin, X-chromosome inactivation and, 126
Transgenic mice, 13-22
  albumin and α-fetoprotein and, 134, 135, 138, 141
  c-fos expression and, 235, 241
    AP-1 expression, 240, 241
    H2, 238-240
    metallothionein, 236-238
    proliferation, 242-244
    specificity, 241, 242
  insertional mutation in, 159-161, 168
    analysis, 165-168
    identification, 161-165
Translation
  Drosophila, hobo element and, 40
  intracisternal A-particle genes and, 181, 185
Translocation
  chromosome breakpoints and, 281, 295, 296
    ABL, 286
    chronic myelocytic leukemia, 288-292
    DNA sequences, 293
  lymphoid tumorigenesis and, 279
    Burkitt's lymphoma, 271-273
    chromosome, 277, 278
    cytogenetics, 269, 270
    follicular lymphomas, 273, 274
    t(11;14), 274-276
  methylation mosaic model and, 147

transgenic mice, insertional mutation and, 161
variable number of tandem repeat sequences and, 199
X-chromosome inactivation and, 119, 120, 123, 125, 127
Transposition
  endogenous retroviral genes and, 262
  molecularly tagged P elements and, 59-67
  P-transposable elements, see Drosophila, P-transposable elements and
Transposons
  Drosophila and, 325
    gene cloning, 99, 100, 102, 104, 105
    molecular lesions, 111, 114, 115
  Drosophila, suppressible insertion-induced mutations and, 87
    alleles, 90-94, 97
    parasites, 88-90
  endogenous retroviral genes and, 247
  X-chromosome inactivation and, 123
Trisomy, variable number of tandem repeat sequences and, 194, 195
Triton X-100, intracisternal A-particle genes and, 182
tRNA
  endogenous retroviral genes and, 250
  homologous recombination and, 307
  intracisternal A-particle genes and, 173
Tumor
  endogenous retroviral genes and, 247, 263
  gene transfer into primates and, 318
  intracisternal A-particle genes and, 174, 176
  methylation mosaic model and, 153
  transgenic mice, c-fos expression and, 236, 238, 242-244
  variable number of tandem repeat sequences and, 187, 193, 195, 197-199, 201, 202
Tumorigenesis, lymphoid, see Lymphoid tumorigenesis
Tyrosine kinase, chromosome breakpoints and, 282, 284, 294, 296

V

V-abl, chromosome breakpoints and, 282, 283, 293, 296, 298

INDEX

Variable number of tandem repeat
sequences, 187, 199, 201, 202
  cell culture models of neoplasia,
    193-195
  DNA fingerprinting, 188-194
  neoplastic process, 195-200
Virus
  chromosome breakpoints and,
    283, 284
  gene transfer into primates and,
    318-321
  homologous recombination and,
    301, 302
  intracisternal A-particle genes and,
    178, 179
  murine leukemia virus and
    ecotropic proviruses, 224, 226, 229,
      230
    future prospects, 233
    gene susceptibility determinants,
      231, 232
    spontaneous germ-line virus
      infection, 232, 233
  retroviral insertion in dilute locus
    and, 211-214
  transgenic mice, insertional mutation
    and, 159

X

X chromosome, methylation mosaic
  model and, 154

X-chromosome inactivation, 119
  center, 119, 120
  signals, 120, 121
    allelic state, 121, 122
    developmental changes, 122
    dosage, 124-126
    evolution, 128
    homology, 123, 124
    parental chromosome imprinting,
      121
    resistance, 126-128
    roles, 126
    spreading, 122, 123

Y

Y chromosomes, X-chromosome
  inactivation and, 123, 124, 128
Yeast
  Drosophila, transposable elements
    and, 26, 28
  homologous recombination and,
    301, 304
  molecularly tagged P elements
    and, 59
  retroviral insertion in dilute locus
    and, 216
  transgenic mice, *c-fos* expression and,
    235